"Working memory is a central concept in contemporary cognitive psychology, and this collection of articles by Alan Baddeley traces the evolution of the construct from the seminal chapter by Baddeley and Hitch in 1974 to current notions of the central executive and the episodic buffer. Each group of papers is preceded by an introduction written in Baddeleyís lucid and informal style, nicely setting the articles in a personal scientific context. This is an excellent selection of important papers, essential reading for students and researchers in the area of memory and cognition."

Fergus Craik, Senior Scientist, Rotman Research Institute, Toronto, Canada

"In cognitive psychology, Alan Baddeley is considered as the *father* of working memory, and his work has revolutionized nearly all fields of psychology, establishing working memory as one of the leading concepts of psychology. This book presents his most influential papers in a comprehensive collection, but it goes far beyond a simple assemblage. The introductory description of each section sheds a new light on the creation and the development of what is now considered as the heart of human cognition, leading to a new and enlightening understanding of Alan Baddeley's work."

Valérie Camos, Université de Fribourg, Switzerland

Exploring Working Memory

In the *World Library of Psychologists* series, international experts themselves present career-long collections of what they judge to be their finest pieces – extracts from books, key articles, salient research findings, and their major theoretical and practical contributions.

Alan Baddeley has an international reputation as an eminent scholar and pioneer in the field of human memory, and is principally known for the theory of working memory, devised with Graham Hitch. This model continues to be valuable today in recognising the functions of short-term memory. This volume includes a specially written introduction by Alan Baddeley which gives an overview of the start of his career and his entry into the field of Psychology. Throughout the book he also provides introductions to the selection of works included and contextualises them in relation to changes in the field during this time. *Exploring Working Memory* includes the author's most influential publications on topics including short-term memory, the distinctions between short and long-term memory, the theory of working memory, the phonological loop, the concept of the central executive, and the episodic buffer. This exceptional selection concludes with an article giving a broad overview of the author's current views on working memory and its relation to other theories in the field.

Through his outstanding work Alan Baddeley has become known as a world-leading expert on human memory. *Exploring Working Memory* is a unique collection which will be of great interest to both students and researchers interested in human memory from psychology backgrounds.

Alan Baddeley is Professor of Psychology at the University of York and one of the world's leading authorities on Human Memory. He is celebrated for devising the ground-breaking and highly influential working memory model with Graham Hitch in the early 1970s, a model which still proves valuable today in recognising the functions of short-term memory. He was awarded a CBE for his contributions to the study of memory, is a Fellow of the Royal Society, the British Academy, the Academy of Medical Sciences, and the American Academy of Arts and Sciences. In 2012 he received the BPS Research Board's Lifetime Achievement Award and in 2016 the International Union of Psychological Sciences Award for Major Advancement in Psychological Science.

World Library of Psychologists

The *World Library of Psychologists* series celebrates the important contributions to psychology made by leading experts in their individual fields of study. Each scholar has compiled a career-long collection of what they consider to be their finest pieces: extracts from books, journals, articles, major theoretical and practical contributions, and salient research findings.

For the first time ever the work of each contributor is presented in a single volume so readers can follow the themes and progress of their work and identify the contributions made to, and the development of, the fields themselves.

Each book in the series features a specially written introduction by the contributor giving an overview of their career, contextualising their selection within the development of the field, and showing how their thinking developed over time.

Discovering the Social Mind
Selected Works of Christopher D. Frith
Christopher D. Frith

Towards a Deeper Understanding of Consciousness
Selected Works of Max Velmans
Max Velmans

Thinking Developmentally from Constructivism to Neuroconstructivism
Selected Works of Annette Karmiloff-Smith
Annette Karmiloff-Smith

Acquired Language Disorders in Adulthood and Childhood
Selected Works of Elaine Funnell
Edited by Nicola Pitchford and Andrew W. Ellis

Exploring Working Memory
Selected Works of Alan Baddeley
Alan Baddeley

Exploring Working Memory
Selected Works of Alan Baddeley

Alan Baddeley

LONDON AND NEW YORK

First published 2018
by Routledge
2 Park Square, Milton Park, Abingdon, Oxon OX14 4RN

and by Routledge
605 Third Avenue, New York, NY 10017

First issued in paperback 2020

Routledge is an imprint of the Taylor & Francis Group, an informa business

Copyright © 2018 Alan Baddeley

The right of Alan Baddeley to be identified as author of this work has been asserted by him in accordance with sections 77 and 78 of the Copyright, Designs and Patents Act 1988.

All rights reserved. No part of this book may be reprinted or reproduced or utilised in any form or by any electronic, mechanical, or other means, now known or hereafter invented, including photocopying and recording, or in any information storage or retrieval system, without permission in writing from the publishers.

Trademark notice: Product or corporate names may be trademarks or registered trademarks, and are used only for identification and explanation without intent to infringe.

British Library Cataloguing in Publication Data
A catalogue record for this book is available from the British Library

Library of Congress Cataloging in Publication Data
Names: Baddeley, Alan D., 1934– author.
Title: Exploring working memory : selected works of Alan Baddeley / Alan Baddeley.
Description: Abingdon, Oxon ; New York, NY : Routledge, 2017. | Series: World library of psychologists
Identifiers: LCCN 2017013070 (print) | LCCN 2017031883 (ebook) | ISBN 9781315111261 (ebook) | ISBN 9781138066908 (hardback : alk. paper)
Subjects: LCSH: Short-term memory. | Memory.Classification: LCC BF378.S54 (ebook) | LCC BF378.S54 .B326 2017 (print) | DDC 153.1/2–dc23
LC record available at https://lccn.loc.gov/2017013070

ISBN 13: 978-0-367-73578-4 (pbk)
ISBN 13: 978-1-138-06690-8 (hbk)

Typeset in Bembo
by Wearset Ltd, Boldon, Tyne and Wear

Contents

Permissions acknowledgements   x

Introduction   1
A. D. BADDELEY

PART I
How many kinds of memory?   3

1 **Short-term memory for word sequences as a function of acoustic, semantic and formal similarity**   9
A. D. BADDELEY, 1966

2 **Simultaneous acoustic and semantic coding in short-term memory**   15
A. D. BADDELEY AND J. R. ECOB, 1970

3 **Amnesia and the distinction between long- and short-term memory**   18
A. D. BADDELEY AND E. K. WARRINGTON, 1970

PART II
A multicomponent model   39

4 **Working memory**   43
A. D. BADDELEY AND G. HITCH, 1974

5 **The recency effect: implicit learning with explicit retrieval?**   80
A. D. BADDELEY AND G. HITCH, 1993

6 The concept of working memory: a view of its current state and probable future development 99
A. D. BADDELEY, 1981

PART III
The phonological loop 107

7 Word length and the structure of short-term memory 109
A. D. BADDELEY, N. THOMSON AND M. BUCHANAN, 1975

8 Exploring the articulatory loop 130
A. D. BADDELEY, V. J. LEWIS AND G. VALLAR, 1984

9 When long-term learning depends on short-term storage 150
A. D. BADDELEY, C. PAPAGNO AND G. VALLAR, 1988

10 The phonological loop as a language learning device 164
A. D. BADDELEY, S. E. GATHERCOLE AND C. PAPAGNO, 1998

PART IV
The visuo-spatial sketchpad 199

11 Reaction time and short-term visual memory 201
W. A. PHILLIPS AND A. D. BADDELEY, 1971

12 Spatial working memory 206
A. D. BADDELEY AND K. LIEBERMAN, 1980

13 Interference with visual short-term memory 224
R. H. LOGIE, G. M. ZUCCO AND A. D. BADDELEY, 1990

PART V
The central executive 243

14 The central executive: a concept and some misconceptions 247
A. D. BADDELEY, 1998

15 **Exploring the central executive** 253
A. D. BADDELEY, 1996

16 **Dementia and working memory** 280
A. D. BADDELEY, R. LOGIE, S. BRESSI,
S. DELLA SALA AND H. SPINNLER, 1986

PART VI
The episodic buffer 295

17 **The episodic buffer: a new component of working memory?** 297
A. D. BADDELEY, 2000

18 **Binding in visual working memory: the role of the episodic buffer** 312
A. D. BADDELEY, R. J. ALLEN AND G. J. HITCH, 2011

19 **Working memory: theories, models, and controversies** 332
A. D. BADDELEY, 2012

Index 370

Permissions acknowledgements

I would like to thank Taylor & Francis for permission to reproduce the following papers:

Baddeley, A. D. (1966) Short-term memory for word sequences as a function of acoustic, semantic and formal similarity. *Quarterly Journal of Experimental Psychology*, 18, 362–365.

Baddeley, A. D. (1996) Exploring the central executive. *Quarterly Journal of Experimental Psychology*, 49A, 5–28.

Baddeley, A. D., Lewis, V. J., & Vallar, G. (1984) Exploring the articulatory loop. *Quarterly Journal of Experimental Psychology*, 36, 233–252.

Baddeley, A. D., Logie, R., Bressi, S., Della Sala, S., & Spinnler, H. (1986) Dementia and working memory. *Quarterly Journal of Experimental Psychology*, 38A, 603–618.

I would like to thank Nature Publishing Group for permission to reproduce the following paper:

Baddeley, A. D., & Ecob, J. R. (1970) Simultaneous acoustic and semantic coding in short-term memory. *Nature*, 227, 288–289.

I would like to thank Springer for permission to reproduce the following papers:

Baddeley, A. D., & Hitch, G. J. (1993) The recency effect: implicit learning with explicit retrieval? *Memory and Cognition*, 21, 146–155.

Phillips, W. A., & Baddeley, A. D. (1971) Reaction time and short-term visual memory. *Psychonomic Science*, 22, 73–74.

I would like to thank Elsevier for permission to reproduce the following papers:

Baddeley, A. D. (1981) The concept of working memory: a view of its current state and probable future development. *Cognition*, 10, 17–23.

Baddeley, A. D. (2000) The episodic buffer: a new component of working memory? *Trends in Cognitive Sciences*, 4, 417–423.

Baddeley, A. D., & Hitch, G. (1974) Working memory. G. H. Bower (Ed.), *Psychology of Learning and Motivation*, Vol. 8. Academic Press, pp. 47–89.

Baddeley, A. D., & Warrington, E. K. (1970) Amnesia and the distinction between long- and short-term memory. *Journal of Verbal Learning and Verbal Behavior*, 9, 176–189.

Baddeley, A. D., Allen, R. J., & Hitch, G. J. (2011) Binding in visual working memory: the role of the episodic buffer. *Neuropsychologia*, 49, 1393–1400.

Baddeley, A. D., Papagno, C., & Vallar, G. (1988) When long-term learning depends on short-term storage. *Journal of Memory and Language*, 27, 586–595.

Baddeley, A. D., Thomson, N., & Buchanan, M. (1975) Word length and the structure of short-term memory. *Journal of Verbal Learning and Verbal Behavior*, 14, 575–589.

Logie, R. H., Zucco, G. M., & Baddeley, A. D. (1990) Interference with visual short-term memory. *Acta Psychologica*, 75, 55–74.

I would like to thank Lawrence Erlbaum Associates for permission to reproduce the following paper:

Baddeley, A. D., & Lieberman, K. (1980) Spatial working memory. In R. S. Nickerson (Ed.), *Attention and Performance VIII*. Hillsdale, N.J.: Lawrence Erlbaum Associates, pp. 521–539.

I would like to thank the American Psychological Association for permission to reproduce the following paper:

Baddeley, A. D., Gathercole, S. E., & Papagno, C. (1998) The phonological loop as a language learning device. *Psychological Review*, 105, 1, 158–173.

I would like to thank Cambridge University Press for permission to reproduce the following paper:

Baddeley, A. D. (1998) The central executive: a concept and some misconceptions. *Journal of the International Neuropsychological Society*, 4, 523–526.

I would like to thank Annual Reviews for permission to reproduce the following paper:

Baddeley, A. D. (2012) Working memory, theories models and controversy. *The Annual Review of Psychology*, 63, 12.1–12.29.

Introduction

I grew up in Leeds in the north of England, finding science rather boring and preferring geography and history, and latterly acquiring an interest in philosophy. Earning a living as a philosopher seemed unlikely so I opted for a degree in psychology at University College London where I took enthusiastically to experimental psychology, excited by the fact that theories could be tested experimentally rather than relying on disputation. After graduating I spent a year at Princeton, returning, and after a number of temporary jobs, was fortunate to be offered a job at the Medical Research Council Applied Psychology Unit in Cambridge where Donald Broadbent (1958) had just published his seminal book *Perception and Communication*, a book that was to play an important role in the cognitive revolution.

My post allowed me to register for a Cambridge PhD, based on my assigned project concerned with the development of memorable postal codes. This led to an interest in applying the newly developing concepts of information theory to memory and language, and to the creation of a set of British postal codes generated using the statistical structure of English words to enhance memorability, a list that was promptly ignored since the post office had already decided on its own solution. At a theoretical level I was able to show that when applied to the learning of lists of nonsense syllables, still a dominant approach to human memory at the time, my information-based measure was a better predictor of performance than standard measures of associative meaningfulness. Pressure to publish at the time, however, at the Unit was not great and my ideas on applying my measure based on the structure of language to memory was overshadowed by a book using a similar approach by Underwood and Schulz (1960), two major figures in verbal learning at the time.

My four years of research on post codes did, however, provide a valuable training in the newly developing field of cognitive psychology. The Unit was buzzing with new ideas both from within the Unit and from the numerous visitors from North America and Europe. I had virtually unlimited access to experimental participants who could be tested either individually or in groups with the result that my PhD thesis contained no fewer than 17 experiments, at least some of which were eventually published. I was therefore well prepared to engage in the controversy that dominated much of cognitive psychology in the 1960s, the question of how many kinds of memory it was necessary to assume.

References

Broadbent, D. E. (1958). *Perception and communication.* London: Pergamon Press.
Underwood, B. J., & Schulz, R. W. (1960). *Meaningfulness and verbal learning.* Chicago: Lippincott Company.

Part I
How many kinds of memory?

My work on postal coding was supervised by Conrad (who like a certain fictional detective preferred to avoid his first name). Conrad was a brilliant applied psychologist, much valued by the government postal and telecommunications department. It was he who optimised the postal code that was eventually adopted and that subsequently proved highly successful. Our next assignment was to undertake a further quite different research project concerned with improving measures of the sound quality of telephone lines. This was typically assessed by speaking a series of potentially confusable words and checking the accuracy with which they were detected by a listener. It was suggested that requiring the listener to then *process* the words in some way might impose an additional cognitive load which could make the task more sensitive to the quality of the telephone line. I was assigned this project while Conrad left for a year's sabbatical in the United States.

As part of the project, the Post Office were due to provide a system that would take speech and distort it in a way that resembled the loss of information during telephone transmission. This equipment, however, took some time to develop, and meanwhile I settled for a simple white noise generator, rather like the sound of a waterfall, where the intensity could be adjusted appropriately. I began by inventing games on topics that required conversation between participants under quiet or noisy conditions. This left me with large amounts of recorded verbiage and few ideas as to how to analyse it. I decided instead to simplify the task, presenting lists of five words for immediate serial recall. I could check the accuracy of simply hearing the words and see if errors increased when, in addition, they had to be remembered. To make the task a little harder, I had one condition in which the words were similar in sound (e.g. *man, can, cat, mat, map* versus *pit, cow, sun, day, top*). Since it was only a pilot experiment, I did not worry about the niceties of psychoacoustics, simply assembling a group of about 20 people, playing the words on a tape recorder either with or without concurrent background noise.

There proved to be a few more errors on the similar lists, but this tendency was not substantially increased by the requirement to remember, suggesting that was not the answer to the Post Office problem. What I did find however was that the similar sequences were hugely more difficult to *remember* than the dissimilar items, whether in noise or quiet. This was expected since Conrad had already demonstrated that similar sounding letters (*b, g, t, p, c*) were harder to recall in the correct order than dissimilar (*f, k, m, r, l*). Conrad (1964) related his findings to the proposal of a short-term memory system that differed from long-term memory, arguing that the system was based on an acoustic code. Having absorbed a good deal of standard verbal learning theory during my time at University College London and Princeton, I realised that verbal learning theorists typically did not discriminate among types of similarity and hence might reasonably object that this did not prove that the store was acoustic as the result might be the same, regardless of whether the similarity was acoustic or not. I therefore created a set of parallel conditions, again involving remembering sequences of five words, but this time varying the degree of semantic similarity comparing sequences like *huge, big, wide, long, tall* with *old, late, wet, thin, hot*. The results were clear, with a tiny though still significant drop in performance based on semantic similarity.

I duly submitted my paper to the *Journal of Experimental Psychology* whose referees failed to be convinced, although Arthur Melton the Editor and leading figure in the verbal learning community at the time was sufficiently intrigued to replicate our results and report his successful replication at a meeting in Cambridge a year or so later. They were duly published in a UK journal, followed by a paper in which we studied the effect of similarity on long-term memory by doubling the length of the lists and presented for several trials, finding exactly the opposite pattern. This time *semantic* similarity was crucial (Baddeley, 1966). In a series of experiments with my friend and colleague Harold Dale we subsequently re-demonstrated our dissociation using paired-associate learning in a series of papers that were indeed accepted by North American journals.

Showing a clear difference between a short-term system that appeared to rely on acoustic cues and a long-term system that reflected semantic coding, had implications for the hot topic of the moment, namely whether memory comprised a single unitary system as was widely assumed at the time, or whether it was necessary to assume more than one type of memory. A challenge to the unitary view first came in the late 1950s with the demonstration by John Brown (1958) in the UK and Peterson and Peterson (1959) in Indiana that quite small amounts of information such as a consonant triplet could be forgotten over a matter of seconds, unless some form of rehearsal was possible. Both studies were interpreted in terms of a fading short-term memory trace, while accepting the general view that forgetting in long-term memory depended on interference rather than decay. Other studies provided evidence that certain tasks might involve both a long-term and a short-term component, again suggesting separate long-term memory and short-term memory systems. A good example of this was free recall in which a list of unrelated words is presented and must be recalled in any order. When recalled immediately, the most recent items are very well recalled, the recency effect. This recency advantage completely disappears, however, after a filled interval of as little as 5–10 seconds suggesting that the recency effect may be based on a more fragile short-term memory system (Glanzer & Cunitz, 1966).

Perhaps the most convincing evidence for separate systems, however, came from neuropsychology with the demonstration by Milner (1966) that a densely amnesic patient HM appeared to have excellent immediate memory as demonstrated, for example, by digit span, coupled with grossly impaired performance on a wide range of long-term memory tests. Other patients showed exactly the opposite pattern of impaired short-term memory with preserved long-term memory (Shallice & Warrrington, 1970).

The three papers chosen in this section reflect my own introduction to the long-term memory/short-term memory debate. The first describes the initial discovery of a clear acoustic-semantic difference in verbal short-term memory. The second illustrates further complications in our interpretation of this distinction while the third describes my introduction to neuropsychology and its implications for the long-term memory/short-term memory controversy.

Although our semantic-acoustic similarity differences between short-term memory and long-term memory proved very replicable, our initial simple assumption of separate long and short-term systems based on different codes came under increasing pressure as we expanded our research on coding across a wider range of situations. It is clearly the case that phonological long-term memory must exist, otherwise how could we learn new words or unfamiliar names? The fact that memory span for sentences extends to around fifteen words rather than five unrelated words, provides strong evidence for the potential influence of meaning in immediate serial recall. The third paper in this section describes an experiment to tease apart the influences of sound and meaning, showing that people *will* use semantic coding, if it is easy to do so, as is the case for meaningful sentences. With unrelated words; however, it is typically too difficult to create the necessary meaningful bonds between the words on a single brief trial with the result that semantic encoding that captures the order in which they are presented will only develop over successive attempts. The relationship between semantic and acoustic coding is complex with both codes potentially operating at the same time; although, the acoustic code appears to be less durable. Chapter 2 shows that this can lead to a paradoxical tendency for performance to *improve* after a delay when people switch from a defective phonological to a more stable semantic code. The link between memory systems and coding was proving rather more complex than we at first thought.

The third chapter stems from an invitation from Elizabeth Warrington, a neuropsychologist at the National Hospital in London, to study a group of densely amnesic patients with otherwise well-preserved cognitive abilities. I was initially reluctant, arguing that it was unlikely that patients would have their lesions in a sufficiently convenient location to allow theoretically convincing experiments to be run. I did, however, accept the invitation and was immediately struck by the apparent purity of the deficit in the first patient I encountered. The paper illustrates our attempt to combine the methods that were evolving in mainstream cognitive psychology with the study of carefully selected patients. It presents evidence for a distinction between impaired long-term memory and preserved short-term memory, although with one or two anomalies. One concerns the preserved performance on the Peterson short-term forgetting task, even after the longest delay. Another surprise came from the Hebb repeated digit test which showed normal long-term learning of a sequence of digits that is embedded in a task that ostensibly involves different digit sequences on every trial. Although we did not realise it at the time, these anomalies foreshadowed the later discovery of preserved implicit learning in amnesic patients (Brooks & Baddeley, 1976; Schacter & Tulving, 1994).

An obvious extrapolation from my previous work might be to suspect that the amnesic patients have a deficit in semantic coding. We did not find this in our group who, as we later showed, could use clusters of semantically related words in a free recall task to enhance their long-term performance (Baddeley & Warrington, 1973). An apparent semantic coding deficit was, however, found by Laird Cermak and Nelson Butters in their group of patients whose amnesia resulted from alcoholic Korsakoff syndrome. Eventually, however, it proved to be the case that their patients had additional, previously undetected, frontal lobe damage that also interfered with their performance on the Peterson task. This was confirmed by their later study of a patient with a purer amnesic deficit who behaved broadly like own patient group.

References

Baddeley, A. D. (1966). The influence of acoustic and semantic similarity on long-term memory for word sequences. *Quarterly Journal of Experimental Psychology, 18*, 302–309.

Baddeley, A. D., & Warrington, E. K. (1973). Memory coding and amnesia. *Neuropsychologia, 11*(2), 159–165.

Brooks, D. N., & Baddeley, A. D. (1976). What can amnesic patients learn? *Neuropsychologia, 14*, 111–122.

Brown, J. (1958). Some tests of the decay theory of immediate memory. *Quarterly Journal of Experimental Psychology, 10*, 12–21.

Conrad, R. (1964). Acoustic confusion in immediate memory. *British Journal of Psychology, 55*, 75–84.

Glanzer, M., & Cunitz, A. R. (1966). Two storage mechanisms in free recall. *Journal of Verbal Learning and Verbal Behavior, 5*, 351–360.

Milner, B. (1966). Amnesia following operation on the temporal lobes. In C. W. M. Whitty & O. L. Zangwill (Eds), *Amnesia* (pp. 109–133). London: Butterworths.

Peterson, L. R., & Peterson, M. J. (1959). Short-term retention of individual verbal items. *Journal of Experimental Psychology, 58*, 193–198.

Schacter, D. L., & Tulving, E. (1994). *Memory systems.* Cambridge, MA: MIT Press.

Shallice, T., & Warrington, E. K. (1970). Independent functioning of verbal memory stores: A neuropsychological study. *Quarterly Journal of Experimental Psychology, 22*, 261–273.

Warrington, E. K., & Baddeley, A. D. (1974). Amnesia and memory for visual location. *Neuropsychologia, 12*, 257–263.

Papers

1 Baddeley, A. D. (1966). Short-term memory for word sequences as a function of acoustic, semantic and formal similarity. *Quarterly Journal of Experimental Psychology, 18*, 362–365.
2 Baddeley, A. D., & Ecob, J. R. (1970). Simultaneous acoustic and semantic coding in short-term memory. *Nature, 227*, 288–289.
3 Baddeley, A. D., & Warrington, E. K. (1970). Amnesia and the distinction between long- and short-term memory. *Journal of Verbal Learning and Verbal Behavior, 9*, 176–189.

1 Short-term memory for word sequences as a function of acoustic, semantic and formal similarity

A. D. Baddeley

Experiment I studied short-term memory (STM) for auditorily presented five-word sequences as a function of acoustic and semantic similarity. There was a large adverse effect of acoustic similarity on STM (72·5 per cent.) which was significantly greater ($p<0.001$) than the small (6·3 per cent.) but reliable effect ($p<0.05$) of semantic similarity.

Experiment II compared STM for sequences of words which had a similar letter structure (formal similarity) but were pronounced differently, with acoustically similar but formally dissimilar words and with control sequences. There was a significant effect of acoustic but not of formal similarity.

Experiment III replicated the acoustic similarity effect found in Experiment I using visual instead of auditory presentation. Again a large and significant effect of acoustic similarity was shown.

Introduction

In a series of short-term memory (STM) experiments Conrad (1963, 1964) has shown that sequences of items which are hard to discriminate in noise are also hard to remember, even though presented visually. Analogous effects of intra-list similarity have also been shown in long-term memory (LTM) where several types of similarity have proved to be relevant including similarity of letter structure (Horowitz, 1961) and of meaning (Underwood and Goad, 1951; Baddeley and Dale, 1966). However, Baddeley and Dale (1966) using paired-associate learning failed to show an equivalent effect of semantic similarity in STM and suggested that STM may differ from LTM in relying more on acoustic cues and much less on the meaning of the material to be retained. The present study uses the method of serial recall to explore further the role of similarity in STM.

Experiment I compares the influence of acoustic similarity on ordered STM for word sequences with that of semantic similarity.

Experiment I
Method

Design

A separate group of subjects did each of two conditions, A and B. Both groups attempted to recall 24 sequences of five words. In condition A these comprised 12 drawn from a set of eight acoustically similar words (mad, man, mat, cap, cad, can, cat, cap) and 12 from a control set of acoustically different words of equal Thorndike-Lorge frequency (Thorndike and Lorge, 1944) (cow, day, bar, few, hot, pen, sup, pit). In Condition B, 12 sequences were drawn from a set of eight adjectives with similar meanings (big, long, broad, great, high, tall, large, wide) and 12 from a set of eight semantically different words of equal Thorndike-Lorge frequency (old, deep, foul, late, safe, hot, strong, thin). All sequences were drawn at random with the constraint that no word appeared more than once in the same sequence. Similar and different sequences were presented in the same random order in both conditions.

Procedure

Subjects were tested in groups of about 20. Word sequences were presented by tape recorder at a rate of one word per sec. and subjects were allowed 20 sec. to write their ordered responses. To maximize response availability the relevant sets of words were written on cards and both sets were visible throughout the test session. Subjects were instructed that no sequence would contain words from more than one set but were not told before each sequence which set would be involved. A listening test was given both before and after the memory test to ensure that subjects were hearing words correctly. The 16 relevant words were presented in random order and subjects were allowed 5 sec. per word to copy them. Two Condition A subjects did not score perfectly and were discarded, leaving 20 subjects in Condition A and 21 in Condition B. Housewives from the A.P.R.U. subject panel served as subjects. These were paid for participation and were assigned haphazardly to one of the two conditions.

Results

Performance was scored in terms of percentage correct sequences, and since scores were not normally distributed they were analysed using non-parametric tests.

Acoustic similarity. A mean of 9·6 per cent. of the acoustically similar sequences were correctly reproduced (range 0–33·3), compared with 82·1 per cent, of the control sequences (range 58·3–100). Since there is no overlap between the two distributions, the difference is clearly highly statistically significant, $p < 0.001$.

Semantic similarity. The mean score for semantically similar sequences was 64·7 per cent, correct (range 16·7–100), and that for control sequences was 71·0 per cent. (range 16·7–100). Although the mean difference is only 6·3 per cent, a Wilcoxon test indicates that it is statistically significant, $p < 0.05$.

Comparing acoustic and semantic similarity. The mean difference between acoustically similar and control sequences was 72·5 per cent. (range 50·0–91·7). The equivalent effect of semantic similarity was 6·3 per cent. (range 0–41·7). Since there is no overlap between the two distributions, the greater effect of acoustic similarity is clearly statistically significant, $p < 0.001$.

These results suggest that STM for word sequences shows a massive effect of intrasequence acoustic similarity compared with only a slight effect of semantic similarity. In fact, however, Experiment I confounds acoustic and formal similarity, since the words which sound alike do so because they have letters in common. Since formal similarity has been claimed to have a marked effect on verbal learning (Horowitz, 1961) it is clearly desirable to separate its effect on STM from that of acoustic similarity. Experiment II attempts to do this.

Experiment II
Method

Design

All subjects attempted to recall 24 five-word sequences, comprising eight sequences drawn at random by sampling without replacement from each of three sets of words. Set A comprised five words which were acoustically similar but relatively dissimilar in letter-structure (bought, sort, taut, caught, wart), Set B comprised five words with similar letter structure but relatively dissimilar pronunciation (rough, cough, through, dough, bough) and Set C comprised five words of approximately equal Thorndike-Lorge frequency to sets A and B, presenting a roughly equivalent degree of spelling difficulty due to unusual letter structure or occurrence of homophones (e.g. caught–court, dough–doe), but which both sound and look relatively dissimilar (plea, friend, sleigh, row, board).

Procedure

The 24 sequences were presented in random order to a group of 17 housewives. As in Experiment I presentation was auditory, the rate was one word per sec. and subjects were allowed 20 sec. to write their responses. To maximize response availability the three sets of words were written on cards which were visible throughout the test session. To prevent the use of position on the card as a cue, four cards were used, each with a different order and were interchanged frequently. Again listening tests were given before and after the main experiment. In each test, the 15 words were read out in random order and the subject was allowed 5 sec. to write down what she heard. All 17 subjects scored perfectly on both tests.

Results

Mean recall scores were as follows, acoustically similar words 36·5 per cent. correct sequences (range 0–100), formally similar 55·8 per cent. (range 25·0–100), control words 63·5 per cent. (range 12·5–100). Comparison using the Wilcoxon test showed that performance on acoustically similar sequences was significantly poorer than performance on either control sequences, $p<0.001$, or formally similar sequences, $p<0.01$. There was no significant difference between performance on formally similar and control sequences, $p>0.05$.

The general level of performance differs from Experiment I in being higher for acoustically similar sequences but lower for control sequences. The acoustically similar sequences are probably easier for two reasons. First, they are selected from a set of five instead of eight items, which seems likely to reduce both degree of inter-item confusion and information load. Secondly the items differ only in terms of the initial sound, so that if the subject remembers only the initial letter, she can easily reconstruct the sequence. This latter point also holds for the other two word sets since neither has more than one word starting with the same letter, although reconstructing the control words might take slightly longer since the latter part of these words have neither a common sound nor a similar letter structure. The fact that the words used in this study were less frequent than those used in Experiment I and presented more difficulties due to spelling and competition from homophones probably accounts for the rather lower performance on control sequences.

Although the set of acoustically similar words are more formally similar than would be ideal, and the formally similar words are probably not as acoustically distinctive as the control list, it nevertheless seems fairly clear that acoustic rather than formal similarity was the principal source of difficulty.

However, both Experiment I and Experiment II have used auditory presentation so that both results are open to the objection that some of the acoustic effect may be due to perceptual error. The listening tests used suggest that it is unlikely that very much of the acoustic similarity effect is due to mishearing, but the possibility exists that some of the effect may be due to the interaction of perceptual and memory loads. The following experiment therefore attempts to replicate the acoustic similarity effect using visual presentation.

Experiment III
Method

Material

Twenty-four five-word sequences were prepared comprising 12 sequences drawn at random from a set of 10 acoustically similar words (mad, man, map, mat, max, can, cad, cap, cat, cab), and 12 from a control set (pen, rig, day, bar, cow, sup, pit, hot, few, bun). Each word was typed on a 4½ × 3½ in. card.

Procedure

Subjects were tested individually; the experimenter presented the material manually at a rate of one word per sec. after which the subject attempted to write down the sequence in the appropriate order. The two sets of words were visible throughout the experiment to maximize response availability, and again several different arrangements of each set were used. Ten young enlisted men served as subjects.

Results

The mean recall score for acoustically similar sequences was 1·7 per cent. (range 0–8·3) and for control sequences was 58·3 per cent. (range 8·3–91·7). The clear difference between the two types of sequence was shown by all 10 subjects and is thus highly statistically significant, $p < 0.001$.

The overall level of performance was lower than that shown in Experiment I. The question of whether this is due to the different method of presentation, the selection of sequences from 10-word instead of eight-word sets or to the different type of subject used is, however, beyond the scope of the present study.

Discussion

All three experiments agree in showing a large and consistent adverse effect of acoustic similarity on ordered STM for words, and Experiments I and II show that neither semantic nor formal similarity has an effect of comparable magnitude. The relative unimportance of semantic similarity shown in Experiment I together with the failure of Baddeley and Dale (1966) to find an effect of semantic similarity among stimuli on STM for paired associates suggests that subjects show remarkable consistency and uniformity in using an almost exclusively acoustic coding system for the short-term remembering of disconnected words. There is abundant evidence that this is not true of LTM (Underwood, 1951; Underwood and Goad, 1951; Baddeley, 1966; Baddeley and Dale, 1966).

References

Baddeley, A. D. (1966). The influence of acoustic and semantic similarity on long-term memory for word sequences. *Quart. J. exp. Psychol.*, **18**, 302–9.

Baddeley, A. D., and Dale, H. C. A. (1966). The effect of semantic similarity on retro-active interference in long- and short-term memory. *J. verb. Learn, verb. Behav.* (in press).

Conrad, R. (1963). Acoustic confusions and memory span for words. *Nature*, **197**, 1029–30.

Conrad, R. (1964). Acoustic confusion and immediate memory. *Brit. J. Psychol.*, **55**, 75–84.

Horowitz, L. M. (1961). Free recall and ordering of trigrams. *J. exp. Psychol.*, **62**, 51–7.

Thorndike, E. L., and Lorge, I. (1944). *The Teacher's Word Book of 30,000 Words.* New York: Teachers' College, Columbia University.

Underwood, B. J. (1951). Studies of distributed practice: II. Learning and retention of paired-adjective lists with two levels of intra-list similarity. *J. exp. Psychol.*, **42**, 153–61.

Underwood, B. J., and Goad, D. (1951). Studies of distributed practice: I. The influence of intra-list similarity in serial learning. *J. exp. Psychol.*, **42**, 125–34.

2 Simultaneous acoustic and semantic coding in short-term memory

A. D. Baddeley and J. R. Ecob

It has been suggested[1,2] that memory for verbal material comprises two components, one of which is labile and depends on the acoustic properties of the words while the other, which is more durable, is based on their meaning. The observed relationship between rate of forgetting and type of coding is explicable in at least the three following ways. (a) Material may pass from a short-term store which uses an acoustic code into a long-term semantically coded store. (b) Material may be encoded on input either acoustically, in which case rapid forgetting occurs, or else semantically, in which case forgetting is relatively slow. (c) Material may be encoded both acoustically and semantically on input, in which case immediate recall will show the effects of both methods of encoding, but because the effects of acoustic coding are short lived, delayed recall will show only semantic effects.

The following experiment aims to decide between the three hypotheses. Subjects tried to remember sequences of three words which were either similar or dissimilar in sound and which either made up a meaningful phrase or were unrelated. The combination of two levels of acoustic similarity with two levels of semantic compatibility thus gave four basic conditions, similar and compatible, for example, "I might fly"; similar and incompatible, for example, "eye fight dry"; dissimilar and compatible, for example, "we could stop"; dissimilar and incompatible, for example, "king wake slit". Subjects attempted recall after either 2 or 20 s during which they were required to perform a distracting task. We assumed that a considerable short term component would be present after 2 s but would have dissipated after 20 s of distraction[3]. Subjects were successively presented with two sequences of three words on each test trial, the sequences being drawn from the same condition, and in the similar conditions they were acoustically similar to each other; for example, "Jude chewed food, lewd dude wooed". Each word was presented visually for 1 s.

The acoustic similarity effect is typically reflected in the difficulty of recalling the order in which items were presented rather than in recall of the items themselves, which may indeed be enhanced by acoustic similarity[4]. Item recall was not therefore required, the six relevant words on each trial being written in random order on a prompt card which was displayed during recall. To ensure that subjects could not respond correctly merely by rearranging the compatible sets to form meaningful triads, sets of three words were selected, which were comparatively meaningful in any of the six possible permutations of order, for example, "I might fly"; "might I fly"; "fly I might" and so on. Six subgroups of subjects were tested so that the six permutations of each triad were used once. For the intervening task, the experimenter read out random digits at a rate of one per second and subjects were required to add one to the digit and write down the answer. After two or twenty such digits they were given a verbal recall signal and shown the relevant prompt card, whereupon they attempted to write down the two word triads in the correct order. After 20s the signal "ready" was given and the next trial began. The order of presentation of the four conditions and two delays was random. Six groups of five subjects were tested.

Performance was scored in terms of the number of words recalled in the appropriate serial position. The results (Table 1) were analysed using the Wilcoxon test with two-tailed significance levels. There was a reliable effect of semantic compatibility at both the 2s ($t=13$, $P<0.01$) and the 20s delay ($t=8$, $P<0.01$). The effect of acoustic similarity, however, occurs only at the short retention interval where it is shown by both compatible ($t=45$, $N=24$, $P<0.01$) and incompatible triads ($t=88$, $N=27$, $P<0.02$). At the longer delay, neither compatible nor incompatible triads show an acoustic similarity effect. Only the incompatible dissimilar triads show significant forgetting during the 20s delay ($t=90$, $N=27$, $P<0.05$), while triads which are both compatible and acoustically similar show a significant improvement in performance after the 20s delay ($t=60$, $N=25$, $P<0.01$).

These results are consistent with the hypothesis of simultaneous semantic and acoustic coding. Recall after 2s uses both sources of information even though this is a sub-optimal strategy for one of the four conditions (acoustically similar compatible). Because acoustically coded information is rapidly forgotten, recall after 20s depends chiefly on semantically coded information and the adverse effects of acoustic similarity are therefore absent.

Table 1 Mean percentage of words recalled as a function of acoustic similarity, semantic compatibility and retention interval

	Condition		Delay	Percent forgotten
		2s	20s	$\left(\frac{2s-20s}{2s} \times 100\right)$
Compatible	Similar	61.0	68.9	−13.0
	Dissimilar	71.8	68.2	5.0
	Difference	10.8	−0.7	
Incompatible	Similar	42.8	40.9	4.4
	Dissimilar	52.2	46.0	11.7
	Difference	9.4	5.1	

Acknowledgment

We thank the Medical Research Council for financial support.

Notes

1 Baddeley, A. D., *Quart. J. Exp. Psychol.*, **18**, 302 (1966).
2 Baddeley, A. D., *Brit. J. Psychol.* (in the press).
3 Glanzer, M., and Cunitz, A. R., *J. Verb. Learn. Verb. Behav.*, **5**, 351 (1966).
4 Wickelgren, W. A., *Amer. J. Psychol.*, **78**, 567 (1965).

3 Amnesia and the distinction between long- and short-term memory[1]

A. D. Baddeley[2] and E. K. Warrington

Evidence for a dichotomy between long-term memory (LTM) and short-term memory (STM) comes from: (*a*) amnesic patients with a normal digit span but defective LTM, and (*b*) tasks comprising two components, one labile (STM) and the other stable (LTM). This study examines the compatibility of (*a*) and (*b*) by comparing the performance of amnesic and control Ss on immediate and delayed free recall, the Peterson short-term forgetting task, development of PI in STM, minimal paired-associate learning, digit span, and the Hebb repeated digit-sequence technique. Results suggest that amnesic Ss have normal STM but defective LTM. There is some evidence of a stable component in certain STM tasks, on which amnesic Ss are also unimpaired. Implications for the dichotomy between STM and LTM are discussed.

There has in recent years been considerable controversy over the question of whether short-term memory (STM), retention of items over a period of seconds, and long-term memory (LTM), retention of items over longer intervals, are explicable in terms of a single unitary memory system. The most cogent defence of the unitary hypothesis was made by Melton (1963) who argued that none of the evidence available at the time was incompatible with a unitary system, and that such a system should therefore be accepted on the grounds of parsimony. A good deal of subsequent evidence, however, cannot be accounted for without assuming more than one type of memory. Such evidence comes from three main sources.

The first source is based on the discovery that many standard techniques for studying verbal memory have two distinct components, one showing rapid forgetting over the first few seconds after presentation, while the second component is relatively stable and may even show improvement over time. Free recall (Glanzer & Cunitz, 1966), minimal paired-associate learning (Peterson, 1966), and the probe-digit technique (Waugh & Norman, 1965) have all been shown to reflect two such components.

A second source of evidence lies in differences in the coding systems used in STM and LTM. Verbal STM depends largely on the sound of the items to be remembered, shows marked effects of acoustic or articulatory similarity among the stimuli (Conrad, 1964; Wickelgren, 1965), and is relatively insensitive to the effects of similarity of meaning among the stimuli (Baddeley, 1966a). For LTM the reverse is true: no effect of acoustic similarity and a marked effect of semantic similarity (Baddeley, 1966b; Baddeley & Dale, 1966).

In general the two sources of evidence for a dual memory system are mutually compatible, with STM showing evidence of acoustic coding and LTM of semantic coding. Thus, Kolers (1966) has shown that free recall depends mainly on semantic coding, with the exception of the short-term component which Craik (1968) has shown to produce acoustic intrusion errors. Using the probe technique, Levy and Murdock (1968) have shown that the latter part of the serial position curve which is assumed to reflect STM is affected by acoustic but not semantic similarity, while the reverse is true of that part of the curve which is assumed to reflect LTM.

There are, however, a number of results which suggest that the assumption that STM is exclusively associated with acoustic coding and LTM with semantic coding may be an over-simplification. Thus, Loess (1967) has shown that semantic factors may enhance performance on the short-term retention task devised by Brown (1958) and Peterson and Peterson (1959), a task which would appear to depend mainly on STM. A problem is also raised by the repeated digit string technique devised by Hebb (1961). Melton (1963) showed that improvement occurs even when the repetitions of the string are separated by five different items, which suggests that LTM is involved. It is, however, difficult to see how one could attribute the effect to semantic coding.

Perhaps the most dramatic evidence for more than one memory system comes from experiments on amnesic patients suffering from temporal lobe damage. They show marked impairment in long-term learning but seem to have a normal immediate memory span (Drachman & Arbit, 1966), a result that is very hard to account for in terms of a single unitary memory system. However, although the evidence is broadly consistent with the assumption that such patients have impaired LTM but normal STM, the evidence is far from clear. Indeed, Talland (1965) refers to the normal digit span of his amnesic Korsakoff patients as "a notable exception to their short-term memory disorder" (p. 268), basing his conclusion mainly on a task involving the reproduction of visual designs. Since it is not clear to what extent performance on such a task depends on LTM rather than STM, interpretation is difficult. Unfortunately, this problem occurs with much of the work in this area, since with the exception of Wickelgren's (1968) study carried out on a single patient, most of the evidence is based on techniques that are not directly comparable with those currently used for studying normal memory. This lack of unanimity is particularly worrying in view of the claim that amnesic Ss forget as soon as their attention is distracted (Milner, 1966, 1968; Talland, 1965). Since many studies of STM involve recall after a delay during which rehearsal is prevented by a secondary, and presumably distracting task, one might expect amnesic patients to show abnormally poor STM. There is clearly a need for further information on this point.

The following experiments therefore compare the performance of amnesic and normal control Ss using a range of standard laboratory memory tasks varying in the extent to which performance is assumed to depend on STM and LTM. It aims first to test the hypothesis that amnesic Ss have a normal STM but impaired LTM. This in turn should provide information about the relative contribution of STM and LTM to the various tasks and should thus help interpretation of the performance of normal Ss.

Amnesic subjects. The experiment was initially performed on N. T., a patient with a unilateral temporal lobectomy who exhibited symptoms closely analogous to H. M., the patient studied extensively by Milner and her co-workers (Milner, 1966). A detailed description of this patient is given by Dimsdale, Logue, and Piercy (1964). Five more amnesic patients were subsequently tested, and since no difference could be detected between their performance and that of N. T., the six patients are treated subsequently as a single group. Criteria used in selecting patients were:

1. A clinically diagnosed amnesia—the patient did not remember such simple pieces of information as the day of the week, where he was, and how long he had been in the hospital.
2. Patients with any sign of intellectual impairment other than a memory defect were excluded. Intelligence test scores, together with S's age and diagnosis are shown in Table 1. It will be noted that four of the six

Table 1 Details of the two groups of patients

S sex	Age	Verbal score	Performance score	Diagnosis
		Amnesics		
1 F	60	10	12	Right temporal lobectomy
2 F	60	11	7	Alcoholic Korsakoff
3 F	47	9	11	Vascular?
4 F	42	8	7	Alcoholic Korsakoff
5 F	55	10	8	Alcoholic Korsakoff
6 M	60	11	11	Alcoholic Korsakoff
Mean	54.8	9.83	9.33	
		Controls		
1 F	53	10	9	
2 F	55	12	10	
3 F	60	—[a]	13	
4 F	44	9	8	All peripheral nerve lesions
5 F	53	8	5	
6 M	58	7	8	
Mean	53.8	8.1	8.8	

Note
a No test data available.

amnesic Ss were diagnosed as alcoholic Korsakoff cases. While such patients may show signs of dementia, this is by no means invariably the case (Victor & Adams, 1953), and provided potential Ss are carefully screened it is quite possible to select amnesic Korsakoff patients who are unimpaired on intellectual tasks.

Control subjects. These were six patients suffering from peripheral nerve lesions who were selected as comparable in age, intelligence, and occupation to the amnesic patients. Control Ss were given two subtests of the WAIS, vocabulary and block design (see Table 1).

Experiment I: free recall

In the free-recall technique S is presented with a list of words which he subsequently tries to recall in any order. Glanzer and Cunitz (1966) have shown that this task comprises two components, a stable long-term component and a more labile short-term component, the latter being reflected mainly in the high probability of recall of the last few items presented (the recency effect). The short-term component disappears if recall is delayed by 20–30 sec. It was predicted that the performance of the amnesic Ss would be impaired on delayed recall, which reflects mainly LTM, but that they would have a normal short-term component which would be reflected in the recency effect with immediate recall.

Method

Twenty lists of ten nouns were selected at random from the Thorndike-Lorge (1944) count. Each word was written in upper case letters on a 3×5-in. index card. The cards were presented manually at a rate of 3 sec per card and S read the words out loud. For ten of the lists the tenth word was followed immediately by a red asterisk, the signal to recall, at which point S attempted to recall verbally as many of the words as possible. When S indicated that he could not remember any more words, the next list was presented. For the other ten lists recall was delayed for 30 sec during which S performed a self-paced counting task to minimize rehearsal. Subjects varied enormously in their counting ability. Since Marcer (1968) has shown that rehearsal is only effectively prevented if S is counting at his optimal rate, it was first necessary to discover for each patient what type of counting task he could manage and at what rate. Two counting tasks were used, counting backwards from a three-digit number, or whenever this proved too difficult, counting backwards from 100. An estimate of counting rate with no memory load was made and every effort was made, first to ensure that S counted at this rate throughout the experiment, and secondly to ensure as far as humanly possible that Ss did not rehearse while counting. Immediate and delayed tests occurred in an ABBA design with all Ss being tested immediately on Trials 1–5 and

16–20, and after a 30-sec delay on Trials 6–15. All Ss began with the immediate condition to ensure that they were not too discouraged by starting with a task that they often found very difficult.

Results

Mean percentages of responses correct for each presentation position are shown for immediate and delayed recall in Figures 1 and 2, respectively. On delayed recall and on the initial part of the immediate recall curve, amnesic Ss clearly recalled less than control Ss, while their performance on the last few items presented was as high as that of the control Ss, provided recall was immediate. This result is consistent with the prediction that amnesic Ss have impaired LTM but normal STM. On the assumption that delayed recall represents only LTM while immediate recall represents both STM and LTM, it is possible to make a quantitative estimate of the two components. Table 2 gives the relevant estimates for the six amnesic and six control patients. Delayed recall is used as an estimate of LTM, the difference between immediate and delayed recall is assumed to reflect the number of items which are

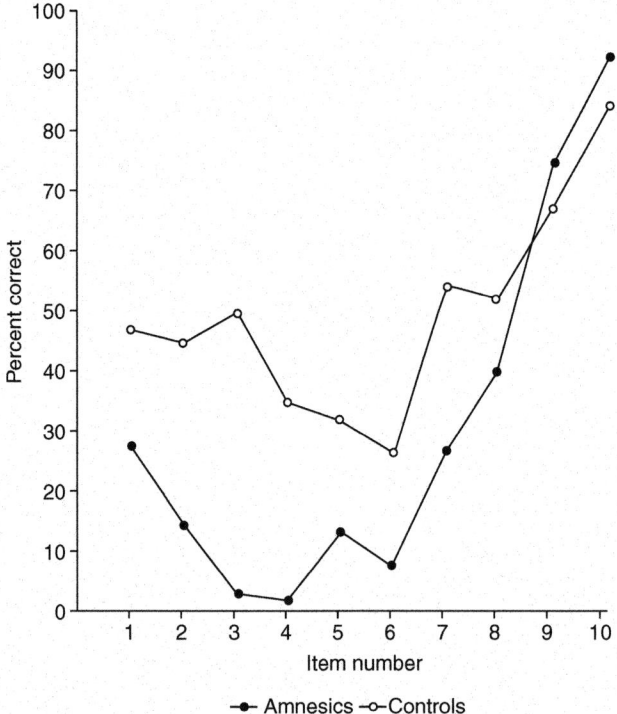

Figure 1 Mean percentage correct recall as a function of order of presentation for amnesic and control Ss with immediate recall.

Table 2 Mean words correct on immediate (I) and delayed (D) test, mean estimated number of words registered in STM but not LTM (I−D), and estimated total content of STM

	I	D	I−D	$\dfrac{N(I-D)}{N-D}$
		Amnesics		
1	3.1	.3	2.8	2.98
2	3.5	1.3	2.2	2.53
3	3.0	.7	2.3	2.47
4	3.0	1.0	2.0	2.22
5	2.9	2.1	.8	1.01
6	2.9	1.3	1.6	1.84
Mean	3.07	1.11	1.95	2.16
		Control Ss		
1	5.4	3.2	2.2	3.23
2	5.4	2.6	2.8	3.78
3	5.2	3.8	1.4	2.26
4	5.9	4.4	1.5	2.68
5	4.1	2.6	1.5	2.03
6	3.7	3.2	.5	.74
Mean	4.95	3.30	1.65	2.45

represented in STM but not in LTM (Baddeley, Scott, Drynan, & Smith, 1969). If we assume (a) that an item may be registered simultaneously in both LTM and STM, and (b) that the probability of being in STM is independent of whether an item is also in LTM, then the total contribution of STM to the free-recall score is given by the expression $N(I-D)/N-D$ where N=number of items in the list, I=immediate recall score, and D=delayed recall score. (For a justification of this estimate see Baddeley, 1970.)

Comparison of amnesic and control Ss using the Mann-Whitney test showed the performance of control Ss to be superior on both immediate and delayed recall, since there is no overlap between the groups in either case ($U=0$, $p<.001$). On the other hand, there is no difference between the groups in the estimates of either number of items in STM but not LTM ($U=12$, $p>.1$) or in the total estimated STM score ($U=14$, $p>.1$). The quantitative estimates therefore support the conclusions based on data in Figures 1 and 2: amnesic Ss have impaired LTM but normal STM.

Table 3 gives the pattern of intrusions. None of the differences between the two groups approaches significance. It is notable that whereas five of the six Ss in each group make acoustic intrusions, only one amnesic and two control Ss make semantic intrusions. One might have expected the amnesic Ss to show fewer prior list intrusions since these are presumably dependent on LTM. Warrington and Weiskrantz (1968) first noted the occurrence of prior list intrusions in amnesic Ss. The present data confirm their findings and suggest that the tendency to make such intrusions is at least as great in amnesic as in control Ss.

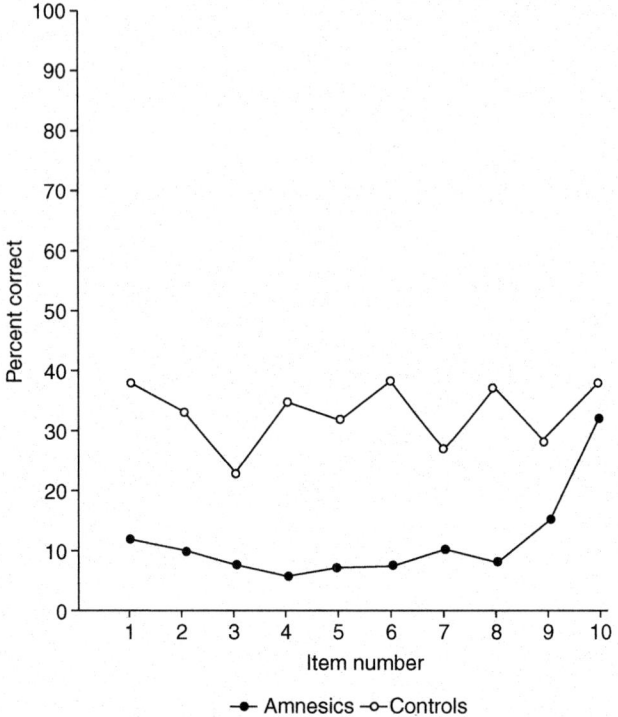

Figure 2 Mean percentage correct recall as a function of order of presentation for amnesic and control Ss with delayed recall.

Experiment II: short-term forgetting

It was shown by Brown (1958) and Peterson and Peterson (1959) that sequences of items well within the memory span show marked forgetting over a matter of seconds if rehearsal is prevented by a task intervening between presentation and test. There is evidence that Ss use acoustic coding (Conrad, 1964), and since this is a classical STM technique, amnesics would be expected to be unimpaired. On the other hand, the interpolated task used to prevent rehearsal is certainly a source of distraction, and on the basis of Milner's observation of the dependence of amnesic memory on continuous rehearsal and the catastrophic effects of distraction (Milner, 1966, 1969), one would predict that amnesics would perform very poorly. Should the prediction based on the distinction between STM and LTM prove successful, with amnesic Ss showing little impairment, there would be considerable theoretical interest in noting the point at which the forgetting curves reach asymptote. One possibility is that the asymptote of forgetting is set by the level of the stable long-term component. If so, amnesic Ss should reach asymptote at a lower level than the control Ss.

Table 3 Distribution of intrusions in free recall

	Prior list		Acoustic confusions	Semantic confusions	Other intrusions	Total
	Previous list	Earlier lists				
Amnesics	34	15	10	2	48	109
Controls	26	20	5	4	12	67

Method

Subjects were presented with sequences of three three-letter words, each triplet printed on a 3×5-in. index card. The S was required to read out the triplet, and after 4 sec the card was covered with a second card containing either the recall sign (a red asterisk), or a three-digit number from which S was required to count backwards, as quickly as possible. On some occasions the counting task had to be modified as described in the section on free recall, but the mean *rate* of counting in the two groups was comparable, being 65.4 items per min for the amnesic group (range 28–78) and 55.6 per min for the control group (range 49–65).

Unpaced recall was tested after delays of 0, 5, 10, 15, 30, and 60 sec. Subjects were tested in blocks of 12 trials, two at each delay in random order, with the constraint that no interval was tested twice until all other intervals had been tested. Ideally each S performed five such blocks making a total of 60 trials. Owing to time limitations the number of blocks completed ranged from three to five, with a median of 4.5 for each group. Performance on this task, however, was sufficiently reliable to give a consistent picture in all cases. Each trial block was preceded by a block of five tests, each with a 15-sec delay used to assess proactive interference effects (see Experiment 3) and was followed by a break of approximately 5 min. Subjects were not given knowledge of results.

Results

Mean percentage of words recalled, regardless of serial position, was calculated for each S at each delay. These means were then averaged for each group. The resultant forgetting curves for the amnesic and control groups are shown in Figure 3.

There is obviously no difference between the two curves at any point. Inasmuch as this task provides an estimate of STM it is clearly consistent with the hypothesis that amnesic patients have a normal STM. It is interesting to note that the curves appear to be reaching a similar asymptote, thus suggesting that the level of asymptote is not determined by LTM as was suggested previously. The rate of forgetting is slower than that commonly

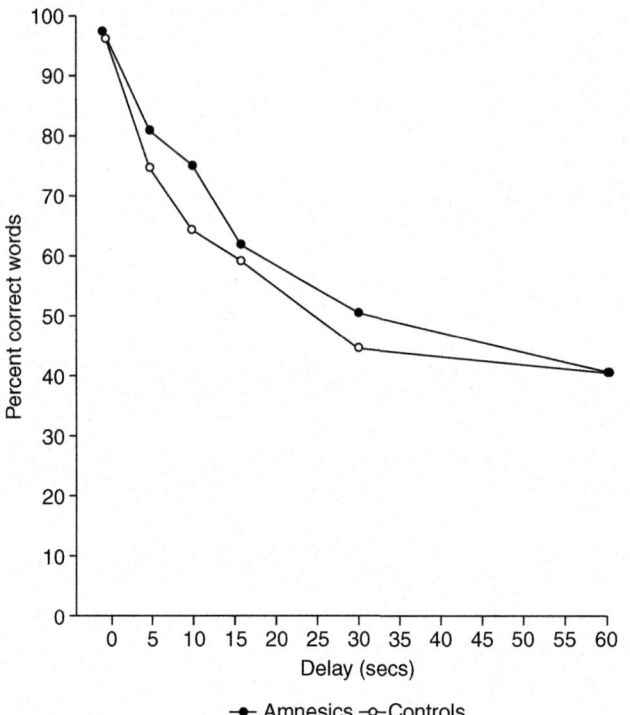

Figure 3 Short-term retention of word triads by amnesic and control Ss.

found: short-term forgetting curves usually approach an asymptote within 20 sec (Peterson & Peterson, 1959). The reason for the slow forgetting is not clear but it may be connected with the relatively long presentation time needed to ensure that Ss of this age and educational level could read out the stimuli.

One striking feature of the data is the considerable differences between Ss on this task. These differences were reliable from day to day and were unrelated to intelligence; in addition we took considerable pains to ensure that Ss were counting at their optimum rate and that differences were not due to rehearsal during counting or to inadequate motivation. The most likely explanation would seem to be in terms of individual differences in STM.

Intrusion distributions for this task are shown in Table 4. Although there is a tendency for amnesic Ss to make more prior list intrusions, this is not significant ($U=10$, $p>.1$). They do, however, make fewer omissions ($U=6$, $p<.05$). Acoustic confusions are made by all Ss, but only three Ss (one amnesic and two control Ss) make any semantic confusions, and even these tend also to be acoustically similar (for instance, *sat* for *sit*, *tip* for *top*, *eat* for *ate*).

Table 4 Distribution of intrusions in the Brown–Peterson short-term forgetting task

	Prior list (separation)[a]				Total	Acoustic	Semantic	Other intrusions	Omissions
	0	1	2	>2					
Amnesics	202	40	12	22	276	46	5	37	104
Controls	125	17	7	8	157	58	2	57	188

Note
[a] Separation refers to number of lists intervening between the previous occurrence of an item and its emissions as an intrusion.

Experiment III: proactive interference

Keppel and Underwood (1962) showed that very little forgetting occurs on the first trial of experiments using the Brown–Peterson technique. Marked forgetting occurs on the second and subsequent trials, being attributed by Keppel and Underwood to proactive interference (PI). Whether or not one accepts the interference theory interpretation, prior items clearly play an important role in producing forgetting and hence the term PI seems appropriate. Loess and Waugh (1967) have shown the PI effect to dissipate if trials are separated by 2 to 3 min, thus suggesting the effect depends on STM. If so, then amnesic Ss and normal Ss should show comparable amounts of PI.

Method

This was as described for Experiment II, with the exception that Ss were given a block of five successive tests, each with a 15-sec delay. There were three, four, or five such blocks, depending on the amount of time for which S was available. Each block was preceded by a rest of approximately 5 min and the instruction that the delay would be constant at 15 sec and was followed by one of the blocks of 12 trials at various delays which comprised Experiment II.

Results

Performance on each of the five successive 15-sec delay items comprising a block is shown in Figure 4. There is a clear drop in performance between the first and subsequent items, an effect shown by all Ss in both groups. However, there is clearly no difference between amnesic and control Ss either in rate of development or amount of PI. This, together with the demonstration by Loess and Waugh (1967) of the transient nature of PI in this situation, suggests that it is based on STM.

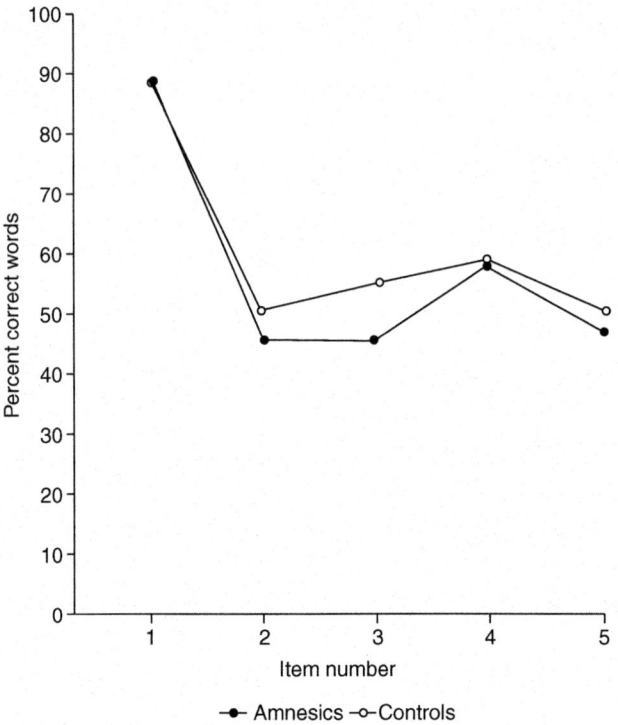

Figure 4 Rate of development of PI for amnesic and control Ss.

Experiment IV: minimal paired-associate learning

With this technique S is presented with a series of pairs of items, one pair at a time, and is then tested by being shown the first word of one of the pairs (S) and attempts to recall the second word (R). Peterson (1966) has shown that this task has two components, a short-term component that influences recall of the last one or two pairs presented and shows rapid forgetting, and a long-term component that is responsible for recall of earlier pairs and shows improvement rather than forgetting if recall is delayed. The task is susceptible to the effects of acoustic similarity (Bruce & Murdock, 1968) but is relatively unaffected by semantic similarity (Baddeley & Dale, 1966). A firm prediction about the performance of amnesic Ss on this task is difficult; Peterson's results suggest a strong LTM component which would indicate impaired performance, but, on the other hand, the evidence for acoustic rather than semantic coding implies that amnesic Ss may perform normally.

Method

In order to maximize the possibility of semantic coding and highlight any differences that might occur, lists were made up with concrete nouns as Ss and high frequency adjectives as Rs, since such pairs have been shown to be particularly easy to learn (Lambert & Paivio, 1956). The adjectives were selected from the list given by Hilgard (1951) and were all disyllabic. Forty lists of four pairs were produced. In order to minimize variations in difficulty, obviously compatible pairs and similar items within the same list were avoided. Each pair of words was printed by hand on a 3×5-in index card. The four pairs were followed by a fifth card containing the S item from one of the pairs. The first, second, third, and fourth pairs were each tested on ten occasions making a total of forty lists. The order in which the four types of list occurred was random. Pairs were presented for 3 sec, during which S read out the pair. Recall was unpaced, and Ss were encouraged to guess if in doubt.

Results

Mean percentage of items recalled for each of the four presentation positions is shown in Figure 5. There was no significant difference in recall score between amnesic and control Ss either overall, or at any one or more positions. There is, however, a suggestion that the two curves are diverging, and since they have clearly not reached an asymptote, it would be premature to claim that the two groups do not differ in minimal paired-associate retention. A further experiment is clearly needed in which the asymptotes of the curves can be estimated by using either longer lists or a filled interval between presentation and recall, or else by testing all four items on each trial.

The distribution of intrusions in this task is shown in Table 5. There were no significant differences between the groups and virtually no semantic intrusions, but unlike in previous tasks, relatively few acoustic confusions occurred.

Experiment V: digit span

In this standard test of immediate memory a random sequence of digits is read out to S who tries to repeat back the sequence immediately afterwards. It is assumed to have a strong short-term component and performance on this test has already been shown to be normal in amnesic patients. It was included to check the comparability of our results with those of Drachman and Arbit (1966).

Table 5 Total intrusions in short-term paired-associate learning

	Within list	*Prior list*	*Acoustic*	*Semantic*	*Others*	*Total*
Amnesics	55	27	8	1	9	100
Controls	56	20	2	1	12	91

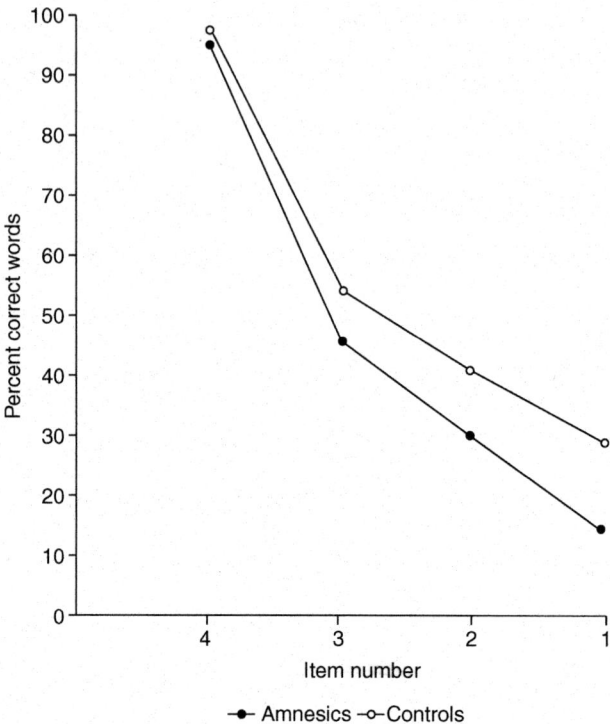

Figure 5 Mean percentage recall of paired-associates as a function of presentation position for amnesic and control Ss.

Method

Digits were read out at a rate of one per second; Ss spoke their unpaced recall. All Ss were tested on six sequences each of length 5, 6, 7, and 8 digits, in that order.

Results

Mean percentage correct strings as a function of length of string is shown in Figure 6. Again there was no significant difference between amnesics and controls. There was, however, a tendency for amnesic Ss to be worse than control Ss on the longest strings; no amnesic S recalled any of the six eight-digit strings perfectly, whereas four of the six control Ss recalled at least one such string. This result is consistent with Drachman and Arbit's (1966) observation that although amnesic patients had unimpaired digit spans their performance broke down more catastrophically when the span was exceeded. This in turn is consistent with the hypothesis that amnesic Ss have normal STM but defective LTM.

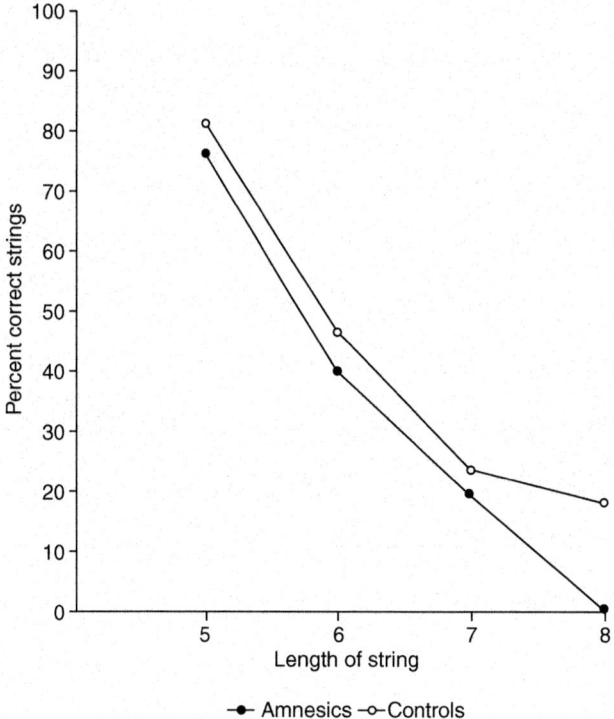

Figure 6 Immediate memory for digits in amnesic and control Ss. Mean percentage of sequences correct as a function of sequence length.

Experiment VI: the Hebb effect

Hebb (1961) has given evidence of a long-term component associated with the traditional digit span technique. He required Ss to recall strings of nine digits, with the modification that unbeknown to the S, the 3rd, 6th, 9th, and 12th strings were identical. Performance on the repeated string improved steadily, indicating a residual long-term component. This presumably reflects LTM, and hence amnesic Ss would not be expected to show improvement on repeated items. In the case of the nonrepeated items, performance would be expected to depend mainly on STM, and amnesic Ss would not be expected to show impairment (Drachman & Arbit, 1966).

Method

This was as in Experiment V, except that Ss were tested on strings of eight digits, of which items 2, 4, 6, 8, 10, 12, 14, 16, 17, 18, 19, and 20 were repetitions of the same string, while items 1, 3, 5, 7, 9, 11, 13, and 15 were all

Results

Performance of the two groups on repeated and nonrepeated strings is shown in Figures 7 and 8. The degree of cumulative learning of the repeated item was assessed in two ways. The first simply involved calculating mean number of digits per string recalled correctly and in the right serial position for repeated sequences and comparing this with the equivalent score for nonrepeated filler sequences. This is shown in the left-hand portion of Table 6. In both amnesic and control groups there was a trend (not significant) for the repeated item to be better recalled than the nonrepeated item. There was, however, no difference between amnesic and normal Ss in the strength of the effect. Indeed, the amnesic Ss tended to show a slightly clearer effect, with five of the six Ss performing better on the repeated item compared with four out of six in the control group. Considering the data for all 12 Ss, the tendency for better recall of the repeated item was significant ($T=7$, $p=.01$).

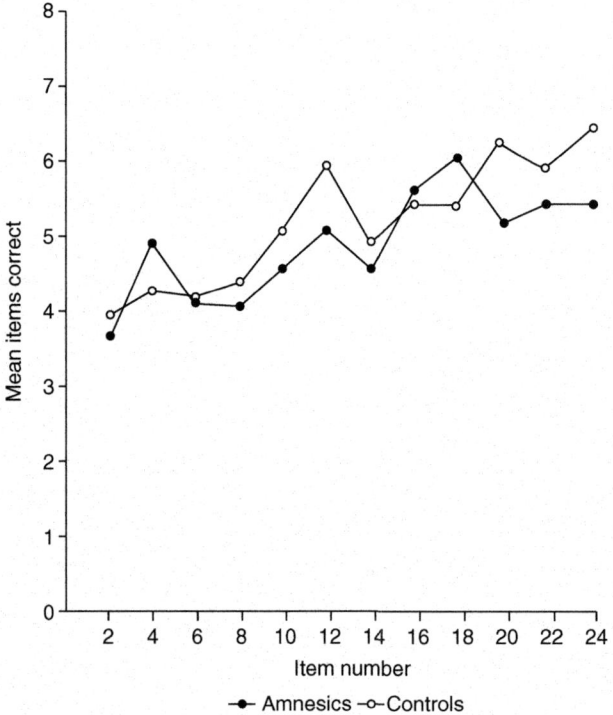

Figure 7 The Hebb effect in amnesic and control Ss. Mean number of digits correct as a function of sequence number for repeated items.

There are, however, certain objections to this simple score. In order to keep conditions as comparable as possible for all Ss, the same repeated item was used throughout. Any tendency for the item to be atypically easy or hard would therefore bias the result either in favor of an effect (if the item was easier than average) or against (if harder). The latter is more likely to be the case, in fact, since care was taken to ensure that the repeated item was not obviously meaningful or easy to code whereas filler items were simply selected at random. A second objection to the simple score used above is that it fails to utilize all the available information since it takes no account of the prediction that performance on repeated strings should show a steady improvement during the experiment. A second score was therefore computed in an attempt to take account of this improvement. Kendall's tau was used to measure the correlation between trial number and number of correct digits. This was calculated separately for repeated and nonrepeated items for each S. Mean correlation scores for the two groups are shown in the right-hand part of Table 6. For all six amnesic and five of the six control Ss improvement on the repeated string was greater than that shown on nonrepeated strings ($T=1$, $N=12$, $p<.01$). There was a slight tendency

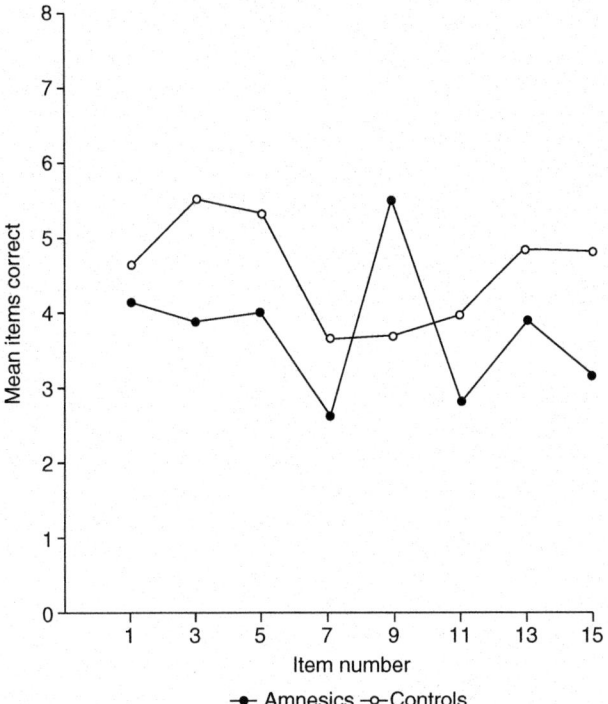

Figure 8 The Hebb effect in amnesic and control Ss. Mean number of digits correct as a function of sequence number for nonrepeated items.

Table 6 Recall of repeated and nonrepeated strings of 8-digits

	Mean digits correct			Mean correlation (τ) between item number and performance		
	Nonrepeated	Repeated	Repeated-Nonrepeated	Nonrepeated	Repeated	Repeated-Nonrepeated
Amnesics	3.75	5.45	1.70	−.191	+.384	+.575
Controls	4.35	5.77	1.42	−.095	+.392	+.487

for performance on nonrepeated sequences to deteriorate, though this was not significant. None of the differences between amnesic and control Ss approached statistical significance.

This result was completely unexpected. Learning of the repeated item is a relatively long-term phenomenon which Melton (1963) has shown can occur with as many as five irrelevant digit strings between repetitions, and hence the effect was assumed to depend on LTM. As Table 6 shows, the amnesic group demonstrated at least as much learning as the control group. From this we must conclude either that amnesic patients are unimpaired on at least one type of LTM, or else that LTM is not responsible for the Hebb effect.

Discussion

We began with a dual hypothesis about amnesic Ss, first that they have normal STM, and secondly that they have impaired LTM. Our results provide strong support for the first part of the hypothesis. Whether STM is assessed in terms of the recency effect in free recall, the rate of short-term forgetting, short-term paired-associate retention, or digit span, the picture is clear: there is no difference in performance between amnesic patients and comparable controls. So far as STM is concerned, amnesic Ss are no more distractable than normal Ss, and no more dependent on rehearsal. It is moreover clearly the case that this group of amnesic patients suffered from defective LTM since their performance was impaired not only on the free-recall task described in the present study but also on a questionnaire involving recall of recent events, and on other experiments requiring long-term learning and recall (Warrington & Weiskrantz, 1968). Our amnesic patients do, therefore, appear to have normal STM and impaired LTM. It thus seems reasonable to use this information to re-examine the standard laboratory techniques used in the present study, using the performance of our two groups of Ss to assess the role of LTM and STM in each task.

Free-recall performance presents no difficulties since the amnesic patients were normal on the labile short-term component but impaired on the more stable long-term component. The Brown-Peterson task, however, does present a problem since forgetting in both groups appears to reach the same asymptote of about 40% correct. After an interval of 60 sec, the labile STM effects should have dissipated, leaving only the contents of LTM. If this residue does indeed represent LTM, however, why does our amnesic group show unimpaired performance? The only formal difference between this task and the delayed free-recall task is in the length of list presented, but why amnesic patients should perform normally with subspan lists, while being grossly impaired with supraspan lists, is far from clear. The PI experiment presents no such problems and is consistent with other evidence (Loess & Waugh, 1967) indicating that the PI effect is itself short-term.

For paired-associate recall Peterson's (1966) evidence for two components would predict impaired performance for amnesic Ss at longer delays, at which point the forgetting curves should have reached asymptote. While we find no significant impairment, the forgetting curves in our study had not reached asymptote, so that any clear conclusion must await further experimentation. Our digit span data were as expected. Since span is assumed to depend almost entirely on STM no difference was predicted, and none was found. On the other hand, the ability of our amnesic group to show learning with the Hebb technique was completely unexpected. Does this imply that the learning shown is based on STM, not on LTM? If so, it raises the problem of whether a memory system showing a gradual increment across a series of trials can reasonably be called short-term. However, since the Hebb effect was not very marked in either group of Ss, further evidence is clearly required.

In summary, then, our results are consistent with predictions, with two exceptions. Short-term forgetting curves for amnesic and normal Ss reach asymptote at the same point, and amnesic Ss appear to show a Hebb effect. How can we explain these discrepancies? One possibility is suggested by the evidence of Mandler (1966) and Tulving (1968) that efficient long-term learning depends on S's ability to organize the material to be learned. The STM tasks we have used have given very little scope for subjective organization since they have all involved unrelated words or digits presented at a rate which was probably too fast to allow very much organization. On the other hand, normal everyday remembering and the LTM tasks used in other studies probably do allow a good deal of subjective organization. On this argument, then, amnesic Ss are normal at tasks involving simple rote memory but are incapable of organizing material when given the chance.

There are, however, two pieces of evidence against this view. Weingartner (1968) showed both clustering and enhanced free-recall learning for associatively structured lists in amnesic Ss. He did not, however, find associative effects in serial learning, although, since he had no control group, it is difficult to interpret this result. Secondly, Zangwill (1946) has shown that immediate memory for sentences is unaffected in amnesic Korsakoff patients. This implies that they are able to make use of the internal constraints found in meaningful prose: if this were not the case they would have shown the much shorter span characteristic of random word sequences. Amnesic Ss therefore do not appear to have a general ability to organize material.

A second hypothesis is that amnesic patients have normal STM and LTM for material coded acoustically. Despite the strong association between acoustic coding and STM there clearly must also be acoustic coding in LTM, otherwise a child could never learn to talk. The occurrence of acoustic confusions in LTM provides further support for this view (Brown & McNeill, 1966). Such an acoustic LTM might be responsible for both the point of asymptote of the Peterson forgetting curve and the Hebb effect. On this argument amnesic patients should show normal long-term learning if all Ss are forced to rely on acoustic coding. This does not seem very likely in view of the gross impairment in the learning of digit sequences shown in amnesic Ss by Drachman and Arbit (1966). While such a task probably depends mainly on acoustic coding, however, the hypothesis cannot be rejected completely without further experimentation.

In conclusion, our results show that amnesic patients combine intact STM with grossly defective LTM. The nature of the LTM impairment, whether it is one of input, storage, or retrieval of information, is beyond the scope of this study.

Notes

1 We thank Dr. R. T. C. Pratt for providing facilities to carry out this work, for permission to study his patients, and for his advice and encouragement. We are grateful to the Medical Research Council for financial support.
2 At the Laboratory of Experimental Psychology, University of Sussex, Brighton BN1 9QY, England.

References

Baddeley, A. D. Short-term memory for word sequences as a function of acoustic, semantic and formal similarity. *Quarterly Journal of Experimental Psychology*, 1966, **18**, 362–365(a).

Baddeley, A. D. The influence of acoustic and semantic similarity on long-term memory for word sequences. *Quarterly Journal of Experimental Psychology*, 1966, **18**, 302–309 (b).

Baddeley, A. D. Estimating the short-term component in free recall. *British Journal of Psychology*, 1970 (In Press).

Baddeley, A. D., & Dale, H. C. A. The effect of semantic similarity on retroactive interference in long- and short-term memory. *Journal of Verbal Learning and Verbal Behavior*, 1966, **5**, 217–220.

Baddeley, A. D., Scott, D., Drynan, R., & Smith, J. C. Short-term memory and the limited capacity hypothesis. *British Journal of Psychology*, 1969, **60**, 51–55.

Brown, J. Some tests of the decay theory of immediate memory. *Quarterly Journal of Experimental Psychology*, 1958, **10**, 12–21.

Brown, R., & McNeill, D. The "tip of the tongue" phenomenon. *Journal of Verbal Learning and Verbal Behavior*, 1966, **5**, 325–337.

Bruce, D., & Murdock, B. B., Jr. Acoustic similarity effects in memory for paired-associates. *Journal of Verbal Learning and Verbal Behavior*, 1968, **7**, 627–631.

Conrad, R. Acoustic confusion and immediate memory. *British Journal of Psychology*, 1964, **55**, 75–84.
Craik, F. I. M. Types of error in free recall. *Psychonomic Science*, 1968, **10**, 353–354.
Dimsdale, H., Logue, V., & Piercy, M. A case of persisting impairment of recent memory following right temporal lobectomy. *Neuropsychologia*, 1964, **1**, 287–298.
Drachman, D. A., & Arbit, J. Memory and the hippocampal complex II. Is memory a multiple process? *Archives of Neurology*, 1966, **15**, 52–61.
Glanzer, M., & Cunitz, A. R. Two storage mechanisms in free recall. *Journal of Verbal Learning and Verbal Behavior*, 1966, **5**, 351–360.
Hebb, D. O. Distinctive features of learning in the higher animal. In J. F. Delafresnaye (Ed.), *Brain mechanisms and learning*. London and New York: Oxford University Press, 1961. Pp. 37–46.
Hilgard, E. R. Methods and procedures in the study of learning. In S. S. Stevens (Ed.), *Handbook of experimental psychology*. New York: Wiley, 1951.
Keppel, G., & Underwood, B. J. Proactive inhibition in short-term retention of individual items. *Journal of Verbal Learning and Verbal Behavior*, 1962, **1**, 153–161.
Kolers, P. A. Interlingual facilitation of short-term memory. *Journal of Verbal Learning and Verbal Behavior*, 1966, **5**, 314–319.
Lambert, W. E., & Paivio, A. The influence of noun-adjective order on learning. *Canadian Journal of Psychology*, 1956, **10**, 9–12.
Levy, B. A., & Murdock, B. B., Jr. The effects of delayed auditory feedback and intra-list similarity in short-term memory. *Journal of Verbal Learning and Verbal Behavior*, 1968, **7**, 887–894.
Loess, H. Short-term memory, word class, and sequence of items. *Journal of Experimental Psychology*, 1967, **74**, 556–561.
Loess, H., & Waugh, N. C. Short-term memory and intertrial interval. *Journal of Verbal Learning and Verbal Behavior*, 1967, **6**, 455–460.
Mandler, G. Organization and memory. In K. W. Spence and J. T. Spence (Eds.), *The psychology of learning and motivation*. New York: Academic Press, 1966.
Marcer, D. Subtraction as interpolated activity in short-term retention. *Psychonomic Science*, 1968, **11**, 359.
Melton, A. W. Implication of short-term memory for a general theory of memory. *Journal of Verbal Learning and Verbal Behavior*, 1963, **2**, 1–21.
Milner, B. Amnesia following operation on the temporal lobes. In C. W. M. Whitty and O. L. Zangwill (Eds.), *Amnesia*. London: Butterworths, 1966.
Milner, B. Disorders of memory after brain lesions in man. Preface: material-specific and generalized memory loss. *Neuropsychologia*, 1968, **6**, 175–179.
Peterson, L. R. Short-term verbal memory and learning. *Psychological Review*, 1966, **73**, 193–207.
Peterson, L. R., & Peterson, M. J. Short-term retention of individual verbal items. *Journal of Experimental Psychology*, 1959, **58**, 193–198.
Talland, G. Deranged memory. New York, Academic Press, 1965.
Thorndike, E. L., & Lorge, I. *The teacher's word book of 30,000 words*. New York: Teachers College Press, 1944.
Tulving, E. Theoretical issues in free recall. In T. R Dixon and D. L. Horton (Eds.), *Verbal behavior and general behavior theory*. Englewood Cliffs N.J.: Prentice Hall, 1968.
Victor, M., & Adams, R. D. The effect of alcohol on the nervous system. In H. Houston Merritt and C. C. Hare (Eds.), *Metabolic and toxic diseases of the nervous system*. Baltimore: Williams and Wilkins, 1953.

Warrington, E. K., & Weiskrantz, L. A study of learning and retention in amnesic patients. *Neuropsychologia*, 1968, **6**, 283–291.

Waugh, N. C., & Norman, D. A. Primary memory. *Psychological Review*, 1965, **72**, 89–104.

Weingartner, H. Verbal learning in patients with temporal lobe lesions. *Journal of Verbal Learning and Verbal Behavior*, 1968, **7**, 520–526.

Weiskrantz, L., & Warrington, E. K. New method of testing long-term retention with special reference to amnesic patients. *Nature*, 1968, **217**, 972–974.

Wickelgren, W. A. Short-term memory for phonemically similar lists. *American Journal of Psychology*, 1965, **78**, 567–574.

Wickelgren, W. A. Sparing of short-term memory in an amnesic patient: implications for strength theory of memory. *Neuropsychologia*, 1968, **6**, 235–244.

Zangwill, O. L. Some qualitative observations on verbal memory in cases of cerebral lesion. *British Journal of Psychology*, 1946, **37**, 8–19.

Part II
A multicomponent model

The 1960s saw a huge increase in research on short-term memory, with a range of different experimental paradigms being invented and their results explained through evermore complex models, often expressed mathematically. The most influential of these was that produced by Atkinson and Shiffrin (1968) which became known as the *modal model* since it summed up the characteristics of many similar although less ambitious approaches. It assumed three storage systems that operated sequentially. Information from the environment flowed first into a bank of sensory memory systems that perhaps can best be regarded as part of the process of perception. These in turn fed into a limited capacity short-term store which held information while it was gradually fed into a long-term storage system. Importantly, the short-term store was assumed to act as a working memory, performing a large number of complex activities such as strategy selection and recoding. This aspect was not, however, explored other than through the simulation of a standard long-term learning task in which verbal material was gradually transferred to the long-term store over successive trials. The model did, however, integrate much of the research carried out in the 1960s and for many years was prominent in most cognitive psychology text books.

There were, however, problems with the model. The first comprised its assumption that simply holding information in the short-term store would guarantee long-term transfer. This proved not to be the case since simply holding on to the relevant information led to little learning. This was highlighted by Craik and Lockhart's (1972) demonstration of the importance for learning of the operations performed on the material to be learned. Superficial visual processing lead to poor recall, acoustic recoding improved performance slightly whereas complex semantic processing led to the most effective learning. These and many similar results were explained via their highly influential Levels of Processing hypothesis.

A second problem for the modal model came from research on patients with a very specific deficit in short-term verbal memory (Shallice & Warrington, 1970). According to the model, impaired short-term memory should result in great difficulty in long-term learning. Furthermore if the short-term system serves as a general working memory such patients should have widespread cognitive problems. In fact, the patients studied had excellent long-term memory and an apparently complete lack of overall cognitive problems.

At this point, I had moved to the new University of Sussex and was urged to apply for a research grant. I chose to investigate the link between short-term and long-term memory, and was fortunate enough to be able to include Graham Hitch as a postdoctoral fellow. Graham had completed a degree in physics in Cambridge before moving to Sussex to complete a Master's degree in experimental psychology which was in turn followed by a PhD under Broadbent back in Cambridge.

The start of our grant coincided with a period when many people were abandoning the study of short-term memory in favour of research on levels of processing or on models of semantic memory. How were we to make progress? We decided to focus on the simple question of what function was served by short-term memory; was it indeed a general working memory or simply a system for the brief storage of verbal information? An obvious way ahead might have been to work with patients with specific short-term memory deficits, but these were very rare and not available to us so instead we opted to simulate such patients using a dual task method. This involved systematically loading the verbal subsystem by requiring memory for sequences of digits. We assumed that the longer the sequence, the more of the storage system it would occupy, the less would be available for other tasks and the greater would be the disruption of a range of other cognitive tasks. The next chapter describes this series of experiments which were, however, almost entirely focused on verbal memory. Visual memory was included in the model, although it did not play a central part in our original paper. A brief but more complete account of the resulting multicomponent model is therefore described in the chapter that follows.

The project seemed to be going well until I received an invitation to contribute a chapter to a prestigious annual series entitled *Recent Advances in Learning and Motivation*. We hesitated since it was clear that the model was far from complete but eventually decided that this was too good an opportunity to miss. It seems to have been the right decision since our paper has since been cited more than 12,000 times across a wide range of disciplines.

The third chapter in this section is somewhat more recent and summarises our views on the recency effect which we, like many others, including Atkinson and Shiffrin, initially regarded as reflecting the contents of short-term memory which were lost over a brief delay as a result of trace decay or interference. This view was challenged by our 1974 studies in which lists of words were presented for free recall while participants were required to remember and continuously repeat sequences of up to six digits. This led to impaired performance on the earlier items in the list, confirming other evidence that occupying the working memory system interfered with long-term learning. However, digit load had no effect on recency, quite inconsistent with the assumption that digit span and recency were utilising the same limited capacity system. At about the same time others were beginning to note the occurrence of recency effects in long-term memory and in certain variants of standard short-term memory tasks, again challenging the classic interpretation. We interpret recency as based on the application of a last-in-first-out strategy that can take advantage of priming in any of a range of different memory systems. This is now a relatively widespread view that has been elaborated and extended by others, notably Brown, Neath, and Chater (2007). Our own views are summarised in the final chapter in this section.

References

Atkinson, R. C., & Shiffrin, R. M. (1968). Human memory: A proposed system and its control processes. In K. W. Spence & J. T. Spence (Eds), *The Psychology of Learning and Motivation: Advances in Research and Theory* (Vol. 2, pp. 89–195). New York: Academic Press.

Brown, G. D. A., Neath, I., & Chater, N. (2007). A temporal ratio model of memory. *Psychological Review, 114,* 539–576.

Craik, F. I. M., & Lockhart, R. S. (1972). Levels of processing: A framework for memory research. *Journal of Verbal Learning & Verbal Behavior, 11,* 671–684.

Shallice, T., & Warrington, E. K. (1970). Independent functioning of verbal memory stores: a neuropsychological study. *Quarterly Journal of Experimental Psychology, 22,* 261–273.

Papers

1 Baddeley, A. D., & Hitch, G. (1974). Working memory. In G. A. Bower (Ed.), *Recent Advances in Learning and Motivation,* Vol. 8. New York: Academic Press, pp. 47–90.

2 Baddeley, A. D. & Hitch, G. J. (1993). The recency effect: Implicit learning with explicit retrieval? *Memory and Cognition, 21,* 146–155.

3 Baddeley, A. D. (1981). The concept of working memory: A view of its current state and probable future development. *Cognition, 10,* 17–23.

4 Working memory

A. D. Baddeley and G. Hitch

I Introduction

Despite more than a decade of intensive research on the topic of short-term memory (STM), we still know virtually nothing about its role in normal human information processing. That is not, of course, to say that the issue has completely been neglected. The short-term store (STS)—the hypothetical memory system which is assumed to be responsible for performance in tasks involving short-term memory paradigms (Atkinson & Shiffrin, 1968)—has been assigned a crucial role in the performance of a wide range of tasks including problem solving (Hunter, 1964), language comprehension (Rumelhart, Lindsay, & Norman, 1972) and most notably, long-term learning (Atkinson & Shiffrin, 1968; Waugh & Norman, 1965). Perhaps the most cogent case for the central importance of STS in general information processing is that of Atkinson and Shiffrin (1971) who attribute to STS the role of a controlling executive system responsible for coordinating and monitoring the many and complex subroutines that are responsible for both acquiring new material and retrieving old. However, despite the frequency with which STS has been assigned this role as an operational or working memory, the empirical evidence for such a view is remarkably sparse.

A number of studies have shown that the process of learning and recall does make demands on the subject's general processing capacity, as reflected by his performance on some simultaneous subsidiary task, such as card sorting (Murdock, 1965), tracking performance (Martin, 1970), or reaction time (Johnston, Griffith & Wagstaff, 1972). However, attempts to show that the limitation stems from the characteristics of the working memory system have proved less successful. Coltheart (1972) attempted to study the role of STS in concept formation by means of the acoustic similarity effect, the tendency for STM to be disrupted when the material to be remembered comprises items that are phonemically similar to each other (Baddeley; 1966b; Conrad, 1962). She contrasted the effect of acoustic similarity on concept formation with that of semantic similarity, which typically effects LTM rather than STM (Baddeley, 1966a). Unfortunately for the working memory hypothesis, her results showed clear evidence of semantic rather than acoustic coding, suggesting that the long-term store (LTS) rather than STS was playing a major role in her concept formation task.

Patterson (1971) tested the hypothesis that STS plays the important role in retrieval of holding the retrieval plan, which is then used to access the material to be recalled (Rumelhart et al., 1972). She attempted to disrupt such retrieval plans by requiring her experimental group to count backwards for 20 seconds following each item recalled. On the basis of the results of Peterson and Peterson (1959), it was assumed that this would effectively erase information from STS after each recall. Despite this rather drastic interference with the normal functioning of STS however, there was no reliable decrement in the number of words recalled.

The most devastating evidence against the hypothesis that STS serves as a crucially important working memory comes from the neuropsychological work of Shallice and Warrington (Shallice & Warrington, 1970; Warrington & Shallice, 1969; Warrington & Weiskrantz, 1972). They have extensively studied a patient who by all normal standards, has a grossly defective STS. He has a digit span of only two items, and shows grossly impaired performance on the Peterson short-term forgetting task. If STS does indeed function as a central working memory, then one would expect this patient to exhibit grossly defective learning, memory, and comprehension. No such evidence of general impairment is found either in this case or in subsequent cases of a similar type (Warrington, Logue, & Pratt, 1971).

It appears then, that STS constitutes a system for which great claims have been made by many workers (including the present authors), for which there is little good evidence.

The experiments which follow attempt to answer two basic questions: first, is there any evidence that the tasks of reasoning, comprehension, and learning share a common working memory system?; and secondly, if such a system exists, how is it related to our current conception of STM? We do not claim to be presenting a novel view of STM in this chapter. Rather, our aim is to present a body of new experimental evidence which provides a firm basis for the working memory hypothesis. The account which follows should therefore be regarded essentially as a progress report on an on-going project. The reader will notice obvious gaps where further experiments clearly need to be performed, and it is more than probable that such experiments will modify to a greater or lesser degree our current tentative theoretical position. We hope, however, that the reader will agree that we do have enough information to draw some reasonably firm conclusions, and will feel that a report of work in progress is not too out of place in a volume of this kind.

II The search for a common working memory system

The section which follows describes a series of experiments on the role of memory in reasoning, language comprehension, and learning. An attempt is made to apply comparable techniques in all three cases in the hope that this will allow a common pattern to emerge, if the same working memory system is operative in all three instances.

In attempting to assess the role of memory in any task, one is faced with a fundamental problem. What is meant by STS? Despite, or perhaps because of, the vast amount of research on the characteristics of STS there is still little general agreement. If our subsequent work were to depend on a generally acceptable definition of STS as a prerequisite for further research, such research would never begin. We suspect that this absence of unanimity stems from the fact that evidence for STS comes from two basically dissimilar paradigms. The first is based on the traditional memory span task. It suggests that STS is limited in capacity, is concerned with the retention of order information, and is closely associated with the processing of speech. The second cluster of evidence derives from the recency effect in free recall. It also suggests that STS is limited in capacity; however, its other dominant feature is its apparent resistance to the effects of other variables, whether semantic or speech-based (Glanzer, 1972). Rather than try to resolve these apparent discrepancies, we decided to begin by studying the one characteristic that both approaches to STS agreed on, namely its limited capacity. The technique adopted was to require S to retain one or more items while performing the task of reasoning, language comprehension, or learning. Such a concurrent memory load might reasonably be expected to absorb some of the storage capacity of a limited capacity working memory system, should such a system exist. The first set of experiments describes the application of this technique to the study of a reasoning task. To anticipate our results, we find a consistent pattern of additional memory load effects on all three tasks that we have studied: reasoning, language comprehension, and free recall. Additionally, all three tasks show evidence of phonemic coding. From this evidence we infer that each of the tasks involves a spanlike component, which we refer to as working memory. Further evidence from the free recall paradigm shows that the recency effect is neither disrupted by an additional memory span task nor particularly associated with phonemic coding. We therefore suggest a dichotomy between working memory and the recency effect, in contrast to the more usual view that both recency and the memory span reflect a single limited capacity short-term buffer store (STS).

A *The role of working memory in reasoning*

The reasoning task selected was that devised by Baddeley (1968) in which *S* is presented with a sentence purporting to describe the order of occurrence of two letters. The sentence is followed by the letters in question, and *S*'s task is to decide as quickly as possible whether the sentence correctly describes the order in which the letters are presented. For example, he may be given the sentence *A is not preceded by B-AB*, in which case he should respond *True*. A range of different sentences can be produced varying as to whether they are active or passive, positive or negative, and whether the word *precedes* or *follows* is used. This task is typical of a wide range of sentence verification tasks studied in recent years (Wason & Johnson-Laird, 1972). Its claim to be a reasoning task of some general validity is supported by the correlation between performance and intelligence (Baddeley, 1968) and its sensitivity to both environmental and speed-load stress (Baddeley, De Figuredo, Hawkswell-Curtis, & Williams, 1968; Brown, Tickner, & Simmonds, 1969). The first experiment requires S to perform this simple reasoning task while holding zero, one, or two items in memory. If the task relies on a limited capacity system, then one might expect the additional load to impair performance.

1 Experiment I: effects of a one- or two-item preload

Subjects were required to process 32 sentences based on all possible combinations of sentence voice (active or passive), affirmation (affirmative or negative), truth value (true or false), verb type (precedes or follows), and letter order (AB or BA). The experiment used a version of the memory preload technique in which S is given one or two items to remember. He is then required to process the sentence and having responded "True" or "False," he is then required to recall the letters. A slide projector was used to present the sentences, each of which remained visible until S pressed the "True" or "False" response key. Twenty-four undergraduate Ss were tested. The order in which the three conditions were presented were determined by a Latin square. For half the Ss the preload was presented visually, while the other half was given an auditory preload. In the zero load condition, S was always presented with a single letter before the presentation of the sentence. However, the letter was the same on all trials, and S was not required to recall it subsequently. With the one- and two-letter loads, the letters differed from trial to trial but were never the same as those used in the reasoning problem. All Ss were informed of this.

The results are shown in Table 1. There was no reliable effect of memory load on solution time regardless of whether the load was one or two letters, and was presented visually or auditorily ($F < 1$ in each case). Since letter recall was almost always perfect, it appears to be the case that Ss can hold up to two additional items with no impairment in their reasoning speed. This result

Table 1 Mean time (sec) to complete verbal reasoning problems as a function of size of additional memory load and method of reading memory items

Method of reading	Memory load		
	Zero	1-letter	2-letters
Silent	3.07	3.35	3.21
Aloud	3.33	3.26	3.41
Means	3.20	3.31	3.31

suggests one of two conclusions; either that the type of memory system involved in retaining the letters is not relevant to the reasoning task, or else that a load of two items is not sufficient to overtax the system. Experiment II attempts to decide between these two hypotheses by increasing the preload from two to six letters, a load which approaches the memory span for many Ss.

2 Experiment II: effects of a six-digit preload

Performance on the 32 sentences was studied with and without a six-letter memory preload. In the preload condition each trial began with a verbal "ready" signal followed by a random sequence of six letters spoken at a rate of one per second. The reasoning problem followed immediately afterwards, details of presentation and method of responding being the same as in Experiment I. After solving the problem, S attempted to recall verbally as many letters as possible in the correct order. In the control condition, the reasoning problem followed immediately after the "ready" signal. After completing the problem, and before being presented with the next problem, S listened to a six-letter sequence and recalled it immediately. This procedure varies the storage load during reasoning, but roughly equates the two conditions for total memorization required during the session.

Separate blocks of 32 trials were used for presenting the two conditions, each block containing the 32 sentences in random order. Half the Ss began with the control condition and half with the preload condition. Two groups of 12 undergraduate Ss were tested. The two groups differed in the instructions they were given. The first group (equal stress) was told to carry out the reasoning task as rapidly as possible, consistent with high accuracy, and to attempt to recall all six letters correctly. The second group (memory stress) was told that only if their recall was completely correct could their reasoning time be scored; subject to this proviso, they were told to reason as rapidly as they could, consistent with high accuracy. All Ss were given a preliminary three-minute practice session on a sheet of reasoning problems, and were tested individually.

Mean reasoning times (for correct solutions) and recall scores for both groups of subjects are shown in Table 2. For the "equal stress" Ss memory

Table 2 Mean reasoning times and recall scores for the "equal stress" and "memory stress" instructional groups

Instructional emphasis	Mean reasoning time (sec)		Mean no. items recalled (max = 6)	
	Control	Memory preload	Control	Memory preload
"Equal stress"	3.27	3.46	5.5	3.7
"Memory stress"	2.73	4.73	5.8	5.0

load produced a slight but nonsignificant slowing down in reasoning time (on a Wilcoxon test, $T=31$, $N=12$, $P>.05$), while for the "memory stress" Ss memory load slowed down reasoning considerably ($T=4$, $N=12$, $P<.01$). There appears to have been a trade-off between reasoning and recall in the memory load condition. The equal stress Ss achieved their unimpaired reasoning at the expense of very poor recall compared with that of the memory stress Ss.

The results show then, that there is an interaction between additional short-term storage load and reasoning performance. In comparison with the results of Experiment I these suggest that the interaction depends on the storage load since, up to two items can be recalled accurately with no detectable effect. Thus the reasoning task does not seem to require all the available short-term storage space. The results show additionally that the form of the interaction depends on the instructional emphasis given to S. It seems likely therefore that interference was the result of the active strategy that Ss employed. One possibility is that the "memory stress" Ss dealt with the memory preload by quickly rehearsing the items, to "consolidate" them in memory before starting the reasoning problem. If this were the case, then reasoning times ought to be slowed by a constant amount (the time spent rehearsing the letters), regardless of problem complexity. Figure 1 shows mean reasoning time for the memory stress group for different types of

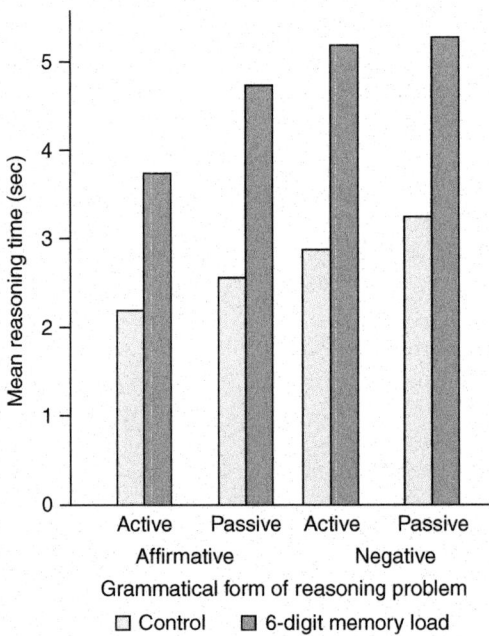

Figure 1 Mean reasoning time for different forms of the problem for the "memory stress" group of subjects.

sentence. Control reaction times (*RT*s) show that problems expressed as passives were more difficult than those expressed as actives, and that negative forms were more difficult than affirmatives. However, the slowing down in reasoning produced by the memory preload was roughly constant regardless of problem difficulty. Analysis of variance showed significant effects of memory load [$F(1,10) = 8.51$, $P<.025$], sentence voice [$F(1,10 = 7.34$, $P<.025$], and negation [$F(1,10 = 34.9$, $P<.001$]. None of the interaction terms involving the load factor approached significance.

The results of this experiment do not adequately demonstrate that the verbal reasoning task involves a short-term storage component. Subjects seem to have adopted a strategy of time-sharing between rehearsal of the memory letters and reasoning. While the time-sharing may have been forced by competition between the tasks for a limited storage capacity, this is not necessarily the case. The tasks may, for example, have competed for use of the articulatory system, without having overlapping storage demands.

Experiment III attempts to prevent the strategy of completely switching attention from the memory task to the reasoning test by changing from a preload to a concurrent load procedure. In the concurrent load procedure, *S* is required to continue to rehearse the memory load items aloud while completing the reasoning task. Since the process of articulation has itself been shown to impair performance in both memory (Levy, 1971; Murray, 1967, 1968) and reasoning (Hammerton, 1969; Peterson, 1969), two additional conditions were included to allow a separation of the effects of memory load and of articulation.

3 Experiment III: effects of a concurrent memory load

All *S*s performed the 32 reasoning problems under each of four conditions, the order in which the conditions were tested being determined by a Latin square. In the control condition, a trial began with a verbal warning signal and the instruction "say nothing." The problem was then presented and solved as quickly and accurately as possible. The second condition used the articulatory suppression procedure devised by Murray (1967). Subjects were instructed to say the word "the" repeatedly, at a rate of between four and five utterances per second. After *S* had begun to articulate, the problem was presented, whereupon he continued the articulation task at the same high rate until he had pressed the "True" or "False" response button. The third condition followed a procedure adopted by Peterson (1969) in which the articulation task consisted of the cyclic repetition of a familiar sequence of responses, namely the counting sequence "one-two-three-four-five-six." Again, a rate of four to five words per second was required. In the fourth condition, *S* was given a random six-digit sequence to repeat cyclically at a four- to five-digit per second rate. In this condition alone, the message to be articulated was changed from trial to trial. The three articulation conditions therefore range from the simple repetition of a single utterance, through the rather more complex articulation involved in counting, up to the digit span repetition task, which presumably makes considerably greater short-term storage demands. Degree of prior practice and method of presentation were as in Experiment II.

Table 3 shows the performance of the 12 undergraduate Ss tested in this study. Concurrent articulation of "the" and counting up to six produced a slight slowing of reasoning time, but by far the greatest slowing occurred with concurrent articulation of random digit sequences. Analysis of variance showed a significant main effect of conditions [$F(3,33) = 14.2$, $P<.01$]. Newman-Keuls tests showed that the effect was mainly due to the difference between the random digit condition and the other three. The slight slowing down in the suppression-only and counting conditions just failed to reach significance.

These results suggest that interference with verbal reasoning is not entirely to be explained in terms of competition for the articulatory system, which may be committed to the rapid production of a well-learned sequence of responses with relatively little impairment of reasoning. A much more important factor appears to be the short-term memory load, with the availability of spare short-term storage capacity determining the rate at which the reasoning processes are carried out. Since difficult problems presumably make greater demands on these processes, one might expect that more difficult problems would show a greater effect of concurrent storage load. Figure 2 shows the mean reasoning times for problems of various kinds. As is typically the case with such tasks (Wason & Johnson-Laird, 1972), passive sentences proved more difficult than active sentences [$F(1,11) = 55.2$, $P<.01$], and negatives were more difficult than affirmatives [$F(1,11) = 38.5$, $P<.01$]. In addition to the main effect of concurrent activity, activity interacted with sentence voice [$F(3,33) = 5.59$, $P<.01$] and with negativity [$F(3,33) = 5.29$, $P<.01$]. Figure 2 shows that these interactions were due largely to performance in the random digit condition. Additional storage load seems to have slowed down solution times to passives more than actives and to negatives more than affirmatives. Thus the greater the problem difficulty, the greater the effect of an additional short-term storage load.

In summary, it has been shown that additional STM loads of more than two items can impair the rate at which reasoning is carried out. Loads of six

Table 3 Mean reasoning times and error rates as a function of concurrent articulatory activity

Concurrent articulation	Mean reasoning RT (sec)	Percent reasoning errors
Control	2.79	8.1
"The-The-The ..."	3.13	10.6
"One-Two-Three ..."	3.22	5.6
Random 6-digit No.	4.27	10.3

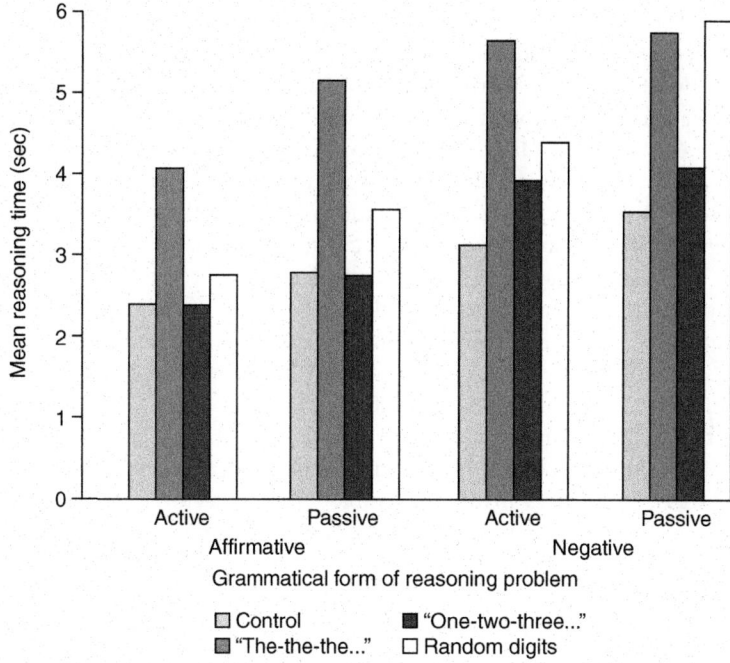

Figure 2 Effects of concurrent articulatory activity on mean reasoning time for different types of problem.

items can produce sizable interference, but the effect may depend on the instructional emphasis given to *Ss* (Experiment II). The interference effects may be partly due to the articulatory activity associated with rehearsal of the memory items, but there is a substantial amount of interference over and above this which is presumably due to storage load (Experiment III). The trade-off between reasoning speed and additional storage load suggests that the interference occurs within a limited capacity "work-space," which can be flexibly allocated either to storage or to processing.

The effect of articulatory suppression in Experiment III was small and did not reach statistical significance. However, Hammerton (1969) has reported evidence that suppression can produce reliable interference in this task. His *Ss* repeated the familiar sentence "Mary had a little lamb" while carrying out the Baddeley reasoning task. Performance was impaired when contrasted with a control group who said nothing when reasoning. This result together with those of Peterson (1969) suggests that reasoning may resemble the memory span task in having an articulatory component. Experiment IV explores the relation between the memory span and working memory further by taking a major feature of the verbal memory span, namely its susceptibility to the effects of phonemic similarity, and testing for similar effects in the verbal reasoning task.

4 Experiment IV: phonemic similarity and verbal reasoning

One of the more striking features of the memory span for verbal materials is its apparent reliance on phonemic (either acoustic or articulatory) coding. This is revealed both by the nature of intrusion errors (Conrad, 1962; Sperling, 1963) and by the impairment in performance shown when sequences of phonemically similar items are recalled (Baddeley, 1966b; Conrad & Hull, 1964). As Wickelgren (1965) has shown, phonemic similarity has its disruptive effect principally on the retention of order information, and since the reasoning task employed depends on the order of the letters concerned, it seems reasonable to suppose that the manipulation of phonemic similarity might prove a suitable way of disrupting any STS component of the task. Experiment IV, therefore, studied the effect of phonemic similarity on the reasoning task and compared this with the effect of visual similarity, a factor which is typically found to have little or no influence on memory span for letters.

A group testing procedure was used in which Ss were given test sheets containing 64 reasoning problems printed in random order and were allowed three minutes to complete as many as possible. A 2×2 factorial design was used with phonemic and visual similarity as factors. There were two replications of the experiment, each using different letter pairs in each of the four conditions. The sets of letter pairs used were as follows: MC, VS (low phonemic similarity, low visual similarity); FS, TD (high phonemic, low visual similarity); OQ, XY (low phonemic, high visual similarity); and BP, MN (high phonemic, high visual similarity). Thirty-two undergraduate Ss were tested, half with one letter set and half with the other. All Ss were first given a preliminary practice session using the letter-pair AB. Each S then completed a three-minute session on each of the four types of problems. Problems were printed on sheets, and Ss responded in writing. The order of presenting the four conditions was determined using a Latin square.

Table 4 shows the mean number of correctly answered questions in the various conditions. Since there were no important differences between results from the two replications, data from the two sets of letter pairs have been pooled. Only the effect of phonemic similarity was significant ($N=32$, $Z=2.91$, $P<.002$), while visual similarity appeared to have no effect ($N=32$, $Z<1$). It appears then that the verbal reasoning task does require the utilization of phonemically coded information, and, although the effect is small, it is highly consistent across Ss.

Table 4 Mean number of reasoning problems correctly solved in three minutes as a function of phonemic and visual similarity of the letters used in the problems

		Phonemic similarity of letters	
Visual similarity of letters		Low	High
	Low	43.2	40.9
	High	42.9	39.8

In summary then, verbal reasoning shows effects of concurrent storage load, of articulatory suppression, and of phonemic similarity. This pattern of results is just what would be expected if the task depended on the use of a short-term store having the characteristics typically shown in the memory span paradigm. However, the magnitude of the effects suggest that the system responsible for the memory span is only part of working memory. We shall return to this point after considering the evidence for the role of working memory in prose comprehension and learning.

B Comprehension and working memory

While it has frequently been asserted that STS plays a crucial role in the comprehension of spoken language (e.g., Baddeley & Patterson, 1971; Norman, 1972), the evidence for such a claim is sparse. There is, of course, abundant evidence that language material may be held in STM (Jarvella, 1971; Sachs, 1967) but we know of no evidence to suggest that such storage is an essential function of comprehension under normal circumstances, and in view of the lack of any obvious defect in comprehension shown by patients with grossly defective STS (Shallice & Warrington, 1970), the importance of STS in comprehension remains to be demonstrated. Experiments V and VI attempt to do so using the memory preload and the concurrent memory load techniques.

1 Experiment V: effects of a memory preload on comprehension

In this experiment, S listened to spoken prose passages under each of two memory load conditions and was subsequently tested for retention of the passages. In the experimental condition, each sentence of the passage was preceded by a sequence of six digits spoken at a rate of one item per second. After listening to the sentence, S attempted to write down the digit sequence in the correct order in time to a metronome beating at a one-second rate. Hence S was required to retain the digit sequence while listening to the sentence. In the control condition the digit sequence followed the sentence and was recalled immediately afterward. Thus both conditions involved the same amount of overt activity, but only in the experimental condition was there a temporal overlap between the retention of the digits and sentence presentation. In both conditions the importance of recalling the digits accurately was emphasized. After each passage, S was allowed three minutes to complete a recall test based on the Cloze technique (Taylor, 1953). Test sheets comprised a typed script of each of the passages, from which every fifth word had been deleted. The passages contained approximately 170 words each, and hence there were about 33 blanks which S was instructed to try to fill with the deleted word. This technique has been shown by Rubenstein and Aborn (1958) to be a reasonably sensitive measure of prose retention. Three different types of passage were included in the experiment: descriptions, narratives, and arguments. Two examples of each type were constructed giving six passages in all, each of which contained ten sentences. Each of 30 Ss was tested on all six passages, comprising one experimental and one control condition for each of the three passage types. Subjects were tested in two separate groups, each receiving a different ordering of the six passages.

Table 5 shows the mean number of correctly completed blanks for the control and experimental conditions together with the mean number of digit sequences correctly reported in the two conditions. The digit preload impaired performance on the comprehension test for all three types of passage. Differences were significant for the descriptions ($Z = 2.81$, $P < .01$, Wilcoxon test) and the narratives ($Z = 2.91$, $P < .01$), but not for the arguments ($Z = 1.14$, $P > .05$). Thus, test performance is impaired when digits have to be held in store during presentation of the passage. Digit recall scores were also poorest in the experimental condition, but this was, of course, to be expected in view of the long filled retention interval in this condition.

While the results can be interpreted as showing that comprehension is impaired by an additional short-term storage load, this conclusion is not unchallengeable. Firstly, the Cloze procedure is probably a test of prompted verbatim recall and may not measure comprehension. Secondly, the control condition of the experiment may not have been entirely satisfactory. If the time *between* sentences is important for comprehension of the meaning of the passage as a whole, the control group itself may have suffered from an appreciable amount of interference. The next experiment goes some way to overcoming both these objections, using the concurrent memory load procedure instead of the preload technique.

2 Experiment VI: *effects of a concurrent memory load on comprehension*

This experiment compared the effects of three levels of concurrent storage load on prose comprehension. In all three conditions, the memory items were presented visually at a rate of one per second using a TV monitor. The

Table 5 Comprehension and digit recall scores with and without additional memory load for three types of passage

Type of passage	Memory load condition	Comprehension score[a]	Digit recall score[b]
Description	No load	16.8	11.8
	Load	13.3	7.7
Narrative	No load	20.1	11.4
	Load	18.0	7.8
Argument	No load	14.5	11.4
	Load	13.6	8.1

Notes
a Mean no. of blanks correctly filled in—max = 33
b Mean no. of digit strings correctly reported—max = 14

concurrent memory load tasks were as follows. In the three-digit load condition, S was always presented with sequences of three digits, each sequence being followed by a 2-second blank interval during which S attempted to recall and write down the three digits he had just seen. In the six-digit condition, the sequences all comprised six items and were followed by a 4-second blank interval. Again S was instructed not to recall the digits until the sequence had been removed. Time intervals were chosen so as to keep S busy with the digit memory task, and were also such that all conditions would require input and output of the same total number of digits. In the control condition, S was presented with sequences of three and six digits in alternation. After each three-digit list there was a 2-second blank interval, and after each six-digit list the blank interval was 4 seconds. In this case, however, S was simply required to copy down the digits while they were being presented. It was hoped that this task would require the minimal memory load consistent with the demand of keeping the amount of digit writing constant across conditions. The main difference between the three conditions was, therefore, the number of digits which S was required to store simultaneously. Instructions emphasized the importance of accuracy on all three digit tasks, and an invigilator checked that Ss were obeying the instructions.

Comprehension was tested using six passages taken from the Neale Analysis of Reading Ability (Neale, 1958). Two passages (those suited for 12- and 13-year-old children) were selected from each of the three parallel test forms. Each passages comprised approximately 120 words and was tested by eight standardized questions. These have the advantage of testing comprehension of the passage without using the specific words used in the original presentation. They can, therefore, be regarded as testing retention of the gist of the passage rather than verbatim recall. Answers were given a score of one if correct, half if judged almost correct, and zero otherwise. At the start of each trial, the experimenter announced which version of the digit task was to be presented before testing began. After a few seconds of the digit processing task, the experimenter began to read out the prose passage at a normal reading rate and with normal intonation. At the end of the passage, the digit task was abandoned and the experimenter read out the comprehension questions. A total of 15 undergraduates were tested in three equal-sized groups, in a design which allowed each passage to be tested once under each of the three memory load conditions.

The mean comprehension scores for the three conditions are shown in Table 6. The Friedman test showed significant overall effects of memory load ($\chi_r^2 = 7.3$, $P < .05$). Wilcoxon tests showed that the six-digit memory load produced lower comprehension scores than either the control condition ($T = 19$, $N = 14$, $P < .05$), or the three-digit condition ($T = 19.5$, $N = 15$, $P < .05$). There was no reliable difference between the three-digit load and control conditions ($T = 44.5$, $N = 14$, $P > .05$). Thus, comprehension is not reliably affected by a three-item memory load, but is depressed by a six-item load, a pattern of results which is very similar to that observed with the verbal reasoning task.

Table 6 Mean comprehension scores as a function of size of concurrent memory load

	Memory load		
	Control (1-digit)	3-digit	6-digit
Mean comprehension score (max=8)	5.9	5.6	4.8

While Experiments IV and V present *prima facie* evidence for the role of working memory in comprehension, it could be argued that we have tested retention rather than comprehension. From what little we know of the process of comprehension, it seems likely that understanding and remembering are very closely related. It is, however, clearly desirable that this work should be extended and an attempt made to separate the factors of comprehension and retention before any final conclusions are drawn.

If comprehension makes use of STM, it should be possible to impair performance on comprehension tasks by introducing phonemic similarity into the test material. To test this hypothesis using the prose comprehension task of the previous experiment would have involved the difficult task of producing passages of phonemically similar words. We chose instead to study the comprehension of single sentences, since the generation of sentences containing a high proportion of phonemically similar words seemed likely to prove less demanding than that of producing a whole passage of such material.

3 Experiment VII: phonemic similarity and sentence comprehension

The task used in this experiment required S to judge whether a single sentence was impossible or possible. Possible sentences were both grammatical and meaningful, while impossible sentences were both ungrammatical and relatively meaningless. Impossible sentences were derived from their possible counterparts by reversing the order of two adjacent words near the middle of the possible sentence. Two sets of possible sentences were constructed, one comprising phonemically dissimilar words and the other one containing a high proportion of phonemically similar words. An example of each type of possible sentence together with its derived impossible sentence is shown in Table 7. In order to equate the materials as closely as possible, each phonemically similar

Table 7 Examples of the sentences used in Experiment VIII

	Possible version	Impossible version
Phonemically dissimilar	Dark skinned Ian thought Harry ate in bed	Dark skinned Ian Harry thought ate in bed
Phonemically similar	Red headed Ned said Ted fed in bed	Red headed Ned Ted said fed in bed

sentence was matched with a phonemically dissimilar sentence for number of words, grammatical form, and general semantic content. There were nine examples of each of the four conditions (phonemically similar possible; phonemically similar impossible; phonemically dissimilar possible, and phonemically dissimilar impossible), giving 36 sentences in all.

Each sentence was typed on a white index card and was exposed to S by the opening of a shutter approximately half a second after a verbal warning signal. The sentence remained visible until S had responded by pressing one of two response keys. Instructions stressed both speed and accuracy. Twenty students served as Ss and were given ten practice sentences before proceeding to the 36 test sentences which were presented in random order.

Since reading speed was a potentially important source of variance, 13 of the 20 Ss were asked to read the sentences aloud at the end of the experiment and their reading times were recorded. The 36 sentences were grouped into four sets of nine, each set corresponding to one of the four experimental conditions and were typed onto four separate sheets of paper. The order of presenting the sheets was randomized across Ss, and the time to read each was measured by a stopwatch.

Table 8 shows mean reaction times for each of the four types of sentence, together with reading rate for each condition. It is clear that phonemic similarity increased the judgment times for both possible and impossible sentences $[F(1,9) = 8.77, P < .01]$, there being no interaction between the effects of similarity and grammaticality $[F(1,9) < 1]$. An interaction between the effects of phonemic similarity and sentence type $[F(8,152) = 4,38, P < .001]$ suggests that the effect does not characterize all the sentences presented. Inspection of the three sentence sets out of nine which show no similarity effect suggests that this is probably because the dissimilar sentences in these sets contained either longer or less frequent words than their phonemically similar counterparts. Clearly, future experiments should control word length and frequency.

Reading times did not vary appreciably with phonemic similarity $[F(1,12) < 1]$. It is, therefore, clear that phonemic similarity interfered with the additional processing over and above that involved in reading, required to make the possible/impossible judgment. As Table 8 suggests, although impossible

Table 8 Results of Experiment VIII

Sentence type	Mean RT for judgment of "possibility" (sec)			Mean reading time (sec)		
	Possible version	Impossible version	Average	Possible version	Impossible version	Average
Phonemically dissimilar	2.84	2.62	2.73	2.93	3.18	3.06
Phonemically similar	3.03	2.83	2.93	2.96	3.19	3.08

sentences took longer to read than possible sentences [$F(1,12) = 41.6$, $P < .001$], they were judged more rapidly [$F(1,19) = 17.3$, $P < .001$]. This contrast suggests that when judging impossible sentences, S was able to make his judgment as soon as an unlikely word was encountered and did not have to read the entire sentence.

To summarize the results of this section: first, comprehension of verbal material is apparently impaired by a concurrent memory load of six items but is relatively unimpaired by a load of three or less. Second, it appears that verbal comprehension is susceptible to disruption by phonemic similarity. It should be noted, however, that use of the term comprehension has necessarily been somewhat loose; it has been used to refer to the retention of the meaning of prose passages on the one hand and to the detection of syntactic or semantic "impossibility" on the other. Even with single-sentence material. Ss can process the information in a number of different ways depending on the task demands (Green, 1973). It should, therefore, be clear that the use of the single term "comprehension" is not meant to imply a single underlying process. Nevertheless, it does seem reasonable to use the term "comprehension" to refer to the class of activities concerned with the understanding of sentence material. Tasks studied under this heading do at least appear to be linked by the common factor of making use of a short-term or working memory system. As in the case of the verbal reasoning studies this system appears to be somewhat disrupted by the demands of a near-span additional memory load and by the presence of phonemic similarity.

It might reasonably be argued that the reasoning task we studied is essentially a measure of sentence comprehension and that we have, therefore, explored the role of working memory in only one class of activity. The next section, therefore, moves away from sentence material and studies the retention and free recall of lists of unrelated words. The free recall technique has the additional advantage of allowing us to study the effects that the variables which appear to have influenced the operation of working memory in the previous experiments have on the recency effect, a phenomenon which has in the past been regarded as giving a particularly clear indication of the operation of STS.

C Working memory and free recall

1 Experiment VIII: memory preload and free recall

This experiment studied the free recall of lists of 16 unrelated words under conditions of a zero-, three-, or six-digit preload. The preload was presented before the list of words and had to be retained throughout input and recall, since S was only told at the end of the recall period whether to write the preload digit sequence on the right- or left-hand side of his response sheet. The experiment had two major aims. The first aim was to study the effect of a preload on the LTM component of the free recall task, hence giving some indication of the possible role of working memory in long-term learning.

The second aim was to study the effect of a preload on the recency effect. Since most current views of STS regard the digit span and the recency effect as both making demands on a common short-term store, one might expect a dramatic reduction of recency when a preload is imposed. However, as was pointed out in the introduction, there does appear to be a good deal of difference between the characteristics of STS revealed by the digit span procedure (suggesting that it is a serially ordered speech-based store) and the characteristics suggested by the recency effect in free recall (which appears to be neither serially ordered nor speech-based).

All lists comprised 16 high-frequency words equated for word length and presented auditorily at a rate of two seconds per word. Subjects were given a preload of zero, three, or six digits and were required to recall the words either immediately or after a delay of 30 seconds during which subjects copied down letters spoken at a one-second rate. In both cases, they had one minute in which to write down as many words as they could remember, after which they were instructed to write down the preload digits at the left-or right-hand side of their response sheets. Instructions emphasized the importance of retaining the preload digits.

The design varied memory load as a within S factor, and delay of recall between Ss. The same set of 15 lists were presented to both immediate and delayed recall groups. Within each group there were three subgroups across which the assignment of particular lists to particular levels of preload was balanced. For each group the 15 trials of the experiment were divided into blocks of three, in which each load condition occurred once. Subject to this constraint, the ordering of conditions was random. Twenty-one undergraduates served in each of the two subgroups. Figure 3 shows the serial position curves for recall as a function of size of preload for the immediate and delayed recall conditions. Analysis of variance showed significant effects due to delay $[F(1,40)=9.85, P<.01]$, serial position $[F(15,600)=49.4, P<.001]$, and the delay $\times$ serial position interaction $[F(15,600)=33.4, P<.001]$. These correspond to the standard finding that delaying free recall abolishes the recency effect. There were also significant effects due to memory load $[F(2,80)=35.8, P<.001]$, to the memory load $\times$ serial position interaction, $[F(30,1200)=1.46, P<.05]$, and to the load $\times$ serial position $\times$ delay interaction $[F(30,1200)=1.80, P<.01]$.

The overall percentage of words recalled declined with increased preload (see Table 9). Comparison between means using the Newman-Keuls procedure

Table 9 Percentages of words recalled in the various conditions of Experiment X

	Memory load		
	Zero	3-digits	6-digits
Immediate recall	43.9	41.6	35.2
Delayed recall	35.5	32.3	24.5

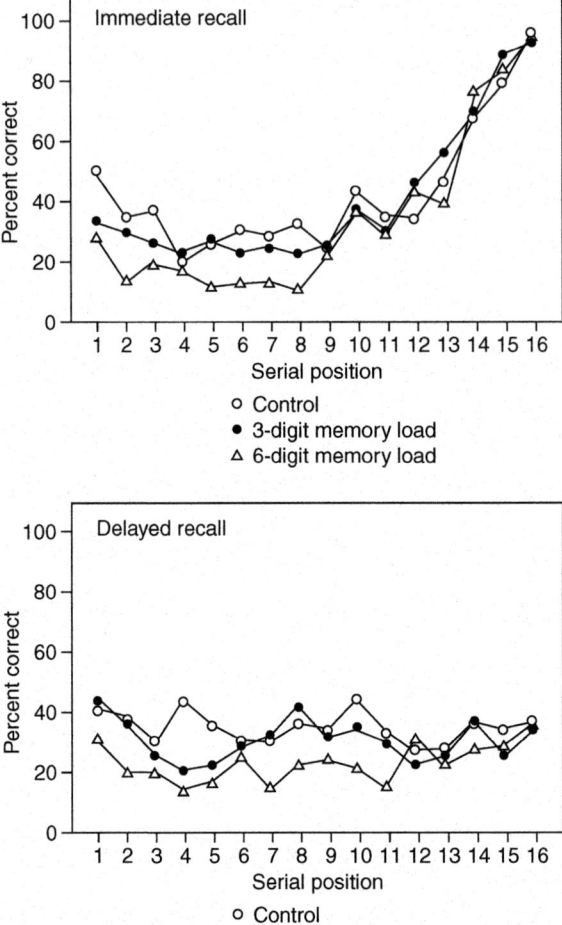

Figure 3 Effects of additional short-term memory load on immediate and delayed free recall.

showed that the impairment due to a three-digit preload was just significant ($P < .05$) while the six-digit preload condition was significantly worse than both the control and the three-digit preload conditions at well beyond the .01 significance level.

It is clear from Figure 3 that the load effect was restricted to the long-term component of recall and did not substantially influence the recency effect.

The first conclusion from this study is that performance on the secondary memory component of free recall is adversely effected by a digit preload, with the size of the decrement being a function of the size of the preload. A somewhat more dramatic finding is the apparent absence of a preload effect on the recency component. There are, however, at least two classes of interpretations of this result. The first is to conclude that an STM preload does not interfere with the mechanism of the recency effect. This would be a striking conclusion, since the "standard" account of recency assumes that the last few items are retrieved from the same store that would be used to hold the preload items. To accept this hypothesis would require a radical change of view concerning the nature of the recency effect. An alternative hypothesis is to assume that S begins to rehearse the preload items at the beginning of the list, and by the end of the list has succeeded in transferring them into LTS, freeing his STS for other tasks. Two lines of evidence support this suggestion: firstly, there was only a marginal effect of preload on recall of the last few items when recall was delayed (see Figure 3). This suggests that the preload effect diminished as the list progressed. Secondly, when questioned after the experiment, 37 out of 39 Ss stated that they carried out some rehearsal of the digits, and 26 of these said that they rehearsed the digits mostly at the beginning of the word list. Clearly, our failure to control Ss, rehearsal strategies prevents our drawing any firm conclusions about the influence of preload on the recency effect. The next experiment, therefore, attempts to replicate the present results under better controlled conditions.

Before passing on to the next experiment, however, it is perhaps worth noting that the delayed recall technique for separating the long- and short-term components in free recall is the only one of the range of current techniques which would have revealed this potential artifact. Techniques which base their estimates of the two components entirely on immediate recall data assume that the LTS component for later items in the list can be estimated from performance on the middle items. In our situation, and possibly in many others, this assumption is clearly not valid.

2 Experiment IX: concurrent memory load and free recall

This experiment again studied the effects of three levels of memory load on immediate and delayed free recall. In general, procedures were identical with Experiment VIII, except that the concurrent load rather than the preload technique was used. This involved the continuous presentation and test of digit sequences throughout the presentation of the memory list. In this way, it was hoped to keep the memory load relatively constant throughout the list and so avoid the difficulties of interpretation encountered in the previous study.

The concurrent load procedure was similar to that described for Experiment VI and involved the visual presentation of digit sequences. In the six-digit concurrent load condition, sequences of six digits were visible for four seconds, followed by a four-second blank interval during which S was required to recall and write down the six digits. The three-digit concurrent load condition was similar except that the three-digit sequences were presented for only two seconds and followed by a two-second blank interval, while in the control condition, S saw alternate sequences of three and six digits, followed, respectively, by gaps of two and four seconds. In this condition, however, he was instructed to copy down the digits as they appeared. The three conditions were thus equal in amount of writing required, but differed in the number of digits that had to be held in memory simultaneously.

The procedure involved switching on the digit display and requiring S to process digits for a few seconds before starting the auditory presentation of the word list. The point at which the word list began was varied randomly from trial to trial. This minimized the chance that a particular component of the digit task (e.g., input or recall) would be always associated with particular serial positions in the word list. After the last word of each list, the visual display was switched off and Ss immediately abandoned the digit task. In the immediate recall condition, they were allowed one minute for written recall of the words, while in the delayed condition they copied a list of 30 letters read out at a one-second rate before beginning the one-minute recall period.

The design exactly paralleled that used in the previous experiment, with 17 undergraduates being tested in the immediate recall condition and 17 in the delayed condition. High accuracy on the digit task was emphasized; each of the three-digit processing procedures was practiced before beginning the experiment, and behavior was closely monitored during the experiment to ensure that instructions were obeyed.

The immediate and delayed recall serial position curves are shown in Figure 4. Because of the scatter in the raw data, scores for adjacent serial positions have been pooled, except for the last four serial positions. Analysis of variance indicated a significant overall effect of memory load [$F(2,64) = 45.2$, $P < .01$], with mean percentage correct scores being 31.8, 31.2, and 24.8 for the zero-, three- and six-item load conditions, respectively. The Newman-Keuls test indicated a significant difference between the six-digit load condition and both other conditions ($P < .01$), which did not differ significantly between themselves.

As Figure 4 suggests, there were highly significant effects of serial position [$F(15,480) = 70.7$, $P < .01$], of delay [$F(1,32) = 26.6$, $P < .01$], and of their interaction [$F(15,480) = 29.1$, $P < .001$], indicating the standard effect of delay on the recency component. The analysis showed no evidence of a two-way interaction between memory load and serial position ($F < 1$) and very weak evidence for a three-way interaction among memory load, serial position, and delay [$F(30,960) = 1.32$, $P > .10$]. The general conclusion, therefore, is that an additional concurrent memory load, even of six items, does not significantly alter the standard recency effect.

This conclusion confirms the result of the previous experiment, but rules out one of the possible interpretations of the earlier data. With the preload technique, the absence of an effect of load on recency might have been due to a progressive decline in the "effort" or "difficulty" associated with the digit task

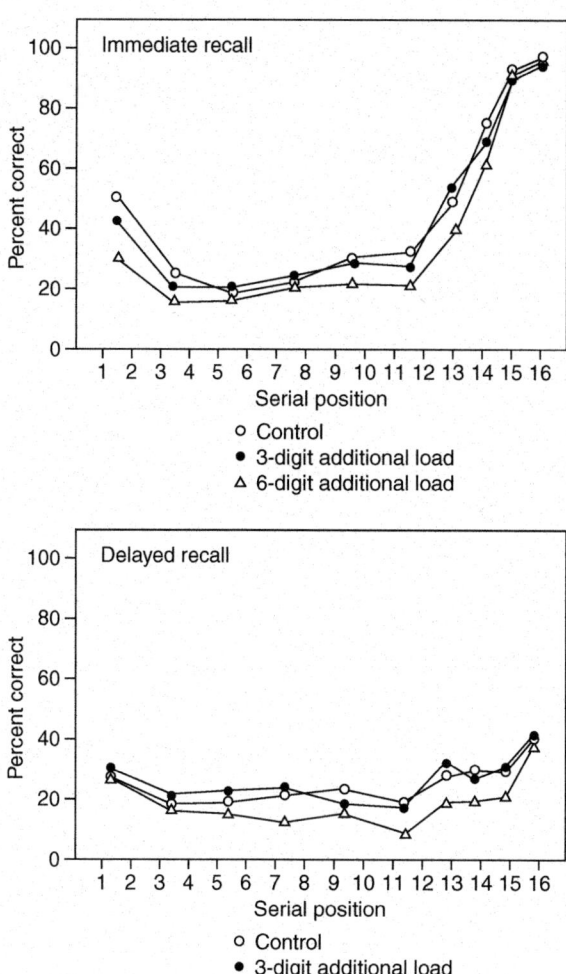

Figure 4 Effects of concurrent short-term storage load on immediate and delayed free recall.

during input of the word list. Such an explanation is not appropriate for the present results since the concurrent load procedure ensured that the digit memory task was carried out right through input of the word lists, a conclusion which is supported by the continued separation of the three- and six-digit load conditions over the last few serial positions in the delayed recall condition (see Figure 3). Thus, even though six digits are concurrently being stored during the input of the final words of the list, the recency effect is unimpaired. To account for this, it must be assumed that the recency mechanism is independent from that involved in the memory span task. According to most dual-store theories, the digit span task ought to keep STS virtually fully occupied. Since the recency effect is commonly supposed to depend on output from this store, the digit span task should seriously reduce the amount of recency observed. It seems, therefore that the buffer-storage account of recency is faced with a major difficulty.

Our data suggest then that a concurrent load of six items does impair the long-term component of free recall. Furthermore, as in the case of our reasoning task and prose comprehension studies, a load of three items has only a marginal effect. These results are consistent with the hypothesis of a working memory, which has some features in common with the memory span task. Since the memory load was present only during input of the words and not during recall, it is reasonable to conclude that working memory is concerned with the processes of transferring information to LTM. The absence of an effect of concurrent storage load on the recency effect suggests that working memory may have little or nothing to do with the recency effect. This hypothesis is discussed more fully in the concluding section of the chapter, when extra evidence against a buffer-storage account of recency is presented and an alternative interpretation suggested.

3 Experiment X: speech coding and free recall

In the case of both verbal reasoning and comprehension, we observed a similar effect of preload to that shown in the last two experiments, together with clear evidence of phonemic coding. This was revealed by effects of both acoustic similarity and articulatory suppression in the reasoning task, and by acoustic similarity effects in comprehension. There already exists evidence that phonemic similarity may be utilized in free recall (Baddeley & Warrington, 1973; Bruce & Crowley, 1970), provided at least that the phonemically similar items are grouped during presentation. The effects observed were positive, but since acoustic similarity is known to impair recall of order while enhancing item recall (Wickelgren, 1965), this would be expected in a free recall task. It is perhaps worth noting in connection with the dichotomy between span-based indicators of STS and evidence based on the recency effect suggested by the results of the last two experiments that attempts to show that the recency effect is particularly susceptible to the effects of phonemic similarity have proved uniformly unsuccessful (Craik & Levy, 1970; Glanzer, Koppenaal, & Nelson, 1972). Although there is abundant evidence that Ss may utilize phonemic similarity in long-term learning, this does not present particularly strong evidence in favor of a phonemically based working memory, since Ss are clearly able to utilize a very wide range of characteristics of the material to be learnt, possibly using processes which lie completely outside the working memory system. The next experiment, therefore, attempts to examine the role of articulatory coding in long-term learning more directly using the articulatory suppression technique. It comprises one of a series of unpublished studies by Richardson and Baddeley and examines the effect of concurrently articulating an irrelevant utterance on free recall for visually and auditorily presented word sequences.

Lists of ten unrelated high-frequency words were presented at a rate of two seconds per word either visually, by memory drum, or auditorily, which involved the experimenter reading out the words from the memory drum, which was screened from S. A total of 40 lists were used, and during half of these S was required to remain silent during presentation, while for the other half he was instructed to whisper "hiya" [an utterance which Levy (1971) found to produce effective suppression] at a rate of two utterances per second throughout the presentation of the word list. Half the Ss articulated for the first 20 lists and were silent for the last 20, while the other half performed in the reverse order. Manipulation of modality was carried out according to an *ABBA* design, with half the Ss receiving visual as the first and last conditions, and half receiving auditory first and last. Each block of ten lists was preceded by a practice list in the appropriate modality and with the same vocalization and recall conditions. Following each list, S was instructed to recall immediately unless the experimenter read out a three-digit number, in which case he was to count backwards from that number by three's. Half the lists in each block of ten were tested immediately and half after the 20-second delay; in each case S was allowed 40 seconds for recall. Sixteen undergraduates served as Ss. The major results of interest are shown in Figure 5, from which it is clear that articulatory suppression impaired retention $[F(1,1185) = 19.6, P < .001]$. The effect is shown particularly clearly with visual presentation and appears to be at least as marked for the earlier serial positions which are generally regarded as dependent on LTS, as for the recency component. This result is consistent with the suggestion of a working memory operating on phonemically coded information and transferring it to LTS. It further supports Glanzer's (1972) conclusion that the recency effect in free recall does not reflect articulatory coding and lends further weight to the suggestion that working memory is probably not responsible for the recency effect.

III A proposed working memory system

We have now studied the effect of factors which might be supposed to influence a working memory system, should it exist, across a range of cognitive tasks. The present section attempts to summarize the results obtained and looks for the type of common pattern which might suggest the same system was involved across the range of tasks.

Table 10 summarizes our results so far. We have studied three types of task: the verbal reasoning test, language comprehension, and the free recall of unrelated words. As Table 10 shows, these have in all three cases shown a substantial impairment in performance when an additional memory load of six items was imposed. In contrast to this, a load of three items appears to have little or no decremental effect, an unexpected finding which is common to all three situations. In the case of phonemic similarity, we have found the type of effect that would be expected on the assumption of a working memory system having characteristics in common with the digit span. Such

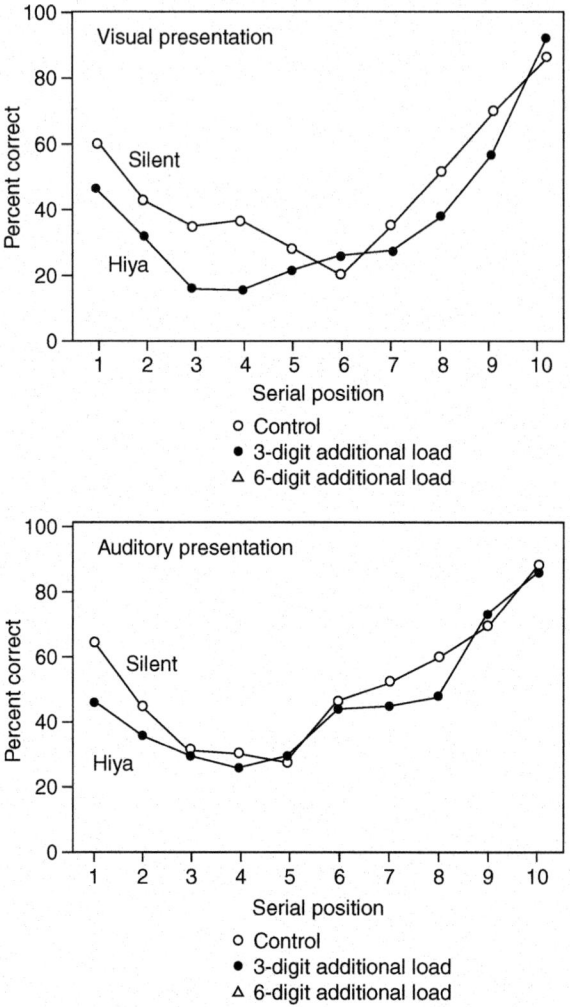

Figure 5 Effect of concurrent articulation on free recall of visually and aurally presented word lists.

Source: Data from Richardson and Baddeley, unpublished.

effects are reflected in a performance decrement in those tasks where the retention of order is important (the reasoning and sentence judging tasks), coupled with a positive effect in the free recall situation for which the recall order is not required. Finally we have found that articulatory suppression, a technique which is known to impair digit span (Baddeley & Thomson, unpublished), has a deleterious effect in the two situations in which we have so far studied it, namely reasoning and free recall learning.

Table 10 Summary of experimental results (paradigm)

		Verbal reasoning	Comprehension	Free recall	
				LTS	Recency
Memory load	1–3 items	No effect	No effect	Small decrement	No effect
	6 items	Decrement	Decrement	Decrement	No effect
Phonemic similarity		Decrement	Decrement	Enhancement	No effect
Articulatory suppression		Decrement	Not studied	Decrement	No effect

There appears then to be a consistent pattern of effects across the three types of task studied, strongly suggesting the operation of a common system such as the working memory initially proposed. This system appears to have something in common with the mechanism responsible for the digit span, being susceptible to disruption by a concurrent digit span task, and like the digit span showing signs of being based at least in part upon phonemic coding. It should be noted, however, that the degree of disruption observed, even with a near-span concurrent memory load, was far from massive. This suggests that although the digit span and working memory overlap, there appears to be a considerable component of working memory which is not taken up by the digit span task. The relatively small effects of phonemic coding and articulatory suppression reinforce this view and suggest that the articulatory component may comprise only one feature of working memory. Coltheart's (1972) failure to find an effect of phonemic similarity on a concept formation task is, therefore, not particularly surprising.

We would like to suggest that the core of the working memory system consists of a limited capacity "work space" which can be divided between storage and control processing demands. The next three sections comprise a tentative attempt to elaborate our view of the working memory system by considering three basic questions: how work space is allocated, how the central processing system and the more peripheral phonemic rehearsal system interact in the memory span task, and, finally, whether different modalities each have their own separate working memory system.

A *Allocation of work space*

Our data suggest that a trade-off exists between the amount of storage required and the rate at which other processes can be carried out. In Experiment III, for example, *S*s solved verbal reasoning problems while either reciting a digit sequence, repeating the word *the*, or saying nothing. It is assumed that reciting a digit sequence requires more short-term storage than either of the other two conditions. Reasoning times, which presumably reflect the rate at which logical operations are carried out, were substantially increased in this condition. Furthermore, problems containing passive and negative sentences were slowed down more than problems posed as active and affirmative sentences. Since grammatically complex sentences presumably require a greater number of processing operations than simple sentences, this result is consistent with the assumed trade-off between storage-load and processing-rate.

The effect of additional memory load on free recall may be used to make a similar point. Experiments on presentation rate and free recall suggest that "transfer" to LTS proceeds at a limited rate. Since increasing memory load reduced transfer to LTS, it is arguable that this may result from a decrease in the rate at which the control processes necessary for transfer could be executed.

However, although our evidence suggests some degree of trade-off between storage-load and processing-rate, it would probably be unwise to regard working memory as an entirely flexible system of which any part may be allocated either to storage or processing. There are two reasons for this. In the first place, there may ultimately be no clear theoretical grounds for distinguishing processing and storage: they may always go together. Secondly, at the empirical level, a number of results show that it is difficult to produce appreciable interference with additional memory loads below the size of the span. This may mean that a part of the system that may be used for storage is not available for general processing. When the capacity of this component is exceeded, then some of the general-purpose work space must be devoted to storage, with the result that less space is available for processing. We shall discuss this possibility in more detail in the next section.

The final point concerns the factors which control the trade-off between the amount of work space allocated to two competing tasks. Results show that instructional emphasis is at least one determinant. In Experiment II, for example, Ss for whom the memory task was emphasized showed a very much greater effect of a six-digit preload on reasoning time than was shown by a second group who were instructed that both tasks were equally important. Evidence for a similar effect in free recall learning has been presented by Murdock (1965). He showed that a concurrent card-sorting task interfered with the long-term component of free recall and that the trade-off between performance on the two tasks was determined by the particular payoff specified in the instructions.

B *The role of working memory in the memory span*

We have suggested that the working memory system may contain both flexible work space and also a component that is dedicated to storage. This view is illustrated by the following suggested interpretation of the role of working memory in the memory span task. It is suggested that the memory span depends on both a phonemic response buffer which is able to store a limited amount of speechlike material in the appropriate serial order and the flexible component of working memory. The phonemic component is relatively passive and makes few demands on the central processing space, provided its capacity is not exceeded. The more flexible and executive component of the system is responsible for setting up the appropriate phonemic "rehearsal" routines, i.e., of loading up the phonemic buffer and of retrieving information from the buffer when necessary. Provided the memory load does not exceed the capacity of the phonemic buffer, little demand is placed upon the central executive, other than the routine recycling of the presumably familiar subroutines necessary for rehearsing digits. When the capacity of the phonemic buffer is exceeded, then the executive component of working memory must devote more of its time to the problem of storage. This probably involves both recoding in such a way as to reduce the length or complexity of the phonemic subroutine involved in rehearsal and also devoting more attention to the problem of retrieval. It is, for example, probably at this stage that retrieval rules become useful in allowing S to utilize his knowledge of the experimental situation in order to interpret the deteriorated traces emerging from an overloaded phonemic buffer (Baddeley, 1972).

According to this account, the span of immediate memory is set by two major factors: the capacity of the phonemic loop, which is presumably relatively invariant, and the ability of the central executive component to supplement this, both by recoding at input and reconstruction at the recall stage. We have begun to study the first of these factors by varying word length in the memory span situation. Figure 6 shows the results of an experiment in

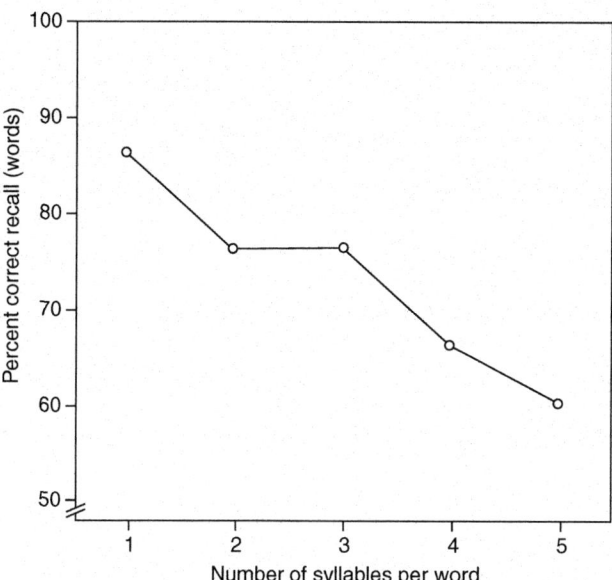

Figure 6 Effect of word length on short-term serial recall.
Source: Data from Baddeley and Thomson, unpublished.

which eight Ss were presented with sequences of five words from each of five sets. Each set comprised ten words of equal frequency of occurrence, but sets varied in word length, ranging in number of syllables from one through five. There is a clear tendency for performance to decline as word length increases. A similar result was independently obtained by Standing (personal communication) who observed a negative correlation between the memory span for a given type of material and the speed at which that material can be articulated. It is perhaps worth noting at this point that Craik (1968) reports that the recency effect in free recall is unaffected by the word length, suggesting once again a clear distinction between factors influencing the recency effect in free recall and those affecting the memory span. Watkins (1972) has further observed that word length does not influence the modality effect, but does impair the long-term component of verbal free recall. The former result would tend to suggest that the precategorical acoustic store on which the modality effect is generally assumed to rely (Crowder & Morton, 1969) lies outside the working memory system.

We, therefore, appear to have at least tentative evidence for the existence of a phonemic buffer, together with techniques such as articulatory suppression and the manipulation of word length which hopefully will provide tools for investigating this component in greater depth. It is possible that this component plays a major role in determining the occurrence of both acoustic similarity effects in memory and perhaps also of such speech errors as tongue twisters and spoonerisms. It seems likely that although it does not form the central core of working memory the phonemic component will probably justify considerably more investigation.

The operation of the central component of working memory seems likely to prove considerably more complex. It seems probable that it is this component that is responsible for the "chunking" of material which was first pointed out by Miller (1956) and has subsequently been studied in greater detail by Slak (1970), who taught subjects to recode digit sequences into a letter code which ensured an alternation between consonants and vowels. This allowed a dramatic reduction in the number of phonemes required to encode the sequence and resulted not only in a marked increase in the digit span, but also in a clear improvement in the performance of a range of tasks involving the long-term learning of digit sequences. A similar recoding procedure, this time based on prior language habits, is probably responsible for the observed increase in span for letter sequences as they approximate more closely to the structure of English words. This, together with the decreased importance of phonemic similarity, suggests that S is chunking several letters into one speech sound rather than simply rehearsing the name of the letter (Baddeley, 1971).

During retrieval, the executive component of the working memory system is probably responsible for interpreting the phonemic trace; it is probably at this level that retrieval rules (Baddeley, 1972) are applied. These ensure that a trace is interpreted within the constraints of the experiment, with the result that Ss virtually never produce completely inappropriate responses such as letters in an experiment using digits. We have unfortunately, however, so far done little to investigate this crucial central executive component; techniques aimed at blocking this central processor while leaving the peripheral components free should clearly be developed if possible.

C One or many working memories?

Our work so far has concentrated exclusively on verbal tasks, and the question obviously arises as to how general are our conclusions. It seems probable that a comparable system exists for visual memory which is different at least in part from the system we have been discussing.

Brooks (1967, 1968) studied a number of tasks in which S is induced to form a visual image and use this in an immediate memory situation. He has shown that performance in such a situation is impaired by concurrent visual processing, in contrast to equivalent phonemically based tasks, which are much more susceptible to concurrent verbal activity. We have confirmed and extended Brooks' results using visual pursuit tracking (Baddeley, Grant, Wight, & Thomson, 1974) which was found to cause a dramatic impairment in performance on a span task based on visual imagery, while producing no decrement in performance on an equivalent phonemically based task. Further evidence for the existence of a visual memory system which may be unaffected by heavy phonemic processing demands comes from the study by Kroll, Parks, Parkinson, Bieber, and Johnson (1970), who showed that Ss could retain a visually presented letter over a period of many seconds of shadowing auditory material.

From these and many other studies, it is clear that visual and auditory short-term storage do employ different subsystems. What is less clear is whether we need to assume completely separate parallel systems for different modalities, or whether the different modalities may share a common central processor. Preliminary evidence for the latter view comes from an unpublished study by R. Lee at the University of St. Andrews. He studied memory for pictures in a situation where Ss were first familiarized with sets of pictures of a number of local scenes, for which they were taught an appropriate name. Several slightly different views of each scene were used although only half of the variants of each scene were presented during the pretraining stage. Subjects were then tested on the full set of pictures and were required in each case to name the scene, saying whether the particular version shown was an "old" view which they had seen before or a "new" one. Subjects' performance was compared both while doing this task alone and while doing a concurrent mental arithmetic task (e.g., multiplying 27 and 42). Subjects were able to name the scenes without error in both conditions, but made a number of errors in deciding whether or not they had seen any given specific view of that scene; these errors were markedly more frequent in the mental arithmetic condition, suggesting that the visual recognition process was competing for limited processing capacity with the arithmetic task. One obvious interpretation of this result is to suggest that the central processor which we have assumed forms the core of working memory in our verbal situations plays a similar role in visual memory, although this time with a separate peripheral memory component, based on the visual system. What little evidence there exists, therefore, suggests that the possibility of a single common central processor should be investigated further, before assuming completely separate working memories for different modalities.

D Working memory and the recency effect in free recall

A major distinction between the working memory system we propose and STS (Atkinson & Shiffrin, 1968) centers on the recency effect in free recall. Most theories of STM assume that retrieval from a temporary buffer store accounts for the recency effect, whereas our own results argue against this view. It is suggested that working memory, which in other respects can be regarded as a modified STS, does not provide the basis for recency.

Experiment IX studied the effect of a concurrent digit memory task on the retention of lists of unrelated words. The results showed that when Ss were concurrently retaining six digits, the LTS component of recall was low, but recency was virtually unaffected. Since six digits is very near the memory span, the STS model would have to assume that STS is full almost to capacity for an appreciable part of the time during the learning of the words for free recall. On this model, both recency and LTS transfer should be lowered by the additional short-term storage load. As there was no loss of recency, it seems that an STS account of recency is inappropriate. Instead, it seems that recency reflects retrieval from a store which is different from that used for the digit span task. Perhaps the most important aspect of this interpretation is that the limited memory span and limited rate of transfer of information to LTS must be regarded as having a common origin which is different from that of the recency effect. It would be useful to consider briefly what further evidence there is for this point of view.

E The memory span, transfer to LTS, and recency

There is a wide range of variables which appear to affect the memory span (or short-term serial recall) and the LTS component of free recall in the same way, but which do not affect the recency component of free recall. In addition to the effects of word length and articulatory suppression which we have already discussed, which probably reflect the limited storage capacity of the working memory system or of one of its components, there are a number of variables which have been shown to affect the second limitation of the STS system, namely the rate at which it is able to transfer information to LTS. Several sets of experimental results show that the recency effect is not influenced by factors which interfere with LTS transfer. Murdock (1965),

Baddeley, Scott, Drynan, and Smith (1969), and Bartz and Salehi (1970) have all shown that the LTS component of free recall is reduced when Ss are required to perform a subsidiary card-sorting task during presentation of the items for free recall. The effect is roughly proportional to the difficulty of the subsidiary task. However, there is no effect on the recency component of recall. Similar results have been reported by Silverstein and Glanzer (1971) using arithmetic varying in level of difficulty as the subsidiary task. As most of these authors concluded, the results suggest that there is a limited capacity system mediating LTS registration which is not responsible for the recency effect. On the present hypothesis, the subsidiary task is viewed as interfering with working memory and does not necessarily, therefore, interfere with recency as well. Hence the crucial difference in emphasis between the two theories (working memory-LTS, and STS-LTS) is that working memory is supposed to have both buffer-storage and control-processing functions, with recency explained by a separate mechanism.

IV The nature of the recency effect

So far, the most compelling argument for rejecting the buffer-storage hypothesis for recency has been the data from Experiment XI, in which a concurrent memory span task did not abolish recency in free recall. Clearly the argument needs strengthening.

Tzeng (1973) presented words for free recall in such a way that before and after each word, S was engaged in a 20-second period of counting backwards by three's. Under these conditions the serial position curve showed a strong recency effect. After learning four such lists, S was asked to recall as many words as possible from all four lists. Even on this final recall, the last items from each of the lists were recalled markedly better than items from earlier positions. Neither of these two recency effects is easily attributable to retrieval from a short-term buffer store. With the initial recall, the counting task ought to have displaced words from the buffer. In the case of the final recall the amount of interpolated activity was even greater. Tzeng's results, therefore, suggest at the very least that the recency effect is not always attributable to output from buffer storage. Tzeng cites further evidence (unpublished at the time of writing) from Dalezman and from Bjork and Whitten, in both cases suggesting that recency may occur under conditions which preclude the operation of STS.

Baddeley (1963) carried out an experiment in which Ss were given a list of 12 anagrams to solve. Anagrams were presented one at a time for as long as it took for a solution to be found, up to a limit of one minute, at which time the experimenter presented the solution. After the final anagram, S was questioned about his strategy and was then asked to freely recall as many of the solution words as possible. The results of the recall test are shown in Figure 7 since they were not reported in the original paper. They show that despite the unexpected nature of the recall request and the delay while S discussed

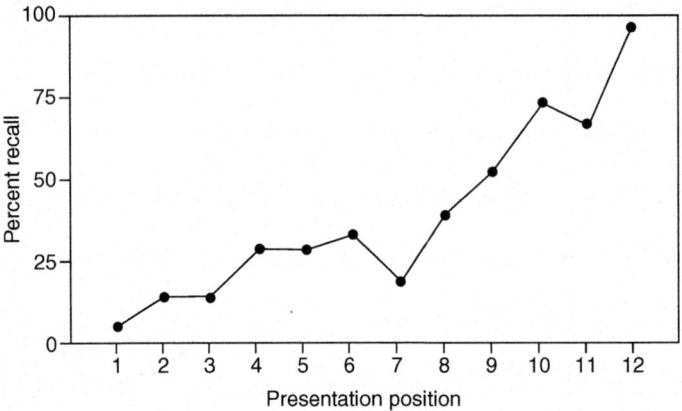

Figure 7 Recall of anagram solutions as a function of order of presentation of the problems.
Source: Data from Baddeley, 1963.

his strategy, a pronounced recency effect occurs. Since each item except the last was followed by up to a minute of problem-solving activity and the last item was followed by a period of question-answering, it is difficult to explain this recency effect in terms of a temporary buffer store.

An experiment by Glanzer (1972) which we have successfully replicated (Baddeley & Thomson, unpublished) presents further problems for a simple buffer-store interpretation of the recency effect. Instead of unrelated words, Glanzer used proverbs as the material to be recalled. His results showed two striking phenomena: first, the recency effect extended over the last few proverbs rather than the last few words; and second, a filled delay reduced, but by no means eliminated, the marked recency effect observed. The extent of recency, therefore, seems to be defined in terms of "semantic units" rather than words. This is not, of course, incompatible with a buffer-storage account, although in this experiment, a good deal of semantic processing would presumably have to occur before entry of a proverb into this buffer. The assumption of a more central store does have the additional advantage of "explaining" the durable recency effect observed in this study, in terms of the suggestion by Craik and Lockhart (1972) that greater depth of processing is associated with greater durability. However, it is clearly the case that such a depth of processing is by no means essential to the recency effect. Indeed, the effect appears to be completely unaffected by factors such as presentation rate (Glanzer & Cunitz, 1966), concurrent processing load (Murdock, 1965), and type of material (Glanzer, 1972), all of which would be expected to have a pronounced influence on depth of processing.

A more promising alternative explanation of recency might be to elaborate the proposal made by Tulving (1968) that recency reflects the operation of a retrieval strategy, rather than the output of a specific store. Provided one assumes that ordinal recency may be one accessible feature of a memory trace, then it is plausible to assume that Ss may frequently access items on the basis of this cue. The limited size of the recency effect, suggesting that recency is only an effective cue for the last few items, might reasonably be attributed to limitations on the discriminability of recency cues. One might assume, following Weber's Law, that with the newest item as a reference point, discriminability of ordinal position ought to decrease with increasing "oldness." The advantage of assuming an ordinal retrieval strategy of this kind is that it can presumably be applied to any available store and possibly also to any subset of items within that store, provided the subset can be adequately categorized. Thus, when an interpolated activity is classed in the same category as the learned items, the interpolated events will be stored in the same dimension as the to-be-remembered items and will hence supersede them as the most recent events. When the interpolated activity is classed in a different category from the learned items, recency will be unaffected. This presumably occurred in the case of proverbs and the anagram solutions. It also seems intuitively plausible to assume that a similar type of recency is reflected in one's own memory for clearly specified classes of events, for example, football games, parties, or meals at restaurants, all of which introspectively at least appear to exhibit their own recency effect. It is clearly necessary to attempt to collect more objective information on this point, however.

The preceding account of recency is highly tentative, and although it does possess the advantage of being able to deal with evidence which presents considerable difficulties for the buffer-store interpretation, it does leave two very basic questions unanswered. The first of these concerns the question of what factors influence the categorization of different types of events; it seems intuitively unlikely that backward-counting activity should be categorized in the same way for example as visually presented words, and yet counting effectively destroys the recency effect in this situation. This is, of course, a difficult problem, but it is no less a problem for the buffer-store interpretation which must also account for the discrepancy between Tzeng's results and the standard effect of a filled delay on recency.

The second basic question is that of how ordinal recency is stored, whether in terms of trace-strength, in terms of ordinal "tags" of some kind, or in some as yet unspecified way. Once again, this problem is not peculiar to the retrieval cue interpretation of recency; it is clearly the case that we are able to access ordinal information in some way. How we do this, and whether ordinal cues can be used to retrieve other information, is an empirical question which remains unanswered.

V Concluding remarks

We would like to suggest that we have presented *prima facie* evidence for the existence of a working memory system which plays a central role in human information processing. The system we propose is very much in the spirit of similar proposals by such authors as Posner and Rossman (1965) and Atkinson and Shiffrin (1971). However, whereas earlier work concentrated principally on the memory system *per se*, with the result that the implications of the system for nonmemory tasks were largely speculative, our own work has been focused on the information processing tasks rather than the system itself. As a consequence of this, we have had to change our views of both working memory and of the explanation of certain STM phenomena.

To sum up, we have tried to make a case for postulating the working memory-LTS system as a modification of the current STS-LTS view. We would like to suggest that working memory represents a control system with limits on both its storage and processing capabilities. We suggest that it has access to phonemically coded information (possibly by controlling a rehearsal buffer), that it is responsible for the limited memory span, but does not underly the recency effect in free recall. Perhaps the most specific function which has so far been identified with working memory is the transfer of information to LTS. We have not yet explored its role in retrieval, so that the implications of Patterson's (1971) results for the nature of working memory are still unclear. Our experiments suggest that the phonemic rehearsal buffer plays a limited role in this process, but is by no means essential. The patient K.F., whom Shallice and Warrington (1970) showed to have grossly impaired digit span together with normal long-term learning ability, presents great difficulty for the current LTS-STS view, since despite his defective STS, his long-term learning ability is unimpaired. His case can, however, be handled quite easily by the view of working memory proposed, if it is assumed that only the phonemic rehearsal-buffer component of his working memory is impaired, while the central executive component is intact. Our experiments also suggest that working memory plays a part in verbal reasoning and in prose comprehension. Understanding the detailed role of working memory in these tasks, however, must proceed hand-in-hand with an understanding of the tasks themselves.

We began with a very simple question: *what is short-term memory for?* We hope that our preliminary attempts to begin answering the question will convince the reader, not necessarily that our views are correct, but that the question was and is well worth asking.

Acknowledgments

Part of the experimental work described in this chapter was carried out at the University of Sussex. We are grateful to colleagues at both Sussex and Stirling for valuable discussion and in particular to Neil Thomson whose contribution was both practical and theoretical. The financial support of both the British Medical Research Council and the Social Sciences Research Council is gratefully acknowledged.

References

Atkinson, R. C., & Shiffrin, R. M. Human memory: A proposed system and its control processes. In K. W. Spence & J. T. Spence (Eds.), *The psychology of learning and motivation: Advances in research and theory.* Vol. 2. New York: Academic Press, 1968. Pp. 89–195.

Atkinson, R. C., & Shiffrin, R. M. The control of short-term memory. *Scientific American,* 1971, **225,** 82–90.

Baddeley, A. D. A Zeigarnik-like effect in the recall of anagram solutions. *Quarterly Journal of Experimental Psychology,* 1963, **15,** 63–64.

Baddeley, A. D. The influence of acoustic and semantic similarity on long-term memory for word sequences. *Quarterly Journal of Experimental Psychology,* 1966, **18,** 302–309. (a)

Baddeley, A. D. Short-term memory for word sequences as a function of acoustic, semantic and formal similarity. *Quarterly Journal of Experimental Psychology,* 1966, **18,** 362–366. (b)

Baddeley, A. D. A three-minute reasoning test based on grammatical transformation. *Psychonomic Science,* 1968, **10,** 341–342.

Baddeley, A. D. Language habits, acoustic confusability and immediate memory for redundant letter sequences. *Psychonomic Science,* 1971, **22,** 120–121.

Baddeley, A. D. Retrieval rules and semantic coding in short-term memory. *Psychological Bulletin,* 1972, **78,** 379–385.

Baddeley, A. D., De Figuredo, J. W., Hawkswell-Curtis, J. W., & Williams, A. N. Nitrogen narcosis and performance under water. *Ergonomics,* 1968, **11,** 157–164.

Baddeley, A. D., Grant, S., Wight, E., & Thomson, N. Imagery and visual working memory. In P. M. Rabbitt & S. Dornic (Eds.), *Attention and performance.* Vol. 5. New York: Academic Press, 1974, in press.

Baddeley, A. D., & Patterson, K. The relationship between long-term and short-term memory. *British Medical Bulletin,* 1971, **27,** 237–242.

Baddeley, A. D., Scott, D., Drynan, R., & Smith, J. C. Short-term memory and the limited capacity hypothesis. *British Journal of Psychology,* 1969, **60,** 51–55.

Baddeley, A. D., & Warrington, E. K. Memory coding and amnesia. *Neuropsychologia,* 1973, **11,** 159–165.

Bartz, W. H., & Salehi, M. Interference in short- and long-term memory. *Journal of Experimental Psychology,* 1970, **84,** 380–382.

Brooks, L. R. The suppression of visualization in reading. *Quarterly Journal of Experimental Psychology,* 1967, **19,** 289–299.

Brooks, L. R. Spatial and verbal components in the act of recall. *Canadian Journal of Psychology,* 1968, **22,** 349–368.

Brown, I. D., Tickner, A. H., & Simmonds, D. C. V. Interference between concurrent tasks of driving and telephoning. *Journal of Applied Psychology,* 1969, **53,** 419–424.

Bruce, D., & Crowley, J. J. Acoustic similarity effects on retrieval from secondary memory. *Journal of Verbal Learning and Verbal Behavior,* 1970, **9,** 190–196.

Coltheart, V. The effects of acoustic and semantic similarity on concept identification. *Quarterly Journal of Experimental Psychology,* 1972, **24,** 55–65.

Conrad, R. An association between memory errors and errors due to acoustic masking of speech. *Nature* (London), 1962, **193,** 1314–1315.

Conrad, R., & Hull, A. J. Information, acoustic confusion and memory span. *British Journal of Psychology,* 1964, **55,** 429–432.

Craik, F. I. M. Two components in free recall. *Journal of Verbal Learning and Verbal Behavior*, 1968, **7**, 996–1004.
Craik, F. I. M., & Levy, B. A. Semantic and acoustic information in primary memory. *Journal of Experimental Psychology*, 1970, **86**, 77–82.
Craik, F. I. M., & Lockhart, R. S. Levels of processing: A framework for memory research. *Journal of Verbal Learning and Verbal Behavior*, 1972, **11**, 671–684.
Crowder, R. G., & Morton, J. Precategorical acoustic storage (PAS). *Perception & Psychophysics*, 1969, **5**, 365–373.
Glanzer, M. Storage mechanisms in free recall. In G. H. Bower (Ed.), *The psychology of learning and motivation: Advances in research and theory.* Vol. 5. New York: Academic Press, 1972. Pp. 129–193.
Glanzer, M., & Cunitz, A. R. Two storage mechanisms in free recall. *Journal of Verbal Learning and Verbal Behavior*, 1966, **5**, 351–360.
Glanzer, M., Koppenaal, L., & Nelson, R. Effects of relations between words on short-term storage and long-term storage. *Journal of Verbal Learning and Verbal Behavior*, 1972, **11**, 403–416.
Green, D. W. A psychological investigation into the memory and comprehension of sentences. Unpublished doctoral dissertation, University of London, 1973.
Hammerton, M. Interference between low information verbal output and a cognitive task. *Nature (London)*, 1969, **222**, 196.
Hunter, I. M. L. *Memory.* London: Penguin Books, 1964.
Jarvella, R. J. Syntactic processing of connected speech. *Journal of Verbal Learning and Verbal Behavior*, 1971, **10**, 409–416.
Johnston, W. A., Griffith, D., & Wagstaff, R. R. Information processing analysis of verbal learning. *Journal of Experimental Psychology*, 1972, **96**, 307–314.
Kroll, N. E. A., Parks, T., Parkinson, S. R., Bieber, S. L., & Johnson, A. L. Short-term memory while shadowing: Recall of visually and aurally presented letters. *Journal of Experimental Psychology*, 1970, **85**, 220–224.
Levy, B. A. Role of articulation in auditory and visual short-term memory. *Journal of Verbal Learning and Verbal Behavior*, 1971, **10**, 123–132.
Martin, D. W. Residual processing capacity during verbal organization in memory. *Journal of Verbal Learning and Verbal Behavior*, 1970, **9**, 391–397.
Miller, G. A. The magical number seven plus or minus two: some limits on our capacity for processing information. *Psychological Review*, 1956, **63**, 81–97.
Murdock, B. B., Jr. Effects of a subsidiary task on short-term memory. *British Journal of Psychology*, 1965, **56**, 413–419.
Murray, D. J. The role of speech responses in short-term memory. *Canadian Journal of Psychology*, 1967, **21**, 263–276.
Murray, D. J. Articulation and acoustic confusability in short-term memory. *Journal of Experimental Psychology*, 1968, **78**, 679–684.
Neale, M. D. *Neale Analysis of Reading Ability.* London: Macmillan, 1958.
Norman, D. A. The role of memory in the understanding of language. In J. F. Kavanagh & I. G. Mattingly (Eds.), Cambridge, Mass.: MIT Press, 1972.
Patterson, K. A. Limitations on retrieval from long-term memory. Unpublished doctoral dissertation, University of California, San Diego, 1971.
Peterson, L. R. Concurrent verbal activity. *Psychological Review*, 1969, **76**, 376–386.
Peterson, L. R., & Peterson, M. J. Short-term retention of individual verbal items. *Journal of Experimental Psychology*, 1959, **58**, 193–198.
Posner, M. I., & Rossman, E. Effect of size and location of informational transforms upon short-term retention. *Journal of Experimental Psychology*, 1965, **70**, 496–505.

Rubenstein, H., & Aborn, M. Learning, prediction and readability. *Journal of Applied Psychology*, 1958, **42**, 28–32.

Rumelhart, D. E., Lindsay, P. H., & Norman, D. A. A process model for long-term memory. In E. Tulving & W. Donaldson (Eds.), *Organisation and memory*. New York: Academic Press, 1972.

Sachs, J. D. S. Recognition memory for syntactic and semantic aspects of connected discourse. *Perception & Psychophysics*, 1967, **2**, 437–442.

Shallice, T., & Warrington, E. K. Independent functioning of verbal memory stores: a neuropsychological study. *Quarterly Journal of Experimental Psychology*, 1970, **22**, 261–273.

Silverstein, C., & Glanzer, M. Difficulty of a concurrent task in free recall: differential effects of LTS and STS. *Psychonomic Science*, 1971, **22**, 367–368.

Slak, S. Phonemic recoding of digital information. *Journal of Experimental Psychology*, 1970, **86**, 398–406.

Sperling, G. A model for visual memory tasks. *Human Factors*, 1963, **5**, 19–31.

Taylor, W. L. "Cloze procedure": A new tool for measuring readability. *Journalism Quarterly*, 1953, **30**, 415–433.

Tulving, E. Theoretical issues in free recall. In T. R. Dixon & D. L. Horton (Eds.), *Verbal behaviour and general behaviour theory*. Englewood Cliffs, N.J.: Prentice-Hall, 1968.

Tzeng, O. J. L. Positive recency effect in delayed free recall. *Journal of Verbal Learning and Verbal Behavior*, 1973, **12**, 436–439.

Warrington, E. K., Logue, V., & Pratt, R. T. C. The anatomical localization of selective impairment of auditory verbal short-term memory. *Neuropsychologia*, 1971, **9**, 377–387.

Warrington, E. K., & Shallice, T. The selective impairment of auditory verbal short-term memory. *Brain*, 1969, **92**, 885–896.

Warrington, E. K., & Weiskrantz, L. An analysis of short-term and long-term memory defects in man. In J. A. Deutsch (Ed.), *The physiological basis of memory*. New York: Academic Press, 1973.

Wason, P. C., & Johnson-Laird, P. N. *Psychology of reasoning: Structure and content*. London: Batsford, 1972.

Watkins, M. J. Locus of the modality effect in free recall. *Journal of Verbal Learning and Verbal Behavior*, 1972, **11**, 644–648.

Waugh, N. C., & Norman, D. A. Primary memory. *Psychological Review*, 1965, **72**, 89–104.

Wickelgren, W. A. Short-term memory for phonemically similar lists. *American Journal of Psychology*, 1965, **78**, 567–574.

5 The recency effect
Implicit learning with explicit retrieval?

A. D. Baddeley and G. Hitch

The recency effect in free recall features prominently in 1960s' theorizing about short-term memory, but has since been largely ignored. We argue that this stems from a preoccupation with the role of recency in the concept of primary memory and the neglect of its role in a broader working-memory framework. It is suggested that the recency effect reflects the application of an explicit retrieval strategy to the residue of implicit learning within a range of cognitive systems. When retrieved implicitly, the same residue is assumed to form the basis of priming effects. The various criteria for implicit learning described by Tulving and Schacter (1990) are successfully applied to the recency effect, and a retrieval process is outlined that can account for both long- and short-term recency effects. It is suggested that a framework combining recency, priming, and implicit learning provides a basis for understanding one of the most important features of cognition and memory, namely, that of maintaining orientation in time and place.

When subjects are presented with a list of words for immediate free recall, the last few words presented tend to be recalled very well, the phenomenon being known as the *recency effect*. It was a phenomenon of considerable interest and theoretical concern during the 1960s and features in most basic textbooks, but has in recent years been relatively neglected by both empirical researchers and by theorists. It has, in short, become rather unfashionable. We ourselves have tended to neglect the study of recency, concentrating instead on the more active features of the multicomponent working-memory model that has increasingly displaced the earlier concept of short-term memory (Baddeley, 1992).

Although psychology is often accused of being excessively prone to changes of fashion, there are, of course, good reasons for ignoring particular topics. For example, a phenomenon might be sufficiently well understood that no further work is necessary. We would argue that this is not the case with recency. Alternatively, investigators may find that they are making little progress, and they very reasonably move to problems that may be more tractable. We would argue, on the contrary, that considerable progress has been made in understanding recency, although this tends not to be reflected in the standard textbook view. A third reason is that theoretical progress makes other areas seem more attractive. We believe this is the case with the phenomenon of recency, but we would like to argue that the time has come to attempt to relate what we know about recency to some of the developments in other areas that have occurred since its heyday in the 1960s. More specifically, we wish to argue that the current interest in the concept of implicit and explicit memory is of direct relevance to understanding the recency effect and placing it within a broader context.

Recency as primary memory

Before going on to discuss the proposed reconceptualization of recency, it is necessary to rule out the simple 1960s view that recency represents the contents of a primary memory store, a view proposed most vigorously by Glanzer (1972). This view argues for a short-term store of limited capacity, able to hold about three items, or chunks, of information, and able to retrieve this information relatively rapidly and with minimal cognitive demand. In the standard free-recall paradigm, the last few items are assumed to be in such a store and, hence, able to be readily recalled. However, if a word list is followed by some intervening activity such as counting, then the word list will be displaced from the primary memory store by the counting material, hence removing the recency effect (Glanzer & Cunitz, 1966).

Evidence in favor of a short-term-storage view includes: (1) the demonstration by Glanzer and his colleagues that recency was uninfluenced by variables that have a major effect on earlier items in the list, such as rate of presentation and word frequency (Glanzer, 1972); (2) the demonstration that the recency effect is preserved in amnesic patients with defective long-term memory (Baddeley & Warrington, 1970) and disrupted in patients with impaired short-term memory (Shallice & Warrington, 1970); and (3) the demonstration by Rundus (1971) that subjects induced to rehearse out loud typically include the most recent items in their rehearsal and show a tendency for these to be recalled, regardless of the number of prior rehearsals. (A more detailed account of this literature is given by Baddeley, 1990, and Crowder, 1976.)

As Greene (1986b) has pointed out, the short-term-memory interpretation remains the view favored by many recent textbook writers, who tend to link the phenomenon to the modal model of memory (Atkinson & Shiffrin, 1968), despite the presence of evidence that makes such an interpretation clearly questionable. The evidence will be described briefly, since it is discussed in more detail elsewhere (Baddeley & Hitch, 1977; Greene, 1986b).

The first reason to doubt the primary-memory interpretation of recency comes from the demonstration of recency effects in long-term memory. These have been studied most extensively using the continuous-distractor technique, in which a list of words is presented with each word separated by some distractor activity such as backward counting, which then fills the interval between the final word and free recall (Bjork & Whitten, 1974; Tzeng, 1973). Under these circumstances, a clear recency effect survives the filled interval, which under normal conditions would have been enough to wipe out recency. Other examples of long-term recency include the recall by subjects of anagram solutions, where each solution is followed by the demanding activity of solving the next anagram (Baddeley & Hitch, 1977). Finally, recency effects can be shown to operate over a matter of weeks, as in the recall of rugby games (Baddeley & Hitch, 1977) or parking locations (Pinto & Baddeley, 1991). Subjects who park regularly in a given car park showed a marked recency effect when asked to recall their parking locations over the previous week, whereas subjects who park infrequently at that location were quite accurate in recalling their parking spot when tested 1 month later (Pinto & Baddeley, 1991).

It is, of course, possible to argue that quite separate mechanisms are responsible for long- and short-term recency effects, although it then becomes problematic to explain why both are fitted by exactly the same function (Glenberg et al., 1980; Pinto & Baddeley, 1991). Furthermore, Greene (1986a) has shown that at least two variables, word frequency and list length, have the same effect on both long- and short-term recency. A model that can explain the whole range of data would have advantages over a model that can only explain part of the data, particularly since attempts to demonstrate actively that the two types of recency are different have met with little success (see Greene, 1986b).

A second and related problem for the primary-memory interpretation of recency is presented by Watkins and Peynircioglu (1983), who demonstrated the existence of up to three simultaneous recency effects, each of which was comparable in magnitude to a recency effect under more conventional, single-stimulus conditions. They presented their subjects with lists of 45 items, selected from three highly distinctive categories, for example, riddles, sounds, and objects, presented in such a way that items from the three categories were interleaved. Hence, the 1st, 4th, and 7th items came from one category, the 2nd, 5th, and 8th items from the second category, the 3rd, 6th, and 9th items from the third category, and so forth. After the 45th item, recall was cued with the category name; a marked recency effect was observed regardless of which of the three categories was cued. This is clearly very difficult to fit into an interpretation of recency that is based on a limited-capacity primary memory other than by making the arbitrary post hoc assumption that the subject has a separate and parallel primary-memory system for each category.

The third problem for the primary-memory view of recency comes from the observation that concurrent activity that ought to occupy primary memory does not appear to interfere with recency. This is shown most strongly in two studies by Baddeley and Hitch (1977) in which free recall of lists of words presented visually in one study and auditorily in the second study was accompanied by a concurrent digit-span task presented in the other modality. Since both digit span and the recency effect are assumed in the short-term-memory view to depend upon the same limited-capacity primary memory, then the recency effect should be wiped out by this concurrent task. In fact, recency was quite unaffected, although performance on items earlier in the list was impaired.

It is not easy to see how a primary-memory interpretation of recency can account for this pattern of data, and we know of no attempt to do so. Historically, this was one of the major reasons for the demise of the concept of primary memory as it appeared in the multistore models of the 1960s (e.g., Atkinson & Shiffrin, 1968; Waugh & Norman, 1965). The idea of a passive primary memory was replaced by more dynamic accounts of temporary information storage and information processing, such as the working-memory model (Baddeley, 1986; Baddeley & Hitch, 1974). However, recency itself became somewhat detached from this theoretical development, since the experimental data suggested it was not an aspect of working memory. Although recency has continued to attract interest, it has tended to become a specialized and somewhat isolated research topic.

In the next section, we describe work that suggests that a simple temporal or ordinal discrimination hypothesis fits the data from both long- and short-term recency extremely well. We then go on to describe an attempt to rehabilitate recency by arguing that it reflects registration in implicit memory by means of a process of priming.

The discrimination hypothesis

We argue that a simple temporal or ordinal discrimination hypothesis fits the data from both long- and short-term recency extremely well. The essence of the hypothesis is perhaps best captured by the analogy proposed by Crowder (1976) of recall from a free-recall list as being rather like looking back at telegraph poles along a railway line. The nearest poles will be easy to discriminate, but as the poles become more distant it becomes progressively harder to tell one from the other. Similarly, if one moves further away from the nearest pole, then that pole will still be the most discriminable, and the immediately preceding one will be the next best, and so forth. Such a view has been quantified both by Glenberg and his colleagues and by Hitch and Rejman (Glenberg et al., 1980; Hitch, Rejman, & Turner, 1980). By analogy with Weber's Law, the assumption is made that the discriminability of the memory trace of a particular item will be a joint function of the interval between that item and the next and the interval between that item and the point of recall, the so-called discrimination ratio.

However, although there is a fair amount of agreement on the nature of the function, there is considerably less agreement on the actual mechanism whereby the function is achieved. A number of authors have suggested the possibility that ordinal cues are involved (e.g., Baddeley & Hitch, 1977; Sanders, 1975; Tulving, 1968). Others have emphasized temporal rather than ordinal distinctiveness (Bjork & Whitten, 1974; Hitch et al., 1980), but on the whole, most accounts have left unspecified the manner in which the relevant cue of order or time is encoded and retrieved. There are, in fact, at least two models that attempt to specify a mechanism in slightly more detail, that of Glenberg et al. (1980) and that of Baddeley (1986).

Glenberg et al. (1980) dealt with the problem of time by assuming that it is represented by a continuously changing context; the context is assumed to change gradually so that the closer two events are in time, the more similar will be their contextual cue. Since the temporal context continues to run during a delay interval, the nearer a recall comes to the end of the list, the more similar the encoding and retrieval contexts will be, and the better the recall. A filled interval is assumed to disrupt and change the contextual situation, hence making the temporal retrieval cue ineffective. However, Glenberg et al. (1980) appeared to make no attempt to unconfound time and context, for example by deliberately changing the background context between learning and retrieval. They did, in fact, include a condition in which it is reasonable to assume that such a change occurs, since the level of difficulty of the concurrent task in a continuous distractor paradigm is changed between learning and retrieval (Glenberg et al., 1980). However, there does not appear to be any decrement in performance, as the contextual interpretation would predict.

It is also difficult to see how the contextual interpretation would handle the microstructure of the recency effect obtained in the Watkins and Peynircioglu (1983) study. Consider, for example, the critical comparison between the recency effect when only one category of task is involved and that in which three categories are interleaved. In the single-category case, the last item is substantially better recalled than the penultimate item, which in turn is better than the item before. According to the contextual interpretation, this is because the background context will have changed over this period of time, causing a disruption of the context and leading to poorer retrieval. Consider now the case where three different categories of task are interleaved, and compare the condition in which a subject attempts to recall the category presented third in order with that presented first. The last item of the third category will have been presented most recently, and the last item of the first category will have been presented at a point separated in time and, presumably, involving an even greater switch in context between learning and recall, and yet there is far less difference in probability of recall of this item than of the equivalently positioned item in the single-category list.

The continuous-temporal-context hypothesis would seem to have even more problems in coping with instances of very long-term recency, for which it has to assume a degree of precision of temporal encoding that appears to be inconsistent with what evidence is available for dating of events in long-term memory. For example, dates for public events appear typically to be derived from information concerning the organization of private, autobiographical events rather than via temporal information per se (Brown, Shevell, & Rips, 1986).

A further difficulty is presented by a study by Pinto and Baddeley (1991) in which subjects who parked their cars at the Applied Psychology Unit a month earlier were asked by mail to remember and mark on a parking plan where they had parked. Subjects who had attended the unit only once in recent months recalled their parking locations as well after 1 month as they appeared to do after a week, or indeed on the same day. However, if they had been to the unit on two occasions, separated by a week, then recall a month later was considerably worse. These and other results from the same study fit the discrimination hypothesis very well, but it seems implausible to assume that there is a precise and continuous contextual cue that is still available 1 month later, when recall is tested by mail in a completely different context. Unless the hypothetical contextual change can be demonstrated much more convincingly, then it is hard to escape the view that the Glenberg et al. (1980) position merely relabels time as context and, as such, does not genuinely provide a mechanism whereby the phenomenon of recency can be explained.

The second proposed mechanism is Baddeley's (1986, chap. 7) suggestion that the recency effect is based on a phenomenon analogous to priming in that the memory representations of recent inputs are assumed to have increased levels of activation. This proposal has the drawback that it is speculative and relatively vague, but it does have the advantage that it is, in principle, able to be modeled. It is also timely in that activation levels are a central feature of current connectionist models of memory that are beginning to attempt to simulate recency as one of the more obvious and dramatic features of memory (e.g., Burgess, Shapiro, & Moore, 1991; Parisi, 1986). In such models, learning typically involves changing the weights of connections throughout an associative network, and retrieval involves using these weights to recover the original learning. Recency emerges naturally because the weights tend to be altered by subsequent learning.

Recency as a priming effect

Baddeley (1986) proposed that recency represents the application of a very basic retrieval mechanism to an isolable memory store or memory domain. The term *priming* refers in this instance to the process that is responsible for the greater availability of information about an item or event as a result of a previous presentation. Priming is typically inferred from the empirical phenomenon of the *priming effect*, in which performance on some (non-memory) task is altered as a function of an earlier presentation of the material being processed. We use the term priming to refer to the underlying *process*, which we suggest may influence performance across a range of experimental paradigms, including both memory and non-memory tasks. As Tulving and Schacter (1990) pointed out, priming effects occur across a wide range of systems, both perceptual and conceptual. Given the range of different systems, it is entirely plausible to assume that the mechanism may be somewhat different, in some cases based on a rapidly fading trace and in others possibly based on interference rather than temporal decay. What such situations have in common, however, is the greater availability of previously presented, and therefore of more recent, items.

How can the system take advantage of this? Clearly, in general, retrieval is not simply based on recency, otherwise the whole retrieval process would become locked into a loop whereby the most recent item is retrieved, thereby becoming even more recent and being even more likely to be retrieved. It seems probable that recency supports a special mechanism used as a supplement to more structured retrieval processes.

How might such a system work? An analogy is proposed whereby the items or events to be recalled are regarded as discriminable nodes within a network. These nodes will be treated as local representations of information, but it is to be understood that they could correspond to patterns of activation at some level within the network. Whether they are considered as nodes or distributed representations is essentially a matter of implementational choice. It is assumed that presentation and processing of an item results in the activation of its node and that recall of an item involves reactivating its node. The recency effect occurs because recently activated nodes are easy to reactivate. The analogy can be made with a bank of lights in which the filament of a given light becomes incandescent when a current is passed through. Presentation of a series of items corresponds to illuminating individual lights in turn. After the series has ended, the last few lights to have been illuminated will tend to be still warm, and one way of taking advantage of this is to flood the bank with a current that gradually increases in amplitude. The first lights to glow during this reactivation process will be those that were warmest, having more recently been illuminated.

An interesting feature of the analogy is that it leads naturally to some of the same predictions as the ratio rule. It is necessary only to assume that the probability of retrieving an item depends on the distinctiveness of its brightness from that of other potential retrieval targets. This, in turn, will be a function of the warmth of their filaments at the time of flooding. Provided the lamps cool at a gradually decreasing rate, and do so independently of one another, the distinctiveness of the warmth of a given lamp relative to the others will depend on time intervals in a way similar to the ratio rule. Thus, the more separated the initial lighting of individual lamps, the greater will be the difference in their warmth at the time of flooding. And since the lamps will cool toward a common baseline, their warmth will become more similar as the delay before flooding increases. We can, therefore, see by this analogy that a system involving discriminations among the activation levels of a set of memory representations is capable of producing recency, which conforms to the general pattern normally observed.

Another feature of the analogy is that if too high a current is passed through during flooding, then many lights will be illuminated, some recent, some distant, and some that were probably not illuminated before. A crucial variable is the domain within which the process of flooding can occur. Watkins and Peynircioglu (1983) managed to find sufficiently distinctive categories to allow them to set up three parallel recency effects, but this was not easy to achieve. Presenting subjects with items selected from three successive semantic categories will not produce three recency effects, but rather one (Rejman, 1979). This is not only a problem for the present theory, but it remains a phenomenon that has yet to be explained by any model. A more basic weakness of the flooding analogy, however, is that it essentially represents verbal speculation, and to become a genuine model it would need more detailed quantitative specification and, preferably, simulation.

In the meantime, how plausible is it to assume that the recency effect resembles priming in reflecting implicit learning, albeit accessed by an explicit retrieval process? We will attempt to answer this question by using the criteria applied by Tulving and Schacter (1990), who argued for perceptual priming as an implicit learning system.

Recency as implicit learning

Over the last decade, one of the most active research areas has been that concerned with the question of whether long-term memory can be regarded as comprising two separate systems, one based on explicit acquisition of information through a system often associated with episodic memory and the other concerned with one or more forms of learning that do not appear to depend on episodic memory and that have typically been termed implicit memory (Richardson-Klavehn & Bjork, 1988). The distinction is reflected most clearly in studies of patients suffering from the classic amnesic syndrome, who, despite having no apparent capacity for recollecting new experiences, may nevertheless show an unimpaired capacity to show improvement with practice on a wide range of learning tasks, ranging from classical conditioning through perceptual priming and motor-skill learning to the acquisition of new problem-solving skills (Baddeley, 1982; Squire, 1982).

Although the existence of a valid distinction between two or more types of long-term-memory tasks is widely accepted, there is much less agreement as to how such a distinction should best be conceptualized and as to what is the most appropriate terminology (Richardson-Klavehn & Bjork, 1988). There is, however, general agreement that one class of implicit tasks that appears to be clearly separable from explicit learning is represented by priming effects. The case for this view was presented cogently by Tulving and Schacter (1990), and although it is questionable as to whether the range of perceptual priming phenomena they describe all reflect the operation of a single system, there would, we think, be general agreement that they all represent broadly comparable forms of implicit memory.

If, as suggested above, the recency effect is based on a priming mechanism, then one might reasonably expect that it also would meet the various criteria proposed by Tulving and Schacter (1990) for classifying learning as implicit, even though the recency effect is measured in quite a different way from the typical priming study. Whereas the perceptual priming effect is detected in a non-memory task, as a change in responding to a stimulus as a function of a previous presentation, the recency effect in free recall is indicated by a change in the probability of recalling the stimulus (or event) in the absence of any specific external cue. The question of whether the characteristics of the recency effect resemble those of priming will be examined next; we will consider in turn the five criteria proposed by Tulving and Schacter.

1. *Intact performance in amnesia.* Amnesic patients typically show priming effects that are as marked as those found in control subjects (see, e.g., Graf, Squire, & Mandler, 1984). In the case of recency, Baddeley and Warrington (1970) studied the free recall of lists of 10 words by a group of densely amnesic patients. When compared to controls, the amnesic patients showed grossly impaired performance on the earlier items in the list but revealed preserved recency. This has been replicated in studies with other amnesic patients (e.g., Wilson & Baddeley, 1988) and with patients suffering memory deficit following closed-head injury (Brooks, 1975).

2. *Developmental dissociation.* Recognition memory in children increases with age, whereas priming can be as large in 3-year-olds as in college students (see, e.g., Parkin & Streete, 1988). A similar pattern occurs for the recency effect. Using an immediate-free-recall task, Craik (1968) compared the performance of young subjects, averaging 22 years of age, and 65-year-olds. The older subjects performed more poorly on the earlier items in the list but showed equivalent recency to the younger subjects. Thurm and Glanzer (1971) carried out a broadly equivalent study using children; the younger children performed more poorly on the earlier items but showed as clear a recency effect as the older children.

3. *Drug dissociations.* Drugs such as alcohol and scopolamine impair explicit recall, while typically having no effect on priming (see, e.g., Parker, Schoenberg, Schwartz, & Tulving, 1983). Again, the recency effect shows a similar pattern. Mewaldt, Hinrichs, and Ghoneim (1983) studied the effect of the tranquilizer Valium on free-recall performance, observing the anticipated deficit following drug ingestion in recall of earlier items in a list but detecting no difference in the magnitude of the recency effect. An equivalent pattern has been observed for a range of other drugs that leave recency unchanged, including alcohol (Baddeley, 1981) and marijuana (Darley, Tinklenberg, Roth, Hollister, & Atkinson, 1973).

4. *Functional independence.* This term was used by Tulving and Schacter (1990) to refer to the observation that variables that influence long-term episodic memory are found not to influence priming, and vice versa. Depth of processing (Craik & Lockhart, 1972) is one of the most influential implicit-explicit dissociations, with priming failing to show the standard tendency for orthographic processing to lead to poorer recall than phonological processing, which in turn is less effective than semantic encoding (Jacoby & Dallas, 1981).

As Glanzer (1972) pointed out, there are many examples of the functional independence of the recency component in free recall and performance on earlier items. Thus, earlier items tend to be influenced by variables such as number of repetitions and word frequency, which affect long-term learning, whereas the recency effect is typically resistant to such variables. Glanzer and Koppenaal (1977) studied the influence of depth of processing on the magnitude of the recency effect, observing that while performance on earlier items followed the pattern of deeper processing leading to better recall, the recency effect was uninfluenced by processing depth. Seamon and Murray (1976) obtained broadly similar results. Other evidence of insensitivity to the nature of coding came from a study by Craik and Levy (1970), in which they manipulated semantic similarity, again finding a dissociation between a clear effect on earlier items and the absence of an effect on recency.

5. *Stochastic independence.* This is the most controversial of Tulving and Schacter's (1990) criteria (Hintzman, 1991; Hintzman & Hartry, 1990). For stochastic independence to occur, a given word presented at the end of a list should have a probability of being retrieved on the basis of recency that is independent of its probability of being recalled subsequently from long-term episodic memory. Such an assumption underlies many of the measures devised to estimate the primary-memory component of recency (see, e.g., Baddeley, 1970; Glanzer, 1972; Waugh & Norman, 1965). Although we know of no studies that have explicitly set out to test this assumption at the level of the individual item, it is supported, at least qualitatively, by a study (Baddeley, 1968) concerned with the role of recency in multitrial free-recall learning. When recall probability was plotted as a function of serial position during presentation, words recalled for the first time showed a clear recency effect, whereas words correct on earlier trials showed no clear recency, although in both cases subjects tended to recall more recent items first.

One might possibly argue that the occurrence of the negative recency effect argues against the assumption of stochastic independence. Negative recency is the observation that when immediate free recall of one or more lists is followed by a final free recall, the last one or two items tend to be particularly poorly remembered (Craik, 1970). Given that these are the items that are best remembered on immediate free recall, one might contend that this argues for a negative correlation, rather than stochastic independence. However, it appears to be the case that negative recency reflects the influence of immediate recall on subsequent recall, since it does not occur when immediate free recall is not required (see Greene, 1986a, for a discussion). The fact that immediate recall may influence subsequent delayed recall does, of course, make it very difficult to test for stochastic independence at the level of the individual word. Hence, if Hintzman (1991) is wrong, and stochastic independence is an important criterion of implicit learning, then we may have to be satisfied with the broad evidence of independence that is presented by the fact that serial position over the latter half of a free-recall list has a massive effect on immediate recall and, typically, no effect on delayed recall.

To recapitulate, Tulving and Schacter (1990) specify five criteria of implicit memory, each of which they apply successfully to the phenomenon of priming. When these are applied in turn to the recency effect, the evidence is consistent in suggesting that the recency effect meets all five criteria of implicit memory.

Recency in implicit memory tests

We next consider the evidence that recency effects can be observed under implicit retrieval conditions. The most obvious example of this is the evidence that priming effects gradually dissipate as a function of the lag separating the rest from the priming presentation. The time course of this function can be very rapid (Kirsner & Smith, 1974; Monsell, 1985), although long-lived priming effects that are resistant to forgetting can persist over periods as long as a week (Tulving, Schacter, & Stark, 1982).

A somewhat different demonstration of recency in implicit memory comes from a series of experiments reported by McKenzie and Humphreys (1991). In a typical study, subjects performed a different classification task on each of two successively presented word lists under incidental learning instructions. After a variable delay, they then performed either an explicit test of memory or an implicit test. For example, subjects were presented with a part-word cue containing some of the letters of a word they had seen. In the explicit condition, subjects were asked to complete the part-word cue to form any of the words that they could remember having seen in the first list. In the implicit condition, they were asked simply to produce the first word that came to mind as a completion for the cue. The two lists were designed so that there was always one word on each that matched the cue, and thus the proportion of List 2 to List 1 responses could be used as an index of recency. McKenzie and Humphreys reported a higher proportion of List 2 responses at the shorter delay, and this was true on both the explicit and implicit tests. Subjects appeared to be obeying the two sets of instructions, since the proportion of extralist responses was much higher for the implicit group. McKenzie and Humphreys's results, therefore, argue convincingly that recency effects can occur under both explicit and implicit retrieval conditions.

If we go back to the analogy of a bank of lights, the implicit recency effect can be understood in terms of the gradual cooling of the lamps with increasing lag since their illumination. This in turn suggests the possibility that recency in implicit and explicit memory tests may be systematically related, since they both depend on activation levels but in slightly different ways. As we have seen, recency in tests of explicit memory appears to reflect the ease of discriminations based on differences among the activation levels of the various presented items. We would expect recency in implicit memory to be a function of differences between the activation levels of presented and non-presented items. Such an interpretation leads to some clear predictions. For example, one would expect that the time scale over which recency is observed for a particular type of material would be similar in tests of explicit memory and in measures of priming effects.

Recency clearly satisfies Tulving and Schacter's (1990) criteria for classification as implicit memory. It may, therefore, seem somewhat paradoxical that recency is typically found under conditions where subjects are intentionally recollecting past experience. However, the apparent contradiction disappears if we assume that recency reflects the application of an *explicit* retrieval strategy to *implicit* learning. To examine this suggestion, we need to consider first whether there is evidence that recency is dependent on the application of a retrieval strategy.

The belief that recency involves a particular retrieval strategy was widely held by experimenters studying immediate free recall in the 1960s and 1970s who would typically attempt to maximize the size of the recency effect by instructing subjects to try to recall the last items first. If such instructions were not given, some subjects would typically take a few trials before they recalled in this way. One of the first studies to show the buildup of recency across trials was one using minimal paired-associate learning (Murdock, 1963). The strategic basis of recency is clearly illustrated by Baddeley and Wilson's (1988) study of a frontal amnesic patient, R.J. Patients with frontal damage tend to perseverate, and having begun to retrieve the early words of a list first, R.J. showed a complete absence of recency. However, on the following day, he began with a strategy of recalling the most recent words, and he continued to do so, demonstrating a very clear recency effect.

Further evidence for the importance of output order comes from one of the studies reported by Baddeley and Hitch (1977) in which subjects were presented with a list of nonsense syllables under instructions for either intentional or incidental learning. Within each group, half the subjects were required to perform a supplementary writing task involving copying each item several times as it was presented. Recency was found for three of the four groups and was absent only for subjects who engaged in intentional learning without the supplementary task. In all three conditions that showed recency, there was a clear tendency for subjects to start recalling with the last items, whereas in the one condition that did not show recency, subjects tended to begin by recalling from the start of the list. These data argue strongly for an association between the recency effect and a retrieval strategy involving a last-in-first-out output order. The importance of output order is further confirmed by evidence that when subjects are instructed to start by retrieving items from the beginning, middle, or end of the list in immediate free recall, recency is only found when subjects retrieve the last items first (Dalezman, 1976).

There is, then, some very strong evidence that recency depends on the adoption of a retrieval strategy that involves recalling the last items first. However, most of the evidence concerns immediate recall, and it appears that output order may be less important in the case of long-term recency. For example, Whitten (1978) examined the effects of instructing subjects to begin recalling at the beginning, middle, or end of the list in the continuous-distractor task. He found that the long-term recency effect in this task was not dependent on recalling the last item first. This evidence obviously challenges the view that recency is always dependent on a particular way of ordering the retrieval process. However, the differential importance of output order in short- and long-term recency is consistent with the view that the retrieval process involves some form of internal discrimination. We saw earlier that support for this idea comes from the "ratio rule" (Glenberg, Bradley, Kraus, & Renzaglia, 1983; Glenberg et al., 1980; Hitch, 1985), which is such that the probability of recall reflects the Weber fraction expressing the interval separating events from one another as a proportion of the interval separating them from the present. Differences in output order will add to the second of these two intervals, leaving the first unaffected. Thus, a change in output order will have a large effect on the discrimination ratio in immediate free recall, since the interval between presentation and retrieval is relatively small. However, the same change in output order will have a much smaller effect on the discrimination ratio in the continuous-distractor task, since the interval between presentation and recall is much larger.

To sum up, there is strong evidence that short-term recency depends on the adoption of a retrieval strategy that involves recalling the last items first. The explicitness of this strategy is evident in its amenability to instruction. However, it remains less clear to what extent long-term recency depends on a similar explicit strategy. We know that output order per se is not critical in this case, but we also know that this can be understood in terms of the discrimination ratio. Interestingly, a number of authors have related the discrimination ratio to the phenomenological experience of mentally "looking back" over the time vista of past events (see, e.g., Crowder, 1976; Hitch, 1985); if valid, this would, of course, imply an explicit strategy.

Discussion

The primary-memory account of recency has a number of strengths. It is simple, it gives a good account of the effects of a filled delay and of the neuropsychological evidence, and, perhaps most importantly, it forms part of a coherent body of evidence related to an important theoretical issue, the need to distinguish between long- and short-term memory. To what extent can our alternative view match these virtues?

There is no doubt that our model is more complex. However, it has considerably broader scope and makes a much more serious attempt to tackle the all-important issue of retrieval. Greater complexity is likely to be an inevitable feature of increased scope and greater attempted precision.

At first sight, the revised model appears to be less able to account for the neuropsychological data, since it appears to change the emphasis from a particular store to a mode of learning coupled with a strategy of retrieval. This discrepancy is, however, more apparent than real, since the revised model of recency is entirely compatible with the view that the immediate free recall of verbal material may depend on the registration of material in a phonological store that may be defective in patients suffering from short-term-memory deficit. Such a view would suggest that short-term-memory patients might well combine the absence of a recency effect in immediate free recall with clear evidence of long-term recency. Exactly this pattern was recently reported in the short-term-memory patient P.V., who showed no recency in immediate free recall of word lists but showed marked long-term recency in her capacity to recall anagram solutions (Vallar, Papagno, & Baddeley, 1991).

In addition to its difficulties in accounting for some of the data, the primary-memory approach also has the draw-back that it tends to raise the further question of whether long- and short-term recency are the same. According to our own interpretation, there is no simple answer to this; the two forms of recency are the same, in that they are assumed to involve the application of the same strategy to primed representations, but they are different in the sense that the representations will typically be within different memory and processing systems. Although posing research questions in an either-or mode can help stimulate controversy and encourage research, such an approach frequently has the longer term penalty of leading to a stand-off. This in turn can result in disillusionment and the abandonment of the question, rather than the development of more complex interpretations that can account for all the data. It is, of course, important that such explanations lead to further testing, but provided they do so, such a synthesis can, in fact, be regarded as a positive step, rather than as reflecting the failure of early theorizing.

Ideally, our link between recency and priming would allow the application of priming theory to the study of recency. Unfortunately, however, priming itself tends to have been regarded as a tool for investigating processing, rather than an area of intrinsic importance that itself demands explanation. It therefore seems appropriate to attempt to develop simultaneously theories of priming and theories of recency. One possible line of attack might be to attempt to link the duration of a given type of recency to the duration of the corresponding priming phenomenon, preferably using free recall and priming paradigms on exactly the same material. Another potential area of fruitful overlap concerns the question of what constitutes a category insofar as recency is concerned. Watkins and Peynircioglu (1983) demonstrated that it is possible to produce several recency effects in parallel, given that the categories of material are sufficiently distinct. This ought to have implications for priming and priming interference, with subsequent items from within a category disrupting priming to a greater extent than items from a distinct category.

A potentially important application of recency is to the task of switching attention from one task to another, a problem that is familiar to any busy administrator. Successful switching depends upon the capacity to reaccess the earlier topic rapidly and accurately, a capacity that one might speculate depends on some form of long-term recency effect. It is, indeed, plausible to assume that this capacity is an important component of executive function and of intelligence, a view that receives at least some support from the observation of Cohen and Sandberg (1977) that intelligence correlates more highly with the recency effect in immediate memory than with performance on earlier items.

Taking an even broader view, one might argue that recency performs the crucial function of helping the organism maintain orientation in time and space (Baddeley, 1988). Note that this is an argument for the usefulness of having privileged access to recent information and not a general justification for human beings having memory. Consider the simple situation of discussing an experiment with a colleague; halfway through the discussion the telephone rings, and you need to discuss the desirability of a repair to your car with the garage, before returning to the conversation with the question, "Now where were we?" One might reasonably argue that you would rely on the recency effect to tell you where you were. A rather more dramatic case occurs when one is traveling and wakes up in a strange bedroom from a deep sleep. Although the latter situation is not easy to investigate experimentally, it should in principle prove possible to set up laboratory simulations that investigate the time it takes to reorientate when switching from one state to another.

Finally, the willingness to consider recency in a broader context than that of primary memory is likely to offer novel ways of conceptualizing phenomena that have so far been largely neglected by students of memory. Consider, for example, the phenomenon of phantom limb pain in patients who have suffered an amputation. Melzack (1992) described the case of a Canadian lumberjack who suffered considerable pain from a sliver of wood lodged under his fingernail. While driving to the hospital, he was involved in an accident that resulted in the need to amputate the arm with the injured hand. He was left with a phantom limb and the continuing experience of the sliver under his nail. This is a particularly dramatic instance of a rather common phenomenon described by Katz and Melzack (1990), who surveyed the extensive evidence for such pain memories and carried out a prospective study to investigate them in more detail. They concluded that "when pain is experienced in a limb at or near the time of amputation there is a high probability that it will persist into the phantom limb and continue to cause the patient distress and suffering.... It appears that if there is a discontinuity or a pain-free interval between the experience of pain and amputation, the likelihood of that pain becoming incorporated into the phantom limb is reduced" (Katz & Melzack, 1990, p. 331). They pointed out that it is now common practice to attempt to ensure that a pain-free interval precedes an amputation. Recency is, of course, not the only interpretation of this result, but it does suggest the possibility of an intriguing explanation that might offer a method of treatment, provided the patient's earlier experience of an intact and pain-free limb can in some sense be repeatedly recollected as part of the treatment, hence allowing the recency effect to be disrupted.

In conclusion, we suggest that the recency effect in immediate free recall should be regarded as simply one instantiation of a much broader phenomenon. This can best be conceptualized in terms of implicit learning, coupled with a particular mode of retrieval that may, but need not be, conscious and explicit. Hence, the recency effect can be viewed as reflecting the utilization of automatic activation by an active, multicomponent, working-memory system.

References

Atkinson, R. C., & Shiffrin, R. M. (1968). Human memory: A proposed system and its control processes. In K. W. Spence (Ed.), *The psychology of learning and motivation: Advances in research and theory* (Vol. 2, pp. 89–195). New York: Academic Press.

Baddeley, A. D. (1968). Prior recall of newly learned items and the recency effect in free recall. *Canadian Journal of Psychology*, **22**, 157–163.

Baddeley, A. D. (1970). Estimating the short-term component in free recall. *British Journal of Psychology*, **61**, 13–15.

Baddeley, A. D. (1981). The cognitive psychology of everyday life. *British Journal of Psychology*, **72**, 257–269.

Baddeley, A. D. (1982). Amnesia: A minimal model and an interpretation. In L. S. Cermak (Ed.), *Human memory and amnesia* (pp. 305–336). Hillsdale, NJ: Erlbaum.

Baddeley, A. D. (1986). *Working memory*. Oxford: Oxford University Press.

Baddeley, A. D. (1988). But what the hell is it for? In M. M. Gruneberg, P. E. Morris, & R. N. Sykes (Eds.), *Practical aspects of memory: Current research and issues: Vol. 1. Memory in everyday life* (pp. 3–18). Chichester: Wiley.

Baddeley, A. D. (1990). *Human memory: Theory and practice*. London: Erlbaum.

Baddeley, A. D. (1992). Working memory. *Science*, **255**, 556–559.

Baddeley, A. D., & Hitch, G. (1974). Working memory. In G. A. Bower (Ed.), *The psychology of learning and motivation* (Vol. 8, pp. 47–89). New York: Academic Press.

Baddeley, A. D., & Hitch, G. (1977). Recency re-examined. In S. Dornic (Ed.), *Attention and performance* (Vol. 6, pp. 647–667). Hillsdale, NJ: Erlbaum.

Baddeley, A. D., & Warrington, E. K. (1970). Amnesia and the distinction between long- and short-term memory. *Journal of Verbal Learning & Verbal Behavior*, **9**, 176–189.

Baddeley, A. D., & Wilson, B. A. (1988). Frontal amnesia and the dysexecutive syndrome. *Brain & Cognition*, **7**, 212–230.

Bjork, R. A., & Whitten, W. B. (1974). Recency-sensitive retrieval processes. *Cognitive Psychology*, **6**, 173–189.

Brooks, D. N. (1975). Long- and short-term memory in head injured patients. *Cortex*, **11**, 329–340.

Brown, N. R., Shevell, S. K., & Rips, L. J. (1986). Public memories and their personal context. In D. C. Rubin (Ed.), *Autobiographical memory* (pp. 137–158). Cambridge: Cambridge University Press.

Burgess, N., Shapiro, J. L., & Moore, M. A. (1991). Neural network models of list learning. *Network*, **2**, 399–422.

Cohen, R. L., & Sandberg, T. (1977). Relation between intelligence and short-term memory. *Cognitive Psychology*, **9**, 534–554.

Craik, F. I. M. (1968). Two components in free recall. *Journal of Verbal Learning & Verbal Behavior*, **7**, 996–1004.

Craik, F. I. M. (1970). The fate of primary memory items in free recall. *Journal of Verbal Learning & Verbal Behavior*, **9**, 143–148.

Craik, F. I. M., & Levy, B. A. (1970). Semantic and acoustic information in primary memory. *Journal of Experimental Psychology*, **86**, 77–82.

Craik, F. I. M., & Lockhart, R. S. (1972). Levels of processing: A framework for memory research. *Journal of Verbal Learning & Verbal Behavior*, **11**, 671–684.

Crowder, R. G. (1976). *Principles of learning and memory*. Hillsdale, NJ: Erlbaum.

Dalezman, J. J. (1976). Effects of output order on immediate, delayed and final recall performance. *Journal of Experimental Psychology: Human Learning & Memory*, **2**, 597–608.

Darley, C. F., Tinklenberg, J. R., Roth, W. T., Hollister, L. E., & Atkinson, R. C. (1973). Influence of marihuana on storage and retrieval processes in memory. *Memory & Cognition*, **1**, 196–200.

Glanzer, M. (1972). Storage mechanisms in recall. In G. H. Bower (Ed.), *The psychology of learning and motivation: Advances in research and theory* (Vol. 5, pp. 129–193). New York: Academic Press.

Glanzer, M., & Cunitz, A. R. (1966). Two storage mechanisms in free recall. *Journal of Verbal Learning & Verbal Behavior*, **5**, 351–360.

Glanzer, M., & Koppenaal, L. (1977). The effect of encoding tasks on free recall: Stages and levels. *Journal of Verbal Learning & Verbal Behavior*, **16**, 21–28.

Glenberg, A. M., Bradley, M. M., Kraus, T. A., & Renzaglia, G. J. (1983). Studies of the long-term recency effect: Support for a contextually guided retrieval hypothesis. *Journal of Experimental Psychology: Learning, Memory, & Cognition*, **9**, 231–255.

Glenberg, A. M., Bradley, M. M., Stevenson, J. A., Kraus, T. A., Tkachuk, M. J., Gretz, A. L., Fish, J. H., & Turpin, B. M. (1980). A two-process account of long-term serial position effects. *Journal of Experimental Psychology: Human Learning & Memory*, **6**, 355–369.

Graf, P., Squire, L. R., & Mandler, G. (1984). The information that amnesic patients do not forget. *Journal of Experimental Psychology: Learning, Memory, & Cognition*, **10**, 164–178.

Greene, R. L. (1986a). A common basis for recency effects in immediate and delayed recall. *Journal of Experimental Psychology: Learning, Memory, & Cognition*, **12**, 413–418.

Greene, R. L. (1986b). Sources of recency effects in free recall. *Psychological Bulletin*, **99**, 221–228.

Hintzman, D. L. (1991). Contingency analyses, hypotheses and artifacts: A reply to Flexser and to Gardiner. *Journal of Experimental Psychology: Learning, Memory, & Cognition*, **17**, 341–345.

Hintzman, D. L., & Hartry, A. L. (1990). Item effects in recognition and fragment completion: Contingency relations vary for different subsets of words. *Journal of Experimental Psychology: Learning, Memory, & Cognition*, **16**, 955–969.

Hitch, G. J. (1985). Short-term memory and information processing in humans and animals: Towards an integrated framework. In L.-G. Nilsson & T. Archer (Eds.), *Perspectives on learning and memory* (pp. 119–136). London: Erlbaum.

Hitch, G. J., Rejman, M. H., & Turner, N. C. (1980, July). *A new perspective on the recency effect*. Paper presented at the meeting of the Experimental Psychology Society, Cambridge, England.

Jacoby, L. L., & Dallas, M. (1981). On the relationship between auto-biographical memory and perceptual learning. *Journal of Experimental Psychology: General*, **110**, 306–340.

Katz, J., & Melzack, R. (1990). Pain "memories" in phantom limbs: Review and clinical observations. *Pain*, **43**, 319–336.

Kirsner, K., & Smith, M. C. (1974). Modality effects in word identification. *Memory & Cognition*, **2**, 637–640.

McKenzie, W. A., & Humphreys, M. S. (1991). Recency effects in direct and indirect memory tasks. *Memory & Cognition*, **19**, 321–331.

Melzack, R. (1992, April). *Memory for pain*. Paper presented at the meeting of the British Psychological Society, Scarborough, England.

Mewaldt, S. P., Hinrichs, J. V., & Ghoneim, M. M. (1983). Diazepam and memory: Support for a duplex model of memory. *Memory & Cognition*, **11**, 557–564.

Monsell, S. (1985). Repetition and the lexicon. In A. W. Ellis (Ed.), *Progress in the psychology of language* (pp. 147–196). London: Erlbaum.

Murdock, B. B., Jr. (1963). Short-term retention of single paired associates. *Journal of Experimental Psychology*, **68**, 184–189.

Parisi, G. (1986). A memory which forgets. *Journal of Physics A: Mathematics & General*, **19**, L617.

Parker, E. S., Schoenberg, R., Schwartz, B. L., & Tulving, E. (1983). Memories on the rising and falling blood-alcohol curve. *Bulletin of the Psychonomic Society*, **21**, 363.

Parkin, A. J., & Streete, S. (1988). Implicit and explicit memory in young children and adults. *British Journal of Psychology*, **79**, 361–369.

Pinto, A. da C., & Baddeley, A. D. (1991). Where did you park your car? Analysis of a naturalistic long-term recency effect. *European Journal of Cognitive Psychology*, **3**, 297–313.

Rejman, M. H. (1979). *Recency and primacy effects in short-term memory*. Unpublished doctoral dissertation, University of Stirling, Stirling, Scotland.

Richardson-Klavehn, A., & Bjork, R. A. (1988). Measures of memory. *Annual Review of Psychology*, **39**, 475–543.

Rundus, D. (1971). Analysis of rehearsal processes in free recall. *Journal of Experimental Psychology*, **89**, 63–77.

Sanders, A. F. (1975). The foreperiod effect revisited. *Quarterly Journal of Experimental Psychology*, **27**, 591–598.

Seamon, J. G., & Murray, P. (1976). Depth of processing in recall and recognition memory: Differential effects of stimulus meaningfulness and serial position. *Journal of Experimental Psychology: Human Learning & Memory*, **2**, 680–687.

Shallice, T., & Warrington, E. K. (1970). Independent functioning of verbal memory stores: A neuropsychological study. *Quarterly Journal of Experimental Psychology*, **22**, 261–273.

Squire, L. (1982). The neuropsychology of human memory. *Annual Review of Neuroscience*, **5**, 241–273.

Thurm, A. T., & Glanzer, M. (1971). Free recall in children: Long-term store vs short-term store. *Psychonomic Science*, **23**, 175–176.

Tulving, E. (1968). Theoretical issues in free recall. In T. R. Dixon & D. L. Horton (Eds.), *Verbal behavior and general behavior theory* (pp. 2–36). Englewood Cliffs, NJ: Prentice-Hall.

Tulving, E., & Schacter, D. L. (1990). Priming and human memory systems. *Science*, **247**, 301–306.

Tulving, E., Schacter, D. L., & Stark, H. E. (1982). Priming effects in word fragment completion are independent of recognition memory. *Journal of Experimental Psychology: Learning, Memory, & Cognition*, **8**, 336–342.

Tzeng, O. J. L. (1973). Positive recency effects in delayed free recall. *Journal of Verbal Learning & Verbal Behavior*, **12**, 436–439.

Vallar, G., Papagno, C., & Baddeley, A. D. (1991). Long-term recency effects and phonological short-term memory: A neuropsychological case study. *Cortex*, **27**, 323–326.

Watkins, M. J., & Peynircioglu, Z. F. (1983). Three recency effects at the same time. *Journal of Verbal Learning & Verbal Behavior*, **22**, 375–384.

Waugh, N. C., & Norman, D. A. (1965). Primary memory. *Psychological Review*, **72**, 89–104.

Whitten, W. B. (1978). Output interference and long-term serial position effects. *Journal of Experimental Psychology: Human Learning & Memory*, **4**, 685–692.

Wilson, B. A., & Baddeley, A. D. (1988). Semantic and episodic memory in a post-meningitic amnesic patient. *Brain & Cognition*, **8**, 31–46.

6 The concept of working memory

A view of its current state and probable future development

A. D. Baddeley*

In the early 1970s, Graham Hitch and I began a series of experiments which aimed to answer the apparently simple question 'What function does short-term memory serve?' (Baddeley and Hitch, 1974). We assumed, as had many others before us that STM serves as a working memory, a temporary storage system that plays a crucial role in many information processing tasks ranging from speech comprehension to arithmetic and from learning to complex reasoning. It followed that if we were to load up STM with some secondary task, then there would be little capacity left for comprehending or calculating, learning or reasoning and hence the subject's performance on these tasks would be dramatically disrupted. We tested this view, using immediate memory for digit strings as our secondary task and requiring our subjects to remember strings of six digits at the same time as performing tasks involving verbal reasoning, comprehending prose or learning lists of words. We did obtain performance decrements, but we were impressed by how little performance was disrupted by a near-span digit load that should have almost totally occupied STM.

In attempting to understand our results, we began to modify our view of STM and to replace it with the concept of working memory (WM). In doing so, we moved away from a strategy of exploring the capacities and implications of a single hypothetical structure (STM) and towards a more functional analysis. Working Memory refers to the role of temporary storage in information processing.

The concept of working memory

The concept of WM with its implication of a limited capacity system for holding and manipulating information has motivated research in a number of areas. In my own case these have included the study of reading (Baddeley, 1979; Baddeley and Lewis, in press; Baddeley, Eldridge and Lewis, in press), while Daneman and Carpenter (1980) have shown that a measure of WM capacity provides an excellent predictor of reading ability in college students, a result which we have subsequently replicated over a wider range of age and ability. An as-yet unpublished series of experiments on the role of WM in retrieval from LTM suggests that whereas WM is important for learning, the process of searching and retrieving from LTM may be largely automatic. Hitch has used the concept highly successfully in studying the processes underlying simple arithmetic (Hitch, 1978), while Broadbent (in press) employs the concept in exploring the role of organization in memory. Finally, an impressive and ingenious series of experiments by Rabbitt, Cohen and their co-workers have shown that much of the cognitive deterioration that accompanies normal aging can be attributed to a decrement in the performance of working memory (Rabbitt, in press).

There is no doubt then that the concept of WM is proving fruitful. However, there is a danger that its very success may lead to its abuse. Given almost any poorly understood performance decrement, it is possible to attribute it to the inadequate performance of WM. Hence, if the concept is to continue to be useful, it is important that an attempt is made to ensure that it is not simply used as a label for one's ignorance of the underlying cause of a given decrement. Perhaps the best way of ensuring that it does not become a vacuous catch-all concept is to develop and test specific models of WM which will allow testable and non-trivial predictions.

A model of working memory

In order to account for our various results, Graham Hitch and I formulated a simple model which we hoped would provide a framework for a more detailed analysis of WM. The model subdivided WM into three components. The *Central Executive* which formed the control centre of the system was assumed to select and operate various control processes. It was assumed to have a limited amount of processing capacity, some of which could be devoted to the short-term storage of information. It was able to offload some of the storage demands on to subsidiary slave systems of which two were initially specified, namely the *Articulatory Loop* which was able to maintain verbal material by subvocal rehearsal, and the *Visuo-Spatial Scratch Pad* which performed a similar function through the visualization of spatial material.

The system we proposed was not a predictive model; we were sure that WM would prove to be far more complex than our original conception, and that given the state of our knowledge, any attempt to make a rigidly specified predictive model was bound to fail. What we proposed was much more in the spirit of a tentative map of new terrain, giving broad guidelines and suggesting areas for more detailed exploration. The evaluation of this type of theory rests on its fruitfulness in generating new knowledge and fresh insights. How successful has our model been?

The articulatory loop

This is the most extensively explored component of the WM model. In its original form, the articulatory loop was assumed to resemble a tape loop of limited duration which could be used to store any information that could be articulated. It has the advantage of neatly tying together results indicating that immediate memory span is susceptible to the effects of both phonological similarity and word length and that both these effects vanish when subvocal rehearsal is suppressed by requiring the subject to articulate repeatedly some irrelevant word such as 'the' or 'double' (Baddeley and Hitch, 1974; Baddeley, Thomson and Buchanan, 1975).

The concept of an articulatory loop has proved useful in investigating other areas. For example, Ellis and Hennelly (1980) used the concept to explain why the norms for the digit span in the Welsh language version of the WISC were consistently lower than the English language equivalent. Welsh digits take longer to articulate than English, and when this factor is eliminated by articulatory suppression, or allowed for by using a time-based measure of span, the Welsh-English difference disappears. As predicted by Hitch's work on WM and arithmetic, the longer spoken duration of Welsh digits also leads to an increase in certain errors in mental arithmetic.

Nicholson (in press) explored the hypothesis that the increase in children's memory span with age might stem from the development of the articulatory loop. He showed a very close relationship between age, memory span and the rate at which children were able to articulate digits.

Research on the role of the articulatory loop in fluent reading has shown that it is not essential for comprehension (Baddeley, 1979), but provides a supplementary source of information that may be important when a high level of accuracy is required (Baddeley, Eldridge and Lewis, in press; Levy, 1978). A number of results suggest that the articulatory loop may play an important role in learning to read (Conrad, 1972; Liberman, Shankweiler, Liberman, Fowler and Fischer, 1977) although the evidence is still not conclusive. Taken overall then, the concept of an articulatory loop does appear to be proving a fruitful one.

Attempts to analyse the articulatory loop in more detail however, have thrown up a number of apparent anomalies. For example, although articulatory suppression removes the effects of phonemic similarity and word length on memory span for visually presented items, with auditory presentation the phonemic similarity effect reappears despite suppression, indicating that phonemic coding is not dependent on the process of articulation. It was initially thought that this was also true of the word length effect, but a recent study has shown that provided articulation is prevented throughout both presentation and recall no word-length effect occurs, even with auditory presentation (Baddeley, Lewis and Vallar, unpublished). It appears then that the word-length effect reflects the process of sub-vocal rehearsal—long words lead to poor recall because of the slower rate at which the memory trace can be refreshed by rehearsal. The nature of the phonemic similarity effect remains unclear. It may reflect the speech-based system which holds the codes that feed the articulatory loop. If so, we must conclude that such codes can be set up either auditorily or by articulation, but not purely visually. Alternatively, the phonemic similarity effect may result from either of two separate codes, one articulatory and the other auditory. The similarity between articulatory and auditory features makes it difficult in practice to distinguish between two such codes.

Further evidence for the existence of a non-articulatory phonological code comes from studies of reading (Baddeley and Lewis, in press; Besner, Davies and Daniels, in press). Articulatory suppression was found to have little effect on either the speed or accuracy with which subjects could make rhyme judgments. Try reading the next sentence while repeating the word 'double' to yourself; most people can still 'hear' an 'inner voice' speaking the words they read. It seems likely that this non-articulatory phonological code will be actively investigated in the next few years; the discovery of a technique for suppressing this code would be very helpful at this point, and might indicate whether it should be considered as an auditory image, or as an abstract pre-articulatory code that is sufficiently deep to be unaffected by concurrent articulation.

The visuo-spatial scratch pad

Although the experiments supporting the concept of a visuo-spatial scratch pad (VSSP) were carried out at the same time as the initial articulatory loop studies they have only just appeared in print (Baddeley and Liberman, 1980), and have so far generated little further research. The scratch pad concept is however, consistent with the flood of research on spatial imagery that has appeared in recent years. It further suggests that the dichotomy between propositional and analogical views of imagery may be a false one. It is for example, quite likely that the scratch pad is a device which takes propositional codes from long-term memory and manipulates and displays them *via* an analogical peripheral system.

The concept of a VSSP provides an appropriate framework for integrating work on visual imagery such as that popularized by Shepard and his co-workers (e.g., Shepard and Metzler, 1971) with work on individual differences in imagery (Ernest, 1977) and spatial intelligence tests (Hunt, 1980). The major theoretical puzzle in this area concerns the existence of a second imagery system concerned with non-spatial, pictorial or pattern information. It is clear phenomenologically that such non-spatial factors as colour, texture and shape can be imaged. It is also clear experimentally that a concurrent spatial task will disrupt spatial imagery, but does not diminish the powerful effects of imageability on memory for words (Baddeley and Liberman, 1980). While evidence for the disruption of such a pictorial system has been reported (At-wood, 1971; Janssen, 1976) effects tend to be small and difficult to replicate. A good technique for disrupting visual but not spatial imagery would represent a major breakthrough in this area.

The central executive

This represents the most important but least understood of the three initial components of WM, and presents the most difficult problems both technically and conceptually. An adequate theory of the Central Executive would probably include not only a specification of its method of manipulating control processes and integrating the growing number of peripheral systems, but would also require an understanding of selective attention and probably of the role and function of consciousness.

In order to make *some* progress, we have adopted a policy of beginning with the more peripheral components of WM, gradually attempting to separate out further subcomponents of the executive. The hope here is that the more components that can be separated off as tractable subproblems, the greater the chance of reducing the Central Executive to a problem of manageable proportions. However, it is at present probably fair to regard the Central Executive as the area of our residual ignorance about WM: it is at present unfortunately very large and very important. However, some progress has been made.

We originally assumed that a concurrent digit load impaired performance on tasks such as comprehension and reasoning because any digits that exceeded the capacity of the articulatory loop had to be held in the central executive; each such digit was assumed to demand some of the limited processing capacity available. Two unpublished subsequent results argue against this. First, a parametric study of the effects of digit load on reasoning showed that some subjects were able to hold seven or eight digits while performing a complex verbal reasoning task with no apparent decrement in speed or accuracy. Although all subjects showed a breakdown in performance if the load were large enough, they seemed to be able to handle many more digits than we had assumed could be maintained in the articulatory loop. This pattern of breakdown suggests a separate verbal memory component that does not interfere with reasoning until its capacity is exceeded. Such a view is consistent with the neuropsychological evidence reviewed by Shallice (1979).

A similar conclusion is suggested by a detailed examination of the interaction of the memory and reasoning tasks. Suppose we assume a pool of general processing capacity that must be shared between the two tasks. For a given level of difficulty, any tendency to assign more capacity to reasoning will aid that task but only at the expense of the memory task, and *vice versa*. In short, performance on the two tasks should be negatively correlated. Examination of the data suggests exactly the opposite; fast reasoning goes with good digit recall, and *vice versa*. Such a pattern of results is inconsistent with the assumption of a pool of general processing capacity. It suggests rather two separate though related systems, a memory system and a controller which runs it. Provided the memory system is coping, it will place few demands on the controller, which can therefore use its available capacity for performing the reasoning task. Once the memory system becomes overloaded however, it requires the controller to bring in further resources in an attempt to avoid breakdown and consequent loss of the stored material. The more attention that is devoted to supporting the memory component, the less is available for reasoning, and the slower and more error prone is the reasoning performance.

It seems probable then that we should distinguish between a verbal memory component and an attentional component. Should this memory component continue to be regarded as part of the Central Executive? Probably not, in which case it could be argued that the Central Executive is becoming increasingly like a pure attentional system. Whether or not this proves to be so, it seems likely that any adequate model of WM will also have to be a model of attention.

Note

* This note was written at the University of Guelph; the support of their short-term visiting professorship scheme is gratefully acknowledged. I wish to thank Tim Shallice for his comments on the manuscript. Reprint requests should be sent to: A. Baddeley, MRC Applied Psychology Unit, 15, Chaucer Road, Cambridge, England.

References

Atwood, G. E. (1971) An experimental study of visual imagination and memory. *Cog. Psychol.*, 2, 290–299.

Baddeley, A. D. (1979) Working memory and reading. In P. A. Kolers, M. E. Wrolstad and H. Bouma (eds.), *Processing of Visible Language*. New York, Plenum.

Baddeley, A. D., Eldridge, M., and Lewis, V. J. (In press) The role of subvocalization in reading. *Q. J. exper. Psychol.*

Baddeley, A. D., and Hitch, G. J. (1974) Working memory. In G. A. Bower (ed.), *The Psychology of Learning and Motivation*, Vol. 8, New York, Academic Press. Pp. 47–90.

Baddeley, A. D., and Lewis, V. J. (In press) Inner active processes in reading: The inner voice, the inner ear and the inner eye. In A. M. Lesgold and C. A. Perfetti (eds.), *Interactive Processes in Reading*. Hillsdale, NJ., Erlbaum.

Baddeley, A. D., and Lieberman, K. (1980) Spatial working memory. In R. Nickerson (ed.), *Attention and Performance VIII*, Hillsdale, NJ., Erlbaum.

Baddeley, A. D., Thomson, N., and Buchanan, M. (1975) Word length and the structure of short-term memory. *J. verb. Learn. verb. Behav.*, 14, 575–589.

Besner, D., Davies, J., and Daniels, S. (In press) Phonological processes in reading: The effects of concurrent articulation: *Q. J. exper. Psychol.*

Broadbent, D. E. (In press) From the percept to the cognitive structure. In J. B. Long and A. D. Baddeley (eds.), *Attention and Performance IX*, Hillsdale, NJ., Erlbaum.

Conrad, R. (1972) Speech and reading. In J. F. Kavanagh and I. G. Mattingley (eds.), *Language by Ear and by Eye*. Cambridge, Mass., MIT Press.

Daneman, M., and Carpenter, P. A., (1980) Individual differences in working memory and reading. *J. verb. Learn. verb. Behav.*, 19, 450–466.

Ellis, N. C., and Hennelley, R. A. (1980) A bilingual word-length effect: Implications for intelligence testing and the relative ease of mental calculation in Welsh and English. *Brit. J. Psychol.*, 71, 43–52.

Ernest, C. H. (1977) Imagery ability and cognition: A critical review. *J. of Ment. Imag., 2,* 181–216.

Hitch, G. J. (1978) The role of short-term working memory in mental arithmetic. *Cog. Psychol., 10,* 302–323.

Hunt, E. (1980) Intelligence as an information-processing concept. *Brit. J. Psychol., 71,* 449–474.

Janssen, W. H. (1976) Selective interference during the retrieval of visual images. *Q. J. exper. Psychol., 28,* 535–539.

Levy, B. A. (1978) Speech analysis during the sentence processing: Reading and listening. *Visible language, 12,* 81–101.

Liberman, I. Y., Shankweiler, D., Liberman, A. M., Fowler, C. and Fischer, F. W. (1977) Phonetic segmentation and recoding in the beginning reader. In A. S. Reber and D. Scarborough (eds.), *Towards a Psychology of Reading.* Hillsdale, NJ., Erlbaum.

Nicholson, R. (In press) The relationship between memory span and processing speed. In. M. Friedman, J. P. Das and N. O'Connor (eds.), *Intelligence and Learning,* New York, Plenum Press.

Rabbitt, P. (In press) Cognitive psychology needs models for changes in performance with age. In J. B. Long and A. D. Baddeley (eds.), *Attention and Performance IX,* Hillsdale, NJ., Erlbaum.

Shallice, T. (1979) Neuropsychological research and the fractionation of memory systems. In L-G. Nilsson (ed.), *Perspectives on Memory Research.* Hillsdale, NJ., Erlbaum.

Shepard, R. N., and Metzler, J. (1971) Mental rotation of three-dimensional objects. *Sci., 171,* 701–703.

Part III

The phonological loop[1]

Our approach to working memory was strongly influenced by the many earlier studies of verbal short-term memory. Because of extensive existing knowledge we felt that the loop was likely to be the most tractable component of our model. The central executive was likely to be most important but also the most difficult to investigate, depending as it did on understanding the processes underpinning attentional *control*, a topic on which there was relatively little research at the time, in contrast to extensive literature on the role of attention in *perception*.

We proposed a relatively simple phonological loop system that accepted Conrad's proposal of an acoustic store within which memory traces would fade unless rehearsed. The first chapter in this section attempts to investigate the process of rehearsal by making the assumption that it occurs in real time with the result that longer words will take longer to rehearse allowing more trace decay to occur. We were concerned that long and short words might differ in ways other than length and attempt to cover some of these possible alternative explanations. While the word length effect is substantial and replicable, its proposed interpretation in terms of trace decay continues to be controversial. Our attempt to contrast bisyllabic words that differ in spoken length has proved particularly controversial. Our effect was replicated using our original word sample but other samples failed to show the effect (e.g. Lovatt, Avons & Masterson, 2000). These failures to replicate have in turn been criticised for their method of measuring articulation rate and for failing to adequately control other variables such as phonological similarity (Mueller, Seymour, Kieras & Meyer, 2003). My own view is that a lack of enough words that are controlled for all possible confounding variables constrains this approach to a level from which is difficult to draw firm theoretical conclusions. Fortunately, while the controversy as to whether short-term forgetting depends on trace decay or interference is of basic interest, either explanation can fit into the broad working memory framework proposed.

The second chapter in this section relates to the use of articulatory suppression to disrupt subvocal rehearsal and the implications of this for the phonological loop hypothesis. This is followed by a neuropsychological study of a patient with a very pure deficit in short-term verbal memory, demonstrating the way in which this impacts on long-term memory as reflected in her capacity to learn new phonological material in the form of foreign language vocabulary. The final chapter reviews the evidence that the phonological loop plays a central role in the development of language, surveying its relevance to the acquisition of native vocabulary, to the development of reading and to the acquisition of further languages.

Note

1 We initially referred to this subsystem as the *articulatory loop*, subsequently becoming concerned that this might over emphasise the rehearsal rather than the storage component. The basis of the code underlying the store remained controversial with claims for both its articulatory and acoustic nature. We initially opted for the term *phonemic* but, when told that this had a specific implication within the speech processing community, switched to *phonological*. We later discovered that this too had specific and unwanted implications but, at this point, gave up. The term is intended to be neutral, accepting that this, like the nature of much of the phonological loop remains an open question, to be resolved in due course by those who are more expert in speech perception and production than myself.

References

Lovatt, P., Avons, S. E., & Masterson, J. (2000). The word length effect and disyllabic words. *Quarterly Journal of Experimental Psychology, 53A*, 1–22.

Mueller, S. T., Seymour, T. L., Kieras, D. E., & Meyer, D. E. (2003). Theoretical implications of articulatory duration, phonological similarity, and phonological complexity in verbal working memory. *Journal of Experimental Psychology: Learning, Memory, and Cognition, 29*, 1353–1380.

Papers

1 Baddeley, A. D., Thomson, N., & Buchanan, M. (1975). Word length and the structure of short-term memory. *Journal of Verbal Learning and Verbal Behavior, 14*, 575–589.
2 Baddeley, A. D., Lewis, V. J., & Vallar, G. (1984). Exploring the articulatory loop. *Quarterly Journal of Experimental Psychology, 36*, 233–252.
3 Baddeley, A. D., Papagno, C., & Vallar, G. (1988). When long-term learning depends on short-term storage. *Journal of Memory and Language, 27*, 586–595.
4 Baddeley, A. D., Gathercole, S. E., & Papagno, C. (1998). The phonological loop as a language learning device. *Psychological Review, 105*, 1, 158–173.

7 Word length and the structure of short-term memory

A. D. Baddeley, N. Thomson and M. Buchanan

A number of experiments explored the hypothesis that immediate memory span is not constant, but varies with the length of the words to be recalled. Results showed: (1) Memory span is inversely related to word length across a wide range of materials; (2) When number of syllables and number of phonemes are held constant, words of short temporal duration are better recalled than words of long duration; (3) Span could be predicted on the basis of the number of words which the subject can read in approximately 2 sec; (4) When articulation is suppressed by requiring the subject to articulate an irrelevant sound, the word length effect disappears with visual presentation, but remains when presentation is auditory. The results are interpreted in terms of a phonemically-based store of limited temporal capacity, which may function as an output buffer for speech production, and as a supplement to a more central working memory system.

Miller (1956) has suggested that the capacity of short-term memory is constant when measured in terms of number of chunks, a chunk being a subjectively meaningful unit. Because of the subjective definition of a chunk, this hypothesis is essentially irrefutable unless an independent measure of the nature of a chunk is available. Typically this problem has been avoided by making the simplifying assumption that such experimenter-defined units as words, digits, and letters constitute chunks to the subject. Hence, although Miller's hypothesis is not refutable in the absence of an independent measure of a chunk, it is meaningful to test a weaker version, namely that the capacity of short-term memory is a constant number of items, where items are defined experimental units. Words represent one commonly accepted type of item, and in this case, the chunking hypothesis would predict that the capacity of short-term memory, as measured in words, should be constant regardless of the size or duration of the words used.

A number of studies testing this hypothesis have used the recency effect in free recall as an estimate of short-term memory capacity. Craik (1968) found no reliable effect of word length on performance in the free recall of separate groups of words comprising one to five syllables. This invariance held true whether performance was measured in terms of either raw scores, or estimates of primary memory and secondary memory components. This result was replicated and extended by Glanzer and Razel (1974) who observed a recency effect which was constant when measured in number of items, even when an item comprised a whole proverb rather than a single word. They concluded from their study that short-term or primary memory has a capacity of two items regardless of item duration or complexity.

Miller's generalization, however, was based on the memory span paradigm, and it is questionable whether recency and span depend on the same memory mechanisms. There is indeed a growing body of evidence suggesting that the recency effect in free recall is basically unrelated to short-term memory as measured by memory span. Such evidence includes:

1. Craik's (1970) observation that a subject's memory span correlates more highly with the secondary memory than the primary memory component of free recall.
2. Memory span shows clear evidence of speech coding, being impaired by both phonemic similarity (Conrad, 1964; Baddeley, 1966) and articulatory suppression (Levy, 1971). This is not the case for the recency effect in free recall which is unaffected by either phonemic similarity (Craik & Levy, 1970; Glanzer, Koppenaal, & Nelson, 1972) or articulatory suppression (Richardson & Baddeley, 1975).
3. Baddeley and Hitch (1974) have shown unimpaired recency in free recall for subjects performing a concurrent memory span task involving the retention of a sequence of six digits. Since the memory span task did not interfere with recency, it is difficult to maintain the view that the two tasks are based on the same limited-capacity system.

Studies investigating the effect of word length on memory span do not in general support the weak version of Miller's hypothesis. Thus, unpublished work by Laughery, Lachman, and Dansereau (Note 1) and by Standing, Bond, and Smith (Note 2) have reported poorer performance in a memory span task when longer words are used. Mackworth (1963) found a high correlation between reading rate and memory span for a wide range of materials, including pictures, letters, digits, shapes, and colors. This result could be interpreted in terms of word length as a determinant of memory span, with reading rate providing an indirect measure of word length. The situation is, however, complicated by the fact that subjects in some cases were asked to label pictures, and in others to read words so that it is not clear whether the result is due to articulation time or to difficulty in retrieving the correct verbal label. Watkins and Watkins (1973) present the clearest published evidence for an effect of word length on memory span, in a study primarily concerned with the modality effect. They found evidence for a word length effect on earlier serial positions, but observed that the modality effect (the enhanced recall of auditorily presented items) did not interact with word length. They suggest that the word length effect observed may have been due to the greater difficulty of perceiving their four-syllable words which were presented at a 1/sec rate.

These studies do not support the hypothesis that memory span capacity is a constant number of items. However, it is always possible to save the item-based hypothesis by questioning the assumption that words constitute items. Given evidence that short-term memory is a speech-based system, it could be reasonably argued that its capacity should be measured in more basic speech units such as syllables or phonemes. The experiments that follow aim first to study the influence of word-length on memory span, secondly to explore the relative importance of number of syllables and temporal duration of a word as determinants of span, and thirdly to explore the implications of this for the question of whether the underlying memory system is time-based or item-based.

Experiment I

This study compared the memory span of subjects for sets of long and short words of comparable frequency of occurrence in English. One set comprised eight monosyllables, namely, *sum, hate, harm, wit, bond, yield, worst,* and *twice.* The other set comprised eight five-syllable words, namely *association, opportunity, representative, organization, considerable, immediately, university,* and *individual.*

Method

Five list lengths were used, comprising sequences of four, five, six, seven, and eight words. Eight sequences of each length were made up from the pool of short words, and eight from the pool of long words. In both cases, sequences were generated by sampling at random without replacement from the appropriate pool of words. All subjects were tested on both long and short words, and all received the sequences in ascending order of list length, beginning with sequences of four words and proceeding up to the point at which they failed on all eight sequences, whereupon testing on the pool of words in question was discontinued. Half the subjects began with the pool of long words, and half with the short words.

The words were read to the subject at a 1.5-sec rate, with each list being preceded by the spoken warning "Ready." Subjects were allowed 15 sec to recall the words verbally in the order presented. Subjects were allowed to familiarize themselves with the two pools of words at the beginning of the experiment, and these two pools remained visible to the subjects on prompt cards throughout the experiment. Several different prompt cards with the words in differing orders were used in this and subsequent experiments so as to prevent the subjects from using location on the card as a cue. The subjects were eight undergraduate or post-graduate students from the University of Stirling.

Results and discussion

Performance was scored in terms of number of sequences recalled completely correctly (i.e., all the items correct and in the correct order). Figure 1 shows the level of performance at each sequence length for the long and the short words. There is a very clear advantage to the short word set which occurs at all sequence lengths and is characteristic of all eight subjects tested.

There is little doubt that the sample of short words used results in better memory span performance than the sample of long words. However, it is arguable that polysyllabic words tend to be linguistically different from monosyllables. In particular, our polysyllables tended to be of Latin origin, compared to the monosyllables which seemed to comprise simpler words of Anglo–Saxon origin. Experiment II attempted to avoid this problem by using words from a single category, country names, a sample of material unlikely to come from any single language source.

Experiment II

Method

Sequences of five words were constructed by sampling without replacement from each of two pools. The pool of short words comprised the country

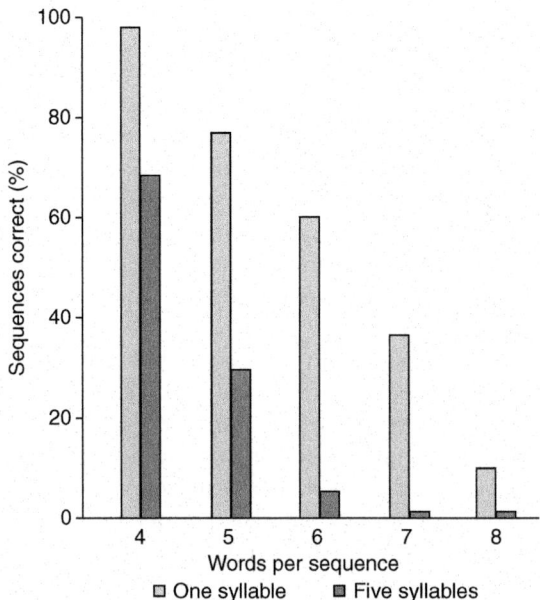

Figure 1 Effect of word length on memory span. Mean percentage recall of long and short words as a function of sequence length.

names *Chad, Chile, Greece, Tonga, Kenya, Burma, Cuba, Malta,* while the long names were *Somaliland, Afghanistan, Venezuela, Czechoslovakia, Yugoslavia, Ethiopia, Nicaragua,* and *Australia*. The names were selected on the basis of their probable familiarity to the subjects, and because they had a similar frequency of repetition of initial and final letters within the pool. Subjects were tested on a total of eight sequences of five short names and eight sequences of five long names. Eight undergraduate subjects were tested using the same presentation procedure as Experiment I.

Results and discussion

Table 1 shows the mean number of sequences recalled completely correctly, and the mean number of items recalled in the appropriate serial position, for long and short names. On both these scores all eight subjects showed a clear word length effect. Since the material in this study was very different at a linguistic level from the material used in the previous study, and since the effect is very large in both cases, it is clear that the word length effect is a robust phenomenon of some generality. However, in these and all previous experiments investigating the effect of word length, two major variables are confounded, namely a word's spoken duration and the number of syllables it contains. The results could therefore indicate either that memory span is limited in the number of items it can hold, with the item being the syllable, or that the temporal duration of the words determines the size of memory span. The latter possibility might be predicted by decay theory (Broadbent, 1958) which assumes that forgetting occurs as a function of time. Many studies have attempted to test the theory by measuring performance as a function of presentation rate, and while some studies report enhanced performance with rapid presentation as predicted by decay theory (Conrad & Hille, 1958), others have found the opposite (Sperling & Speelman, 1970). However, in none of these studies was the subject prevented from rehearsing, and this makes interpretation of the results difficult as the subject is effectively re-presenting the list to himself at a rate of his own choosing. This problem can be avoided by allowing the subject to rehearse while using lists of

Table 1 Mean number of sequences and items correctly recalled as a function of word length in Experiment II

	Short names		Long names	
	Mean	SD	Mean	SD
Sequences correct Max = 8	4.50	2.00	.88	1.27
Items correct Max = 5	4.17	.71	2.80	.24

long- and short-duration words. As less long words than short words can be rehearsed in a given period of time, a word duration effect will be predicted by decay theory (Sperling, 1963). On the other hand, a simple displacement or interference model would predict an effect of number of items, but not duration. Thus, the hypothesis that short-term memory capacity is a constant number of items, where the syllable is the item, predicts no word length effect for words matched for syllable number, but differing in spoken duration. Decay theory, on the other hand, predicts that the amount recalled will be a function of word duration. The next experiment tests these predictions.

Experiment III

Method

Two pools of disyllabic words, matched for frequency, were produced such that one set tended to have a longer duration when spoken normally. The long word set comprised: *Friday, coerce, humane, harpoon, nitrate, cyclone, morphine, tycoon, voodoo*, and *zygote*, and the short words were *bishop, pectin, ember, wicket, wiggle, pewter, tipple, hackle, decor*, and *phallic*. The words were recorded by a female experimenter onto magnetic tape, which was then played through an oscillograph. This plots the wave-form of the signal against time, allowing the duration of the utterance to be measured. The mean duration of the long words was 0.77 sec, and of the short words, 0.46 sec.

From each pool of words, 10 lists of five words were constructed by sampling at random without replacement. The twenty lists were divided into four blocks of five, two comprising lists of short duration words and two of long duration words. A Latin square design was then used to present the blocks in counterbalanced order to each of the 12 subjects. Words were read at a 2-sec rate, and subjects were required to recall verbally at the same rate, paced by a metronome. Recall was paced so as to ensure that the mean delay between input and recall was comparable for long and short words (Conrad & Hille, 1958). Subjects were familiarized with the set of words and with the procedure, and were instructed to commence recall as soon as the last item in each list had been presented. Twelve undergraduates from the University of Stirling served as subjects.

Results and discussion

Figure 2 shows the mean number of words correctly recalled as a function of serial position. A three-way analysis of variance involving subjects, word length, and serial position showed significant effects of word length, $F(1, 11) = 11.33$, $p < .01$, serial position, $F(4, 44) = 36.82$, $p < .001$, and a significant interaction between word length and serial position, $F(4, 44) = 3.28$, $p < .05$. Analysis by t test showed that the word length effect was significant for serial positions 1, 2, and 3, but not for positions 4 and 5.

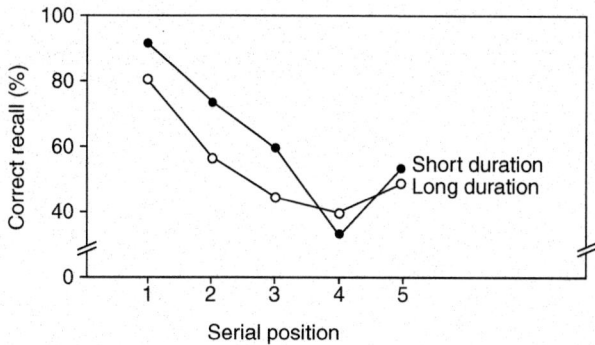

Figure 2 Mean recall of disyllabic words of long and short temporal duration.

These results are very similar to those of Watkins and Watkins (1973) showing a word length effect only for the earlier serial positions; this could reflect the masking of an underlying word length effect by the modality effect. However, the experiment differs from the Watkins and Watkins study in using words which are matched for number of syllables, but differ in spoken duration. As such, the results are consistent with decay theory, and are inconsistent with the hypothesis that short-term memory holds a constant number of syllables.

The last version of Miller's weakened hypothesis to be investigated is that short-term memory holds a constant number of phonemes. In the last experiment, there was a clear tendency for the long words to have more constituent phonemes, thus the result is open to the interpretation that the word length effect represents a limit to the number of phonemes that can be held. Experiment IV compares performance on sets of words which are matched for number of constituent phonemes, but which differ in duration. Decay theory again predicts a difference in performance in favor of the short duration words.

Experiment IV

Two sets of words were generated with the following constraints: They differed in spoken duration; they were equal in number of syllables; they were matched for word frequency; and they were equal in number of phonemes (with Scottish pronunciation). Given all these constraints, the previous sets of words reduced from 10 to five; details are given in Table 2. Sequences of five words were produced, and the experiment performed using a procedure identical to that used in Experiment III, except that the presentation and paced recall rate was increased to 1 sec per word. Eight Scottish undergraduates served as subjects.

Table 2 Details of words used in Experiment IV

	Words	Frequency	Number of phonemes	Duration (sec)
Long	Coerce	1	5	.80
	Harpoon	1	6	.75
	Friday	40	5	.70
	Cyclone	3	6	.88
	Zygote	—	5	.90
Short	Wicket	1	5	.50
	Pectin	1	6	.60
	Bishop	40	5	.28
	Pewter	3	6	.40
	Phallic	—	5	.42

Results and discussion

Subjects recalled a mean of 61.6% of the long words and 72.2% of the short. A three-way analysis of variance showed a significant effect of word length, $F(1, 7) = 18.9$, $p < .01$, and of serial position, $F(4, 28) = 38.06$, $p < .001$, but no interaction between serial position and word length, $F(4, 28) = 0.55$, $p > .05$.

It is clear then that word duration may influence span when the number of both syllables and phonemes is held constant. The absence of an interaction between word length and serial position is puzzling in view of the previous result, it may be due to either the change in material, or more likely, the change in presentation and recall rate. However, despite this minor discrepancy between experiments III and IV, both seem to concur in suggesting that the temporal duration of items is a powerful determinant of memory span. Before finally dismissing the hypothesis that short-term memory capacity is a constant number of items, a procedural point that could have distorted the results should be mentioned. In both experiments the same experimenter read out the words and it is possible that some incidental feature of her mode of delivery produced the observed effect. To avoid this possibility, the experiment was repeated using visual presentation at a 2-sec rate and to our dismay, a statistically reliable word length effect was not observed.

However, a closer examination of the data revealed that most subjects did show the predicted effect, but that two out of eight did substantially better on the long words. On testing a further set of subjects and asking them how they remembered the material, it was found that those who did best on the short words reported using a rehearsal strategy, whilst those who did better on the long words reported using an imagery strategy. Use of this latter strategy was facilitated by the fact that the presentation rate had been reduced to 2 sec/word in order to obviate perceptual difficulties. As the subject of investigation is the articulatory short-term memory system, it is reasonable to instruct subjects to use a rehearsal strategy in order to avoid this difficulty. The next experiment, then, is a replication of the previous one, but using visual presentation with an instruction to the subjects to rehearse.

Experiment V

The same material and design were used as in the previous experiment, except that the material was presented visually on flash cards at a 2-sec rate, and recall was unpaced. The duration of the words was also measured in a different way. The duration of a word is determined by two sets of variables, the acoustic nature of the word and the subjects' articulatory rate. The latter variable has been shown to be very stable over a wide range of conditions within a subject, but to vary considerably between subjects (Goldman–Eisler, 1961), and, as decay theory assumes rehearsal rate determines performance, the subject's rather than the experimenter's pronounciation of the words was used.

Two different estimates of rehearsal rate were made. In the first of these, subjects were timed for reading the 10, five-word lists in each condition, as quickly as they could out loud, the 50 words being typed out in two columns. This was done four times for each word length after the memory task, times being recorded by stopwatch. The times so obtained were transformed into reading rate (RR) scores in units of words per second. The second estimate of rehearsal rate involved requiring the subject to repeat continuously three of the words from one of the pools out loud. Subjects did this as quickly as they could, and were timed by stopwatch for 10 repetitions of the three words. For each condition, they did this four times, always with a different set of three words, and always after the memory task. These times were transformed into articulatory rate scores (AR) in units of words per second. Half the subjects did the reading rate test first, and half the articulatory rate test first. The subjects, who were instructed to remember the lists by repeating the words to themselves, were eight members of the Applied Psychology Unit subject panel who were paid for their services.

Results and discussion

Subjects recalled a mean of 53.4% of the long words correctly and in the right order, and 71.7% of the short words. Analysis of variance showed that there was a significant effect of word length, $F(1, 7) = 15.14$, $p < .01$, indicating that the word duration effect is not dependent on auditory presentation. There was again a significant effect of serial position, $F(4, 28) = 14.79$, $p < .001$, but the interaction between word length and serial position failed to reach significance, $F(4, 28) = 2.43$, $.05 < p < .1$. These results are again inconsistent with the hypothesis that short-term memory capacity is a constant number of items. An alternative view, that short-term memory is a time-based system, will next be explored and the adequacy of decay theory in this context empirically investigated.

Let us assume that the memory system underlying the word length effect exhibits trace decay, but that rehearsal may revive a decaying trace. It then follows that the amount recalled will be a function of rehearsal rate. Thus, if it can be assumed that reading rate (RR) and articulation rate (AR) are good estimates of rehearsal rate, then it should be possible to use them as predictors of memory span. Table 3 shows the ratio of memory span to reading rate and to articulation rate across conditions. A Wilcoxon matched pairs test showed that there was no effect of conditions for either the memory span-reading rate ratio, $T=9$, $N=8$, $p>.05$, or for the memory span-articulation rate ratio, $T=10$, $N=6$, $p>.05$. In short, Table 3 indicates that a subject can recall as many words as he can read in 1.6 sec, or can articulate in 1.3 sec. The next experiment explored this relationship in more detail using five different word lengths rather than two.

Experiment VI

Method

Five pools of 10 words were constructed. Each pool comprised one word from each of 10 semantic categories, the items being matched as closely as possible for familiarity to the subjects. The sets differed in comprising words of either one, two, three, four, or five syllables, as may be seen in Table 4.

From each pool, 10 lists of five words were produced by sampling at random without replacement. The 50 lists were then presented visually on video tape in completely random order; hence subjects were unaware on any given trial what set would be used and so were unlikely to use a different strategy for words of different length. Half the subjects received Lists 1–25 first, and half Lists 26–50 first. Words were written on cards and presented at a 2-sec rate by a card changer which was viewed by a video camera and recorded. A card containing a row of asterisks served as a warning that the list was about to appear. Twelve seconds were allowed for spoken recall.

Table 3 Ratio of memory score to reading rate and to articulation rate for subjects in Experiment V

Subject	Memory score reading rate		Memory score articulation rate	
	K_L^a	K_S^b	K_L^a	K_S^b
1	1.78	1.70	1.48	1.43
2	1.72	1.42	.93	.93
3	1.55	1.80	1.15	1.14
4	1.43	1.83	1.34	1.48
5	1.30	1.46	1.14	1.32
6	1.68	1.95	1.40	1.79
7	1.38	1.59	1.24	1.00
8	2.15	1.63	1.95	1.36
Mean	1.62	1.67	1.33	1.31

Notes
a K_L = Constant for long words.
b K_s = Constant for short words.

Table 4 Pools of words, matched for conceptual class; used in Experiment VI

Number of syllables

1	2	3	4	5
Stoat	Puma	Gorilla	Rhinoceros	Hippopotamus
Mumps	Measles	Leprosy	Diphtheria	Tuberculosis
School	College	Nursery	Academy	University
Greece	Peru	Mexico	Australia	Yugoslavia
Crewe	Blackpool	Exeter	Wolverhampton	Weston-Super-Mare
Switch	Kettle	Radio	Television	Refrigerator
Maths	Physics	Botany	Biology	Physiology
Maine	Utah	Wyoming	Alabama	Louisiana
Scroll	Essay	Bulletin	Dictionary	Periodical
Zinc	Carbon	Calcium	Uranium	Aluminium

Reading rate was measured in this experiment by requiring the subjects to read lists of 50 words comprising five occurrences of each item in a given set. The words were typed in uppercase in random order in two columns on a sheet of paper. Subjects were instructed to read the lists aloud as quickly as they could, consistent with pronouncing each word correctly. Their reading times were measured by stopwatch. Subjects read each list a total of four times, twice before beginning the memory task and twice after completing it. Half the subjects began both tests by reading the one-syllable list and proceeding up to the five-syllable list, while the remainder of the subjects were tested in the reverse order. The subjects, who were tested individually, comprised 14 members of the Applied Psychology Unit's panel who were paid for their services.

Results

Figure 3 shows the effect of word length on mean percentage of words correctly recalled in the appropriate position, and mean reading rate.

Memory scores. Analysis of variance showed a significant effect of conditions, $F(4, 52) = 36.70$, $p < .001$, and of subjects, $F(13,52) = 11.84$, $p < .001$. A Newman–Keuls test between conditions showed that words of one or two syllables were better recalled than words of three or four, which in turn were better than five-syllable words ($p < .05$ in each case).

Reading rate. Analysis of variance showed a significant effect of conditions, $F(4, 52) = 244.02$, $p < .001$. A Newman–Keuls test between conditions showed that each condition was significantly different from every other one ($p < .01$ in each case).

The next set of analyses tested the prediction made by decay theory, that the ratio memory span to reading rate is constant across conditions. Figure 4 shows memory span plotted as a function of reading rate, the line being fitted

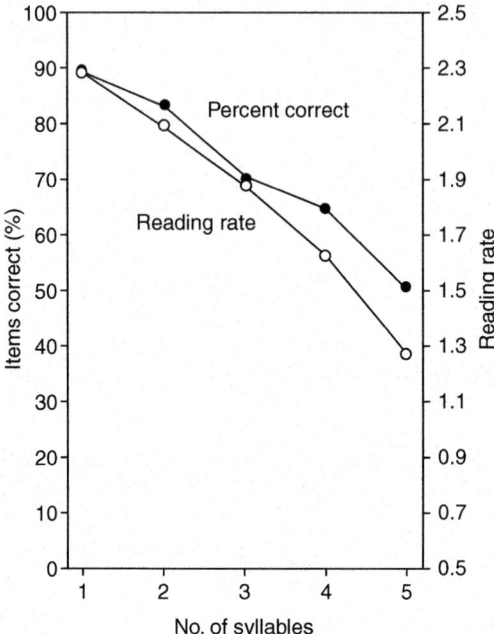

Figure 3 Mean reading rate and percentage correct recall of sequences of five words as a function of word length.

by the method of least squares. The slope of the line is 1.87, and the intercept on the ordinate 0.17. The standard error of the estimate is 0.10. The value of the intercept differs significantly from zero, $t(3) = 3.71$, $p < .05$. Thus the results are well described by the function $S = c + kR$, where S is the memory span, R is reading rate, and k and c are constants.

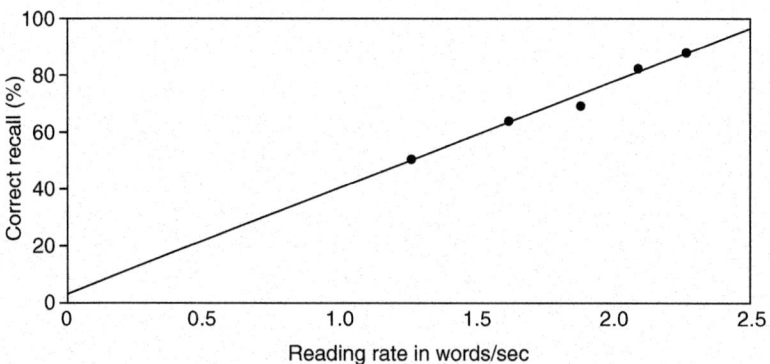

Figure 4 The relationship between reading rate and recall observed in Experiment VI.

One final question of interest is whether such a relationship holds across subjects as well as across word samples, or in other words as to whether fast readers are also good memorizers. This proved to be the case; there was a substantial correlation between memory span and reading rate, $r(13) = .685$, $p < .005$.

Discussion

The results show that the manipulations were effective in producing sets of words of different spoken duration, and that memory score for these words was well predicted by their duration. It has also been shown that fast readers tend to be good memorizers. The relationship between reading rate and memory span thus appears to be remarkably straightforward. Again the ratio of reading rate to span is approximately constant, indicating in the present study that subjects are able to remember as much as they can read out in 1.8 sec. At this stage of research, however, it is probably imprudent to generalize this result too widely. There are many variables which change memory span, but which are unlikely to change reading rate (e.g., list length, word meaningfulness, interpolated delay). It would be of interest to know whether these variables have an effect on the slope, as predicted by decay theory, or on the intercept of the function. Only if the intercept stays consistently near to zero for a variety of conditions can the simple form of decay theory under discussion be accepted. The main result of the experiment however, when seen in conjunction with the previous studies, is that short-term memory capacity, as measured by memory span, is constant when measured in units of *time*, not in units of structure.

The time-based system, which presumably underlies the effects observed, is broadly consistent with a decay theory component of short-term forgetting. Decay theory ascribes to rehearsal the role of reviving a decaying trace, and it is this function of rehearsal that requires the prediction of a word-length effect. It follows that if rehearsal could be prevented then, providing the presentation rate was the same for both long and short words, no word-length effect should occur. The next experiment was designed to test this prediction. The technique used to stop rehearsal was that of articulatory suppression (Murray, 1968) in which the subject is required to articulate an irrelevant item during presentation of the list.

Experiment VII

In this experiment, the recall of visually presented lists of long and short words was compared under two conditions: (1) with the subject remaining silent during list input and being free to rehearse, and (2) with the subject required to articulate an irrelevant sequence of items. The design thus involved four conditions, comprising two word lengths in each of two presentation conditions.

Method

Two pools of 10 words each were produced, one of one-syllable words and one of five-syllable words, matched for word frequency. From each pool, 16 five-word lists were constructed by sampling at random without replacement. Each set of 16 lists was divided into two equal blocks and a Latin square used to determine order of presentation of the blocks. All subjects did all conditions. The lists were presented at a 1.5-sec rate on a memory drum, and subjects were instructed to recall the items in the order presented. In the suppression conditions, subjects counted repeatedly from one to eight, keeping rate of articulation as constant as possible at about three digits per second. They began counting before the list appeared and stopped to recall as soon as the last item had been presented. In the no-suppression condition, subjects were simply told to try to remember the words. The subjects, 12 undergraduate students from the University of Stirling, were familiarized with the pools of words before being tested.

Results and discussion

Figure 5 shows the mean percentage of words recalled in the correct serial position as a function of word length for the two presentation conditions. Analysis of variance showed a significant effect of word length, $F(1, 11) = 17.73$, $p < .005$, of suppression, $F(1, 11) = 67.89$, $p < .001$, and a significant interaction between

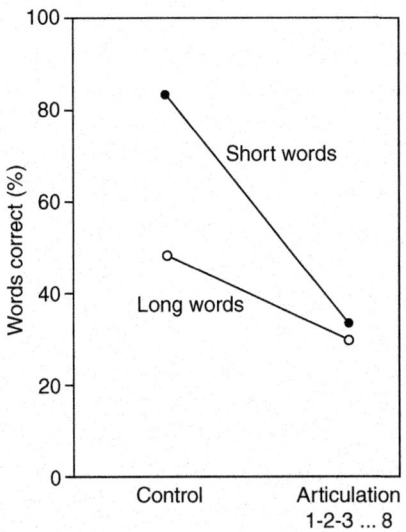

Figure 5 Effect of articulatory suppression on the word length effect. The influence of articulatory suppression on the recall of auditorily presented long and short words.

word length and suppression, $F(1, 11) = 16.30$, $p < .005$. The data may be summarized by saying that the word-length effect disappears under suppression. Thus, these results are consistent with decay theory if it can be assumed that suppression stops rehearsal. Unfortunately, this latter assumption is open to dispute, since the effects of suppression seem to be dependent on presentation modality (Levy, 1971; Peterson & Johnson, 1971). In particular, suppression has been shown to have a large effect on visually presented material, but little effect on auditorily presented material. It would seem unlikely that suppression stops rehearsal with visual presentation, but not with auditory. An alternative explanation might be to assume that suppression stops the transformation of a visual stimulus into a phonemic code. Thus, given that the word-length effect is mediated by a system employing a speech code, and that under suppression, visually presented material does not enter this system, we have an alternative explanation for the above results. Experiment VIII was designed to throw light on this issue.

Experiment VIII

This was essentially a replication of the previous experiment, with the addition of a condition involving auditory presentation. This expanded the design into a $2 \times 2 \times 2$ design, with two levels of word length (one and five syllables), two articulatory conditions (suppression and no suppression), and two presentation modes (auditory and visual). All subjects did all conditions, with the number of replications per condition being reduced to five. The experiment was run in two halves, with the four conditions of one modality in each half. Half the subjects did the visual conditions first, and half the auditory conditions first. Within a modality, the order of the conditions was determined by a Latin square. New pools of words were used, taken from the one- and five-syllable pools of Experiment VI. Presentation rate was slowed to 2 sec; in the auditory condition, the lists were read to the subject, whilst in the visual condition, the lists were presented on a memory drum. In all conditions, recall was verbal. In all other respects the procedure was as for the previous experiment; the subjects were 16 members of the Applied Psychology Unit panel who were paid for their services.

Results and discussion

The mean percentage of words correctly recalled in the appropriate serial position is shown in Figure 6. Analysis of variance showed significant effects of word length, $F(1, 15) = 14.02$, 14.02, $p < .005$, suppression, $F(1, 15) = 85.68$, $p < .001$, and modality, $F(1, 15) = 39.66$, $p < .001$. The interaction between word length and modality was significant, $F(1, 15) = 8.81$, $p < .01$, as was the suppression $\times$ modality interaction, $F(1, 15) = 33.13$, $p < .001$. The remaining two-way interaction, word length $\times$ suppression, just failed to reach significance $F(1, 15) = 4.49$, $.05 < p < .10$. The three-way interaction did reach

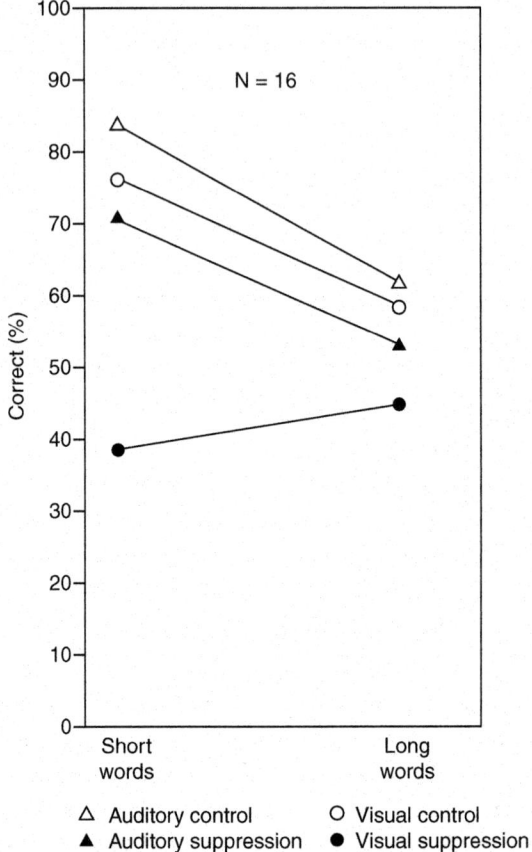

Figure 6 The influence of articulatory suppression on the recall of long and short words as a function of modality of presentation.

significance, $F(1, 15) = 6.23$, $p < .05$. This last result indicates that the change in the word length effect produced by suppression is different in the two presentation modalities. Specifically, the word-length effect is abolished by suppression in the visual modality, but is unchanged in the auditory modality.

The results demonstrate very clearly how the effects of suppression change with presentation modality and provide support for the view that suppression stops the visual to auditory transformation. These results can still be fitted into the simple decay and rehearsal hypothesis, but only if the assumption is made that articulatory suppression does not prevent rehearsal, but simply inhibits the translation of visual material into a phonemic code.

General discussion

The experiments described have shown: (1) That memory span is sensitive to word length across a range of verbal materials. (2) That when number of syllables and number of phonemes are held constant, the word-length effect remains. (3) A systematic relationship between articulation time and memory span, such that memory span is equivalent to the number of words which can be read out in approximately 2 sec. (4) That memory span is correlated with reading rate across subjects. (5) That articulatory suppression abolishes the word length effect when material is presented visually.

We shall discuss the implications of these results for existing empirical generalizations about memory, and will then attempt to fit them into a conceptual framework.

The most obvious implication of these results is for Miller's (1956) suggestion that memory span is limited in terms of number of chunks of information, rather than their duration. It suggests a limit to the generality of the phenomenon which Miller discusses, but does not, of course, completely negate it. The question remains as to how much of the data subsumed under Miller's original generalization can be accounted for in terms of temporal rather than structural limitations. Consider, for example, the tendency for subjects' memory span for letter sequences to vary with order of approximation to English. McNulty (1966) has shown that higher orders of approximation to English lead to higher memory span performance when measured in terms of number of letters recalled, but not when measured in terms of number of adopted chunks. It seems highly probable that sequences which can be reduced to a relatively small number of chunks (e.g., THEMILEAKE) will be not only well remembered, but also spoken much more rapidly than sequences which cannot be reduced in this way (e.g., YVSPCWUECR). Such a view has, of course, been explored in some detail by Glanzer and Clark (1962) in connection with the recall of verbally recodable visual patterns. Both letter sequences and the types of pattern used by Glanzer and Clark are encodable into articulatory sequences which can be produced within the 1- to 2-sec limit implied by our results. It would clearly be desirable to explore the relationship between articulation time and memory for materials such as approximations to English prose, which are broadly consistent with Miller's chunking hypothesis (Tulving & Patkau, 1962), but which would seem likely to involve articulation times considerably in excess of 2 sec.

A second general point arises from the contrast between the recency effect in free recall, which apparently shows no word-length effect (Craik, 1968; Glanzer & Razel, 1974), and the memory span task, which clearly does show such an effect. This fits in with the general pattern of results mentioned in the Introduction, suggesting that memory span relies on phonemic coding, whereas recency does not. This provides further evidence for the suggestion that the two reflect quite different underlying memory processes. Since most current views of the nature of short-term memory assumed a common primary memory system underlying both, it is unclear how they would interpret the word length effect.

One approach which does not have this drawback is the framework suggested by Baddeley and Hitch (1974), who explicitly postulate a working memory system which is responsible for performance on memory span tasks, but is not responsible for the recency effect in free recall. The formulation is based on a range of experiments in which subjects were required to perform reasoning, prose comprehension, or free recall learning tasks while simultaneously holding sequences of up to six random digits in short-term memory. In general, the results suggested that subjects could hold up to three items with virtually no effect on performance, but when required to remember six items, a decrement appeared. A tentative formulation was suggested in terms of a working memory system which acts as a central executive, and a supplementary articulatory rehearsal loop with a capacity of about three items.

Most of the experiments in the present series fit neatly into this broad framework, on the assumption that the word-length effect is the result of the limited capacity of the rehearsal loop. Looked at from this viewpoint, our data suggest that the articulatory loop system is time-based, and hence has a temporally limited capacity. When access to the loop is prevented by articulatory suppression, memory depends entirely on the capacity of the executive working memory system, which is not phonemically based, and does not have the same temporal limitation as the articulatory loop. The tendency for memory span to be impaired by phonemic similarity among the items to be remembered can also be attributed to the operation of the articulatory rehearsal system. As in the case of the word-length effect, the phonemic similarity effect disappears when articulatory suppression occurs with visually presented material (Levy, 1971). Finally, the existence of patients who have drastically impaired digit span, and yet who appear to show none of the general cognitive impairments that might be anticipated from most views of the role of short-term memory (Shallice & Warrington, 1970), can readily be accounted for within this framework if it is assumed that such patients are defective in the operation of the articulatory rehearsal system, while having the executive component of the working memory system intact.

One of our results however does present a problem for such a view. This is raised by the observation that the word-length effect does occur despite articulatory suppression, provided the material is presented auditorally. Levy (1971) has shown a similar pattern of results for phonemic similarity, with the similarity effect disappearing under suppression when visual presentation is used, but not when the presentation is auditory. On the straightforward assumption of an articulatory rehearsal loop which is entirely synonymous with subvocalization, it should follow that suppression effects could not be avoided by auditory presentation. Experiment VIII, however, suggests that although articulatory suppression produces an overall impairment in performance with auditory presentation, it does not influence the word-length effect. This would therefore seem to point to articulation as being a means of converting the visual stimulus into a phonemic code which may be accepted by some form of storage system. With auditory presentation, the material is presumably already encoded in an appropriate form, and can be fed into the supplementary system without the need for articulation. The fact that articulatory suppression still impairs performance, even with auditory presentation, may imply either that there is an additional advantage to be gained by articulation, or simply that the task of suppression provides a secondary task which takes up some of the general processing capacity which might otherwise be devoted to remembering the items presented.

Suppose one tentatively assumes a supplementary phonemically based store, what might its other characteristics be? Could it, for example, be equivalent to the precategorical acoustic store suggested by Crowder and Morton (1969)? This seems unlikely for two reasons: First because the word-length effect occurs with visual presentation, provided suppression is avoided, whereas the precategorical acoustic store does not appear to be operative unless auditory presentation is used. Secondly, Watkins and Watkins (1973) have presented evidence suggesting that the precategorical acoustic store is not sensitive to the effect of word length; if this is so, it can clearly not be used to explain the word-length effect. An alternative is to suggest that the system is an output buffer of some type: A limited-capacity store for holding the motor program necessary for the verbal production of letter names has been suggested by Sperling (1963). It seems plausible to assume that some form of buffer store is necessary for the smooth production of speech; and indeed the existence of the eye–voice span in reading points to some such temporary storage process (Morton, 1964), since what the reader is saying when reading aloud lags consistently behind the point at which he is fixating. Such a buffer system would need to be separate from the act of articulation, since it is presumably necessary to set up new articulatory programs while existing programs are operating. On this interpretation, therefore, articulatory programs can be set up or at least primed, either by the act of overt or covert articulation, or indirectly through auditory stimulation. It is tentatively suggested that such a system may be necessary for fluent speech, and may have the supplementary advantage of providing an additional backup system for the immediate retention of phonemically codable material. Such a view is clearly very tentative and leaves unspecified the complex problem of how such a store might be interfaced with the other components of the system so as to account for even the basic phenomena of the memory span. It does, however, have the advantage of linking together the existing data in a way which is both internally consistent and also likely to generate testable hypotheses.

References

Baddeley, A. D. Short-term memory for word sequences as a function of acoustic, semantic and formal similarity. *Quarterly Journal of Experimental Psychology*, 1966, **18**, 362–365.

Baddeley, A. D., & Hitch, G. Working memory. In G. A. Bower (Ed.), *The psychology of learning and motivation*. New York: Academic Press, 1968, Vol. 8.

Broadbent, D. E. *Perception and communication*. London: Pergamon Press, 1958.

Conrad, R. Acoustic confusion in immediate memory. *British Journal of Psychology*, 1964, **55**, 75–84.

Conrad, R., & Hille, B. A. The decay theory of immediate memory and paced recall. *Canadian Journal of Psychology*, 1958, **12**, 1–6.

Craik, F. I. M. Two components in free recall. *Journal of Verbal Learning and Verbal Behavior*, 1968, **7**, 996–1004.

Craik, F. I. M. The fate of primary memory items in free recall. *Journal of Verbal Learning and Verbal Behavior*, 1970, **9**, 143–148.

Craik, F. I. M., & Levy, B. A. Semantic and acoustic information in primary memory. *Journal of Experimental Psychology*, 1970, **86**, 77–82.

Crowder, R. G., & Morton, J. Precategorical acoustic storage (PAS). *Perception & Psychophysics*, 1969, **5**, 365–373.

Glanzer, M., & Clark, W. H. Accuracy of perceptual recall: An analysis of organization. *Journal of Verbal Learning and Verbal Behavior*, 1962, **1**, 289–299.

Glanzer, M., Koppenaal, L., & Nelson, R. Effects of relations between words on short-term storage and long-term storage. *Journal of Verbal Learning and Verbal Behavior*, 1972, **8**, 435–447.

Glanzer, M., & Razel, M. The size of the unit in short-term storage. *Journal of Verbal Learning and Verbal Behavior*, 1974, **13**, 114–131.

Goldman-Eisler, F. The significance of changes in rate of articulation. *Language and Speech*, 1961, **4**, 171–174.

Levy, B. A. The role of articulation in auditory and visual short-term memory. *Journal of Verbal Learning and Verbal Behavior*, 1971, **10**, 123–132.

Mackworth, J. F. The relation between the visual image and post-perceptual immediate memory. *Journal of Verbal Learning and Verbal Behavior*, 1963, **2**, 75–85.

McNulty, J. A. The measurement of "adopted chunks" in free recall learning. *Psychonomic Science*, 1966, **4**, 71–72.

Miller, G. A. The magical number seven, plus or minus two: Some limits to our capacity for processing information. *Psychological Review*, 1956, **63**, 81–97.

Morton, J. The effects of context upon speed of reading, eye movements and eye-voice span. *Quarterly Journal of Experimental Psychology*, 1964, **16**, 340–354.

Murray, D. J. Articulation and Acoustic confusability in short-term memory. *Journal of Experimental Psychology*, 1968, **78**, 679–684.

Murray, D. J. Articulation and Acoustic confusability in short-term memory. *Journal of Experimental Psychology*, 1968, **78**, 679–684.

Peterson, L. R., & Johnson, S. F. Some effects of minimising articulation on short-term retention. *Journal of Verbal Learning and Verbal Behavior*, 1971, **10**, 346–354.

Richardson, J. T. E., & Baddeley, A. D. The effect of articulatory suppression in free recall. *Journal of Verbal Learning and Verbal Behavior*, 1975, **14**, 623–629.

Shallice, T., & Warrington, E. K. Independent functioning of verbal memory stores: A neuropsychological study. *Quarterly Journal of Experimental Psychology*, 1970, **22**, 261–273.

Sperling, G. A model for visual memory tasks. *Human Factors*, 1963, **5**, 19–31.

Sperling, G., & Speelman, R. G. Acoustic similarity and auditory short-term memory: Experiments and a model. In D. A. Norman (Ed.), *Models of human memory*. New York: Academic Press, 1970.

Tulving, E., & Patkau, J. E. Concurrent effects of contextual constraint and word frequency on immediate recall and learning of verbal material. *Canadian Journal of Psychology*, 1962, **16**, 83–95.

Watkins, M. J., & Watkins, O. C. The postcategorical status of the modality effect in serial recall. *Journal of Experimental Psychology*, 1973, **99**, 226–230.

Reference notes

1 Laughery, K. R., Lachman, R., & Dansereau, D. D. Short-term memory: Effects of item–pronunciation time. (Unpublished.)
2 Standing, L., Bond, B., & Smith, P. The memory span. (Unpublished.)

8 Exploring the articulatory loop

A. D. Baddeley, V. J. Lewis and G. Vallar

A series of five experiments explore the influence of articulatory suppression on immediate memory for auditorily presented items with a view to testing the revised concept of an articulatory loop. Experiments 1, 2 and 3 demonstrate that the phonological similarity effect is not abolished by articulatory suppression, whether this occurs only at input or at both input and recall. Experiments 4 and 5 show that the tendency for long words to be less well remembered than short is abolished by articulatory suppression, even when presentation is auditory, provided suppression occurs during both input and recall. These results are consistent with the concept of a loop comprising a phonological store, which is responsible for the phonological similarity effect, coupled with an articulatory rehearsal process that gives rise to the word length effect.

General introduction

It has been clear for many years that a close relationship exists between speech coding and short-term memory. Conrad (1964) showed that the intrusion errors made by subjects recalling sequences of letters were phonologically similar to the correct item, even though the material was presented visually. Conrad and Hull (1964) showed that immediate memory for sequences of letters was impaired when the letters concerned were phonologically similar, while Baddeley (1966) showed a similar effect for words; immediate memory for sequences of words that sound alike was much poorer than memory for dissimilar words. Visual similarity and semantic similarity did not have comparable effects.

These results were initially taken to imply that short-term memory relies on an acoustic store, with visually presented letters being translated into an acoustic code. The results were, however, equally interpretable in terms of an articulatory code. There is good evidence that short-term memory relies on subvocal rehearsal (Sperling, 1967; Waugh and Norman, 1965; Atkinson and Shiffrin, 1968), and it is equally possible to argue that the code is articulatory rather than acoustic. Further evidence for this view came from a study by Conrad (1970) of memory in congenitally deaf children. Some of these children showed clear evidence of phonologically based errors, while others did not. It subsequently proved to be the case that such errors were shown by those children rated by their teachers as being good speakers. Since the children had never been able to hear, clearly the implication is that the errors were articulatory in origin.

The concept of a unitary short-term memory system was subsequently questioned by Baddeley and Hitch (1974), who suggested that it should be replaced by a multi-component working memory. They proposed a system involving a controlling central executive component supplemented by a number of slave systems. One of these, the articulatory loop, was postulated to account for the role of speech coding in short-term memory. In its initial formulation, the articulatory loop was assumed to function like a tape loop of limited duration. The loop was assumed to hold about 1.5 sec of speech-based material in temporary storage and to be capable of maintaining this by means of articulatory rehearsal. Although a very simple concept, the loop was able to account for a wide range of results. These included:

1. *The phonological similarity effect.* Poor immediate memory for phonologically similar items (Conrad and Hull, 1964; Baddeley, 1966) was assumed to result from confusion among items that had similar articulatory codes.
2. *The word length effect.* Memory span for short words is greater than for long (Baddeley, Thomson and Buchanan, 1975), a phenomenon that can readily be explained by assuming that short words are better remembered simply because they can be spoken more rapidly; 1.5 sec of short words will comprise more items than 1.5 sec of long words.
3. *Articulatory suppression.* Requiring the subject to utter some irrelevant sound—such as the word "the"—during an immediate memory task impairs performance (Murray, 1968). Such an impairment would be expected, since suppressing articulation in this way prevents the use of the articulatory loop for maintaining items in memory.
4. *Suppression, word length and similarity.* Since both the similarity and word length effects were assumed to depend on the articulatory system, blocking the system by means of a suppression task might be expected to remove both of the effects. With visual presentation this was indeed the case (Murray, 1968; Baddeley et al., 1975).

In addition to linking this cluster of findings in a coherent way, the concept of the articulatory loop was applied to a range of a phenomena, from the role of phonological coding in learning to read (Baddeley, 1979) to cultural differences in memory span and arithmetic performance (Ellis and Hennelly, 1980) and the development of digit span in children (Nicholson, 1981).

However, despite its apparent usefulness, one cluster of results does not readily fit into the concept of the articulatory loop as a simple tape loop system. The apparently anomalous results concern the effect of suppression on the phonological similarity and word length effects when presentation is auditory rather than visual. If, as the original concept implies, the word length and phonological similarity effects stem from the articulatory process, then preventing articulation should prevent the occurrence of the effects, regardless of whether the material is presented visually or auditorily. However, clear evidence exists to suggest that with auditory presentation, suppression does not eliminate either the phonological similarity effect (Murray, 1968) or the word length effect (Baddeley et al., 1975). These results could be explained by assuming a parallel but separate acoustic system showing exactly comparable characteristics to that based on articulation. However, the assumption of a further memory system is unparsimonious and, in addition, raises a range of additional problems as to the relationship between the assumed acoustic and articulatory memory systems. The aim of the experiments that follow is to explore this area of apparent inconsistency with a view to producing a single coherent model that will account both for the phenomena previously attributed to the articulatory loop and for the anomalies observed with auditory presentation. Experiment 1 aims to replicate the observation that with auditory presentation, articulatory suppression does not abolish the phonological simiarity effect (Murray, 1968).

Experiment 1

Design

Three variables were manipulated in this study: the phonological similarity of the material, the rate of presentation and articulatory suppression. Presentation rate and suppression were blocked, giving four separate blocks, each of which contained both similar and dissimilar sequences. A Latin square design was used involving the 24 possible combinations of order of presentation of the four conditions, namely control–fast, control–slow, suppression–fast, and suppression–slow. Twenty-four subjects were tested, one in each of the 24 possible orders. Each of the four testing blocks comprised 24 sequences of five words. Twelve of the sequences were from a phonologically similar set and 12 were from a dissimilar set. The first four of the 24 sequences were treated as practice trials and discarded. These always comprised two similar and two dissimilar sequences. The remaining 20 comprised 10 similar and 10 dissimilar sequences in random order.

Material and subjects

The material was that used by Baddeley (1966) and comprised eight phonologically similar words: *can, cad, cat, cap, mad, man, mat* and *map*, and eight dissimilar words: *cow, day, bar, few, hot, pen, sup* and *pit*. The subjects were 24 female members of the Applied Psychology Unit subject panel. Their ages ranged from 21 to 70 years, with a mean of 51.

Procedure

Words were read out at a rate of one word per 0.5 sec or one per 2 sec, followed after a 2-sec delay by the spoken recall signal "now". Subjects then commenced written recall on a response sheet. They were instructed to recall the words serially, beginning with the first item. Under conditions of articulatory suppression, subjects were required continuously to repeat the digits *1, 2, 3, 4* from a "ready" signal that occurred before presentation of the words up to a "recall" signal 2 sec after presentation. Rate of suppression was not strictly monitored, but subjects were encouraged to suppress at a rate of approximately three to four items per second and were cautioned if at any time their rate of suppression showed signs of becoming slower or less regular.

Results

Table 1 shows the mean number of five-word sequences recalled incorrectly or in an incorrect serial order as a function of rate, similarity and suppression. Analysis of variance showed significant effects of rate [$F(1, 23) = 7.73$, $p < 0.025$], of suppression [$F(1, 23) = 47.93$, $p < 0.001$] and of phonological similarity [$F(1, 23) = 232.59$, $p < 0.0001$]. Only one of the possible interactions reached significance, that between suppression and similarity [$F(1, 23) = 15.85$, $p < 0.001$].

These results show the expected influence of articulatory suppression on performance, together with a massive effect of phonological similarity. The interaction between similarity and suppression indicates that preventing articulation did reduce the magnitude of the similarity effect. Nonetheless, it remains massive and certainly does not allow one to conclude that suppression abolishes the similarity effect, as the initial articulatory loop hypothesis would require. We were indeed concerned that the interaction might conceivably be artifactual, since some subjects were performing so poorly in the more difficult conditions that possible floor effects could not confidently be ruled out.

Finally, we observed better performance under conditions of slow than under rapid presentation. Discussion of this will be postponed until after related results have been reported from the next two experiments.

Table 1 Results of Experiment 1: immediate memory for word sequences as a function of phonological similarity, articulatory suppression and rate of presentation (percentage of erroneous sequences)

	Fast		Slow	
	Similar	Dissimilar	Similar	Dissimilar
Control	62.9	24.6	56.6	19.7
Suppression (at input only)	67.1	35.9	61.5	35.7

Experiment 2

While Experiment 1 indicated a strong influence of phonological similarity even when subjects were suppressing articulation, it is conceivable that the similarity effect could have been emerging during the recall process. On this interpretation, the subjects may be holding the material in a temporary store during presentation and at the recall signal transferring it into an articulatory code prior to responding. Experiment 2 attempted to explore this possibility by requiring subjects to suppress articulation throughout both input and written recall. Furthermore, to avoid floor effects, sequence length was reduced to four words. It was hoped that this would allow the interaction between similarity and suppression observed in Experiment 1 to be interpreted less equivocally either as a genuine effect of coding or as a statistical artifact.

Method

Design and procedure

This was identical with Experiment 1, with two exceptions. In the articulatory suppression conditions, subjects continued suppressing until they had completed their recall. Secondly, since we were concerned at the possibility of floor effects in the previous experiment, we reduced the length of the sequences from five to four words. A further 24 female members of the Applied Psychology Unit subject panel served as subjects. Their ages ranged from 20 to 67, with a mean of 42.

Results

Table 2 shows the mean percentage of sequences that were not correctly recalled in each condition.

Analysis of variance again showed a massive effect of phonological similarity [$F(1, 23) = 653.00$, $p < 0.0001$], and suppression again had a clear effect on performance [$F(1, 23) = 82.54$, $p < 0.001$]. On this occasion, however, there was no significant interaction between phonological similarity and

Table 2 Results of Experiment 2: immediate memory for word sequences as a function of phonological similarity, articulatory suppression and rate of presentation (percentage of erroneous sequences)

	Fast		Slow	
	Similar	Dissimilar	Similar	Dissimilar
Control	15.2	2.6	12.2	2.0
Suppression (at input and recall)	19.8	8.5	16.9	8.7

suppression [$F(1, 23) = 1.69$, $p > 0.05$]. There was a significant effect of rate, which again indicated that slow presentation led to better performance than fast [$F(1, 23) = 6.46$, $p < 0.025$], an effect that interacted with phonological similarity [$F(1, 23) = 8.02$, $p < 0.01$]. This seems to reflect a slightly less drastic effect of similarity when rate of presentation is slow. No other interactions approached significance.

Discussion

Experiment 2 resembles the previous experiment in showing clear effects of both phonological similarity and suppression. On this occasion, however, the interaction between the two variables was not significant, indicating that the previously observed interaction might well have been attributable to a floor effect. Unfortunately, it could be argued that interpretation of Experiment 2 was limited by a ceiling effect; although a similarity effect clearly occurs in the control condition, it is possible that the effect would have been substantially larger but for the fact that performance was near ceiling in the dissimilar control conditions. Since five words gave rise to possible floor effects and four to a ceiling effect, it seemed advisable to move away from this particular type of material. We therefore carried out a further experiment, using phonologically similar and dissimilar consonants. Furthermore, in order to ensure that performance was not masked by floor and ceiling effects, we opted for a memory span procedure in which subjects were tested successively with sequences in lengths ranging from two to seven consonants.

Experiment 3
Method

Materials

Consonants were selected at random either from an acoustically similar set comprising the letters *B C D G P T V* or from a dissimilar set, *H J L R S Y Z*. The mean confusability of the two sets was 48.5 and 10.5%, respectively (Hull, 1973). Each condition involved presenting three sequences of from two to seven letters selected at random without replacement from either the similar or dissimilar set. The letter strings were read out crisply by the experimenter at a rate of either one letter per 0.5 sec or one per 1.5 sec. In all conditions the experimenter began by presenting the two-letter condition and systematically increasing sequence length up to seven letters.

Design

This involved three variables, rate of presentation (0.5 or 1.5 sec per letter), phonological similarity (high or low) and articulatory suppression (present or absent). As in Experiment 2, suppression occurred during both input and recall. The eight test conditions were presented in a counterbalanced order such that each condition appeared in the same order for 2 of 16 subjects tested. Within this constraint, the order of the eight conditions was selected at random, except that no two suppression conditions were allowed to occur in succession, since subjects found suppressing for long periods somewhat uncomfortable. During suppression, the subject was required to utter the digits *1 2 1 2* repeatedly, at a rate of approximately 4 per sec. The suppression task was limited to digits 1 and 2 in order to avoid the danger of the digit "three" differentially interfering with the phonologically similar list comprising letters all of which rhyme with "three". The subjects were 16 female members of the Applied Psychology Unit subject panel, ranging in age from 29 to 69, with a mean of 49 years.

Results

Table 3 shows the mean percentage of errors made at each sequence length as a function of phonological similarity, rate of presentation and articulatory suppression. An error was scored whenever the subject failed to reproduce the appropriate item in its correct serial position. Analysis of variance showed significant effects of similarity [$F(1, 15) = 34.1$, $p < 0.001$] and of suppression

Table 3 Memory span for letters as a function of phonological similarity, rate of presentation and articulatory suppression. Percentage error at each sequence length

	Control			
	Fast		Slow	
Length	Similar	Dissimilar	Similar	Dissimilar
2	0	0	0	0
3	0	0	3.5	0
4	5.2	1.0	9.9	0
5	15.8	5.8	23.8	5.4
6	33.0	20.0	26.4	13.5
7	44.0	26.8	42.9	28.0
Mean	16.3	8.9	17.8	7.8
	Suppression			
2	1.0	0	0	0
3	2.8	0	4.9	0
4	10.9	4.7	11.5	4.2
5	31.3	9.2	38.3	15.4
6	34.7	25.3	46.5	29.2
7	51.2	39.3	58.0	49.4
Mean	22.0	13.0	26.5	16.4

[$F(1, 15) = 52.0$, $p < 0.001$] but no interaction between similarity and suppression ($F < 1$). There was no overall effect of rate [$F(1, 15) = 2.59$, $p > 0.05$], but there was a significant interaction between rate and suppression [$F(1, 15) = 6.66$, $p < 0.05$], a result that reflects the greater susceptibility of subjects to suppression when items are presented slowly. No other interactions approached significance.

Discussion

Experiment 3 replicates the very solid effect of phonological similarity observed in the two previous studies. There is, however, no suggestion of an interaction between similarity and suppression, a result that on this occasion can clearly not be attributed to either ceiling or floor effects.

The effects of rate on performance in Experiment 3 differs from that observed in the previous experiments where slow presentation resulted in somewhat better performance. In Experiment 3 no overall difference occurred, but an interaction was observed, with suppression having a more deleterious effect with slow than with fast presentation. One possible interpretation of this pattern of results is as follows: Experiments 1 and 2 both involved presentation of words for which it is plausible to assume that slow presentation would allow the development of a semantic representation that could be used to supplement the less durable phonological code (Craik and Lockhart, 1972). Unrelated consonants such as those used in Experiment 3, on the other hand, are not intrinsically meaningful, and as such they are presumably more difficult to encode semantically and hence might gain less advantage from slow presentation. On the other hand, it is conceivable that with rapid presentation and rapid recall the auditory trace of the items might still have been present in sufficient strength to enhance recall performance, in contrast to slow presentation where the memory trace would have ample time to fade. Such an interpretation is, of course, speculative, and since the effects of rate were small compared with those of similarity and suppression, this aspect of the data will not be discussed further.

The central feature of these experiments is, of course, the relationship between phonological similarity and suppression. Our results indicate a very substantial effect of phonological similarity in all three experiments, regardless of whether the subject is or is not suppressing articulation. A small but significant interaction between these variables was observed in Experiment 1. However, since this could equally well be interpreted as a statistical artifact due to floor effects, and since it did not appear in Experiments 2 or 3, our data suggest that under conditions of auditory presentation the phonological similarity effect is undiminished by articulatory suppression.

Before accepting this conclusion, however, we should consider two possible criticisms. The first of these concerns the relatively small sample of material on which our conclusions are based—eight similar and eight dissimilar words in the first two studies and seven similar and dissimilar consonants in Experiment 3. It has been suggested by a reviewer that our data should be analysed across material using Clarke's (1973) min F test. In the memory task used, however, the probability of recalling any one item will be influenced by the difficulty of others within that sequence. In the absence of independent estimates of item difficulty, Clarke's test is not valid. Indeed, the whole concept of similarity makes no sense when applied to a single item: it is a characteristic of the set of items chosen, not of the individual letters or words. Were we demonstrating the effect of phonological similarity for the first time, the possibility that our results might stem from a chance selection of atypical words might be worrying. However, over the last 20 years the phonological similarity effect has been replicated so frequently on such a wide range of materials that such an interpretation of our results seems quite implausible. In the case of our samples of consonants, Clarke's min F test is clearly inappropriate, since we are using virtually the entire population of phonologically highly similar items, so the question of generalisation does not apply. The effects of similarity, however, are essentially the same as those observed with phonologically similar words, reinforcing our conclusion that our results are not due to a chance sampling of atypical material.

The second potential objection is more serious and concerns the possibility that the effects of phonological similarity with auditory presentation reflect mishearing the items rather than misremembering them. This again is a danger that is more apparent than real. Earlier experiments (Baddeley, 1968) have shown a comparable phonological similarity effect in a series of studies using the same words as those employed in Experiments 1 and 2, but where a listening test indicated that mishearing was not a major factor. Furthermore, when the items were spoken through noise, the resulting increase in difficulty of discrimination did not lead to a substantial, or indeed statistically significant, impairment in immediate memory performance (Baddeley, 1968, Experiment IV). There is then from this and other studies abundant evidence that the phonological similarity effect with spoken presentation is not attributable to mishearing.

This conclusion can, of course, be checked for the present study by examination of the results of Experiment 3. If we examine the probability of responding correctly under conditions where subjects are presented with two items, this will give us an indication of the subject's ability to identify the item when a memory load is relatively low. This should give us a conservative estimate of the probability that an individual item can be detected correctly. On this measure, percent correct recall, and hence detection for the four conditions, are 100, 100, 99 and 100 percent. Explaining our memory result in terms of misperception would require error rates of approximately 10 percent. It is, of course, possible to come up with a more complex discrimination hypothesis by arguing that although the subject was perceiving the items correctly, the discrimination was absorbing enough of her central processing capacity to detract from memory. While there is some evidence that listening through noise may have a general effect on memory span (Rabbitt, 1968), such effects are relatively small and were not observed using this material in an earlier study (Baddeley, 1968). Furthermore, such an interpretation is highly unparsimonious, since it requires a quite separate explanation for the phonological similarity effect with auditory and with visual presentation, together with an explanation of why the two effects are of roughly comparable magnitude and show an essentially similar pattern of performance across a range of manipulations (Baddeley, 1968). In conclusion, then, we do not regard mishearing the auditorily presented items as a plausible interpretation of our data.

If we accept our results at face value, then they appear to replicate those of Murray (1968) but to be inconsistent with the initial version of the articulatory loop model. This suggested that the process of articulatory suppression should preempt the loop, hence abolishing the effects of articulatory coding, of which the phonological similarity effect was assumed to be one example. In fact, however, since the initial formulation of this model it has proved necessary to elaborate it. This elaboration was prompted by experiments on the effect of unattended speech on memory for visually presented digits, a phenomenon explored by Colle and Welsh (1976) and by Salame and Baddeley (1982).

In order to interpret our own results and those of Colle (Colle and Welsh, 1976; Colle, 1980), we were forced to revise the earlier version of the articulatory loop. The revised model makes a distinction between a phonological store and an articulatory rehearsal process. Material can be registered in the store in two ways, either through articulatory rehearsal, an optional process, or via auditory presentation, a process that leads to obligatory registration. Such a reinterpretation was strongly suggested by the evidence obtained by Colle and Welsh (1976) and Salame and Baddeley (1982) on the effects of irrelevant speech on memory and was further supported by evidence from the detailed investigation of a patient suffering from defective short-term memory (Vallar and Baddeley, in press). On this interpretation, the phonological similarity effect reflects the nature of the coding in a phonological input store. When material is presented auditorily, then it is automatically registered in this store, regardless of whether or not the subject is allowed to articulate. With visually presented material, however, the store can be used only if the subject is allowed to articulate the items; hence suppression during visual presentation removes the similarity effect.

However, while the phonological similarity effect is assumed to stem from the phonological store, the word-length effect is assumed to reflect the control process of articulatory rehearsal. Long words are assumed to be remembered more poorly than short simply because the rate at which they can be rehearsed, and hence the rate at which the phonological trace can be refreshed, is less for long than for short words. Since the word length effect is assumed to depend on the process of articulation per se, it should therefore be abolished if subvocal rehearsal is prevented by articulatory suppression, regardless of whether the material is presented visually or auditorily.

A major difficulty for this interpretation is offered by Experiment 8 in the study by Baddeley et al. (1975). This examined the influence of articulatory suppression on the word length effect, using both auditory and visual presentation. The word length effect was found to be abolished with visual, but not with auditory, presentation, a result that is clearly at variance with the prediction of the revised articulatory loop model. The next two experiments explore this point in greater detail. In particular, care is taken to ensure that suppression occurs not only during input, as was the case in the Baddeley et al. (1975) study, but also during recall to prevent subjects rapidly shifting information from an auditory to an articulatory mode before recalling the sequence.

Experiment 4

In this experiment the serial recall of auditorily presented lists of long and short words was compared under two conditions: suppression, in which the subject was required to articulate an irrelevant sequence of items both during input and during recall, and control, in which the subject remained silent.

Method

Two pools of 10 words each from Baddeley et al. (1975) Experiment VI were employed. The two pools comprised one word from each of 10 semantic categories, the items being matched as closely as possible for familiarity to the subjects. The set of short words comprised 10 monosyllables, namely, *Stoat, Mumps, School, Greece, Crewe, Switch, Maths, Maine, Scroll, Zinc*. The set of long words comprised 10 words of five syllables, namely, *Hippopotamus, Tuberculosis, University, Yugoslavia, Weston-super-Mare, Refrigerator, Physiology, Louisiana, Periodical, Aluminium*. From each pool 21 lists of five different words were generated by sampling at random.

In the suppression condition, 16 lists of short words and 16 lists of long words were auditorily presented in a random fixed order. In the control condition five lists for each word length were presented in a random fixed order. The control condition, which was always administered after the suppression condition, was included to ensure that a standard word length effect occurred with this sample of words and subjects. The words were presented by means of a tape-recorder at a rate of 1.5 sec per word. Each list was preceded by a warning tone presented 3 sec before the beginning of the list. Subjects were instructed to write the words in the order of presentation and were allowed 20 sec for writing. In order to avoid differences in the time required to write the two sets of words, subjects were allowed to write the long words in abbreviated form (e.g., *Hippo* for *Hippopotamus*). An acoustic signal warned subjects that writing time was finished. The interval between the end of writing time and the signal announcing the next list was 3 sec. Subjects were instructed to draw a line when they did not remember a word. At the beginning of the experiment, subjects were allowed to familiarise themselves with the two pools of words. In the suppression condition, subjects counted repeatedly from one to eight, keeping rate of articulation as constant as possible at about three digits per second; subjects started counting at the warning signal and stopped when they had completed written recall. In the control condition they remained silent both during input and during recall. The subjects were 18 members of the Applied Psychology Unit subject panel who were paid for their services.

Results

The mean percentage error for long and short words under both control and suppression is shown in Figure 1. Analysis of variance indicated a significant effect of suppression [$F(1, 17) = 249.1$, $p < 0.001$] of word length [$F(1, 17) = 28.7$, $p < 0.001$] and of serial position [$F(4, 68) = 89.9$, $p < 0.001$]. The crucial interaction between condition and word length also reached significance [$F(1, 17) = 11.8$, $p < 0.01$], while there was also a significant interaction between serial position and both suppression [$F(4, 68) = 7.22$, $p < 0.001$] and word length [$F(4, 68) = 3.06$, $p < 0.05$].

Further analysis using the t test indicated a highly significant effect of word length under control conditions $t_{34} = 6.31$, $p < 0.001$), but no overall effect of word length under suppression ($t_{34} = 1.84$, $p > 0.05$). Under control conditions, performance on short words is significantly better than on long at all serial positions other than the last, while under suppression, although there is a suggestion of a difference throughout the curve, this achieves significance only at serial position 2 ($t_{170} = 2.28$, $p < 0.05$).

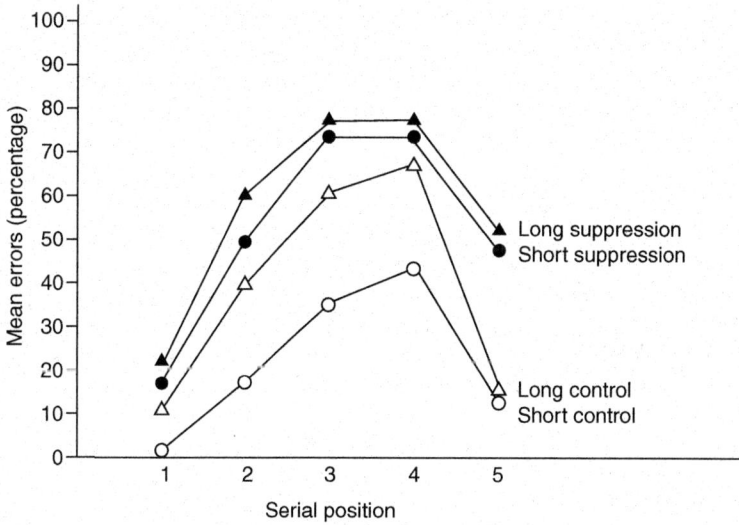

Figure 1 The influence of word length and suppression on immediate memory for auditorily presented words.

Discussion

Although there is a slight trace of a word length effect under suppression, these results are broadly consistent with the prediction that the word length effect would be abolished under suppression, even with auditory presentation. As such, they suggest that if articulatory coding is to be avoided it is crucial that subjects be required to continue to suppress articulation throughout both presentation and recall. It is interesting to note that this precaution apparently is not required when presentation is visual (Baddeley et al., 1975). The most likely interpretation of this difference would seem to stem from the assumption that articulatory repetition of auditory items is a highly compatible skill that can be performed rapidly and with minimal processing demand. Evidence for such a view comes from both the observation by Davis, Moray and Treisman (1961) that repetition of verbal material does not show the standard Hick's law effect of set size, implying that such responses are highly compatible, together with the more recent studies by McLeod and Posner (1984), which indicate that repeating an auditorily presented item demands minimal attentional capacity.

The serial position curves show the absence of a word length effect on the last item. While performance at this point is clearly constrained by a possible ceiling effect, it seems unlikely that this can explain the absence of a difference, since serial position 1, which is similarly constrained, does show a significant difference between long and short words. A similar absence of a word length effect on the last serial position has been observed in earlier studies (e.g. Baddeley et al., 1975, Expt. III) but is not invariably observed (c.f. Baddeley and Hull, 1979, Expt. III), suggesting that performance on the last item may be particularly subject to the strategy adopted. In general, the serial position curve reinforces the suggestion that with auditory presentation, the last item enjoys a special status (Baddeley and Hull, 1979; Crowder and Morton, 1969).

Considered over-all, our results clearly support the revised version of the articulatory loop. However, there remain two causes for caution: the first of these is the clear discrepancy between this result and that of Baddeley et al. (1975). While the requirement to articulate during both input and recall offers a reasonably plausible interpretation of this discrepancy, given the importance of this result within the working memory framework, it would clearly be desirable to replicate Experiment 4 in order to ensure that it does not represent a statistical fluke. Secondly, although no overall effect of word length was observed under suppression, there was a suggestion that an effect might remain, and a significant difference on one of the five serial positions. Under suppression conditions, performance throughout was relatively low, and it seemed possible that a floor effect might have been obscuring a genuine influence of word length. Experiment 5 therefore attempts to replicate the previous finding, including observations of performance on sequences of three, four and five words in order to minimise any danger of obscuring a genuine effect through making the task too difficult. Finally, rather than adding the control condition after the main experiment, as was the case in the previous study, both control and suppression conditions were given equal weight.

Experiment 5
Method

Design

This involved three variables, word length, suppression and sequence length. A counterbalanced within-subjects design was used in which half the subjects began with long words and half with short, and within these groups half were tested initially under suppression while the other half were tested initially under the silent control condition. Half the subjects were tested on the control before suppression for both long and short words, while the other half were tested in the reverse order. There were, therefore, four groups, one of which was tested in the order long–control (LC), long–suppression (LS), short–control (SC) and short–suppression (SS), one tested in the order LS, LC, SS, SC, one tested in the order SC, SS, LC, LS and the fourth tested in the order SS, SC, LS, LC. Each trial block comprised eight sequences of three words followed by eight four-word sequences and finally eight five-word sequences.

Material and subjects

The material comprised the sets of words of one and five syllables used in the previous study. Sequences were drawn at random from the relevant set, with the constraint that no item could occur more than once in the same sequence. The subjects were 24 female members of the Applied Psychology Unit subject panel.

Procedure

Words were read out at a rate of 1.5 sec per word, followed by the verbal signal "recall". The subject then attempted to write the words on a recall sheet. She was instructed to reproduce them in the order presented, but in order to equate the writing time for the two lists she was encouraged to abbreviate her responses to the first three or four letters of each item. As soon as she had completed her recall, the next sequence was presented. Under conditions of articulatory suppression, the subject was required to say softly the numbers *1 2 3 4 5 6 7 8*, at a speed of approximately three to four digits per second. Suppression was required during both input and recall.

Results

Figure 2 shows the mean performance of subjects recalling long and short words with and without articulatory suppression. Analysis of variance showed significant effects of suppression [$F(1, 23) = 112.11$, $p < 0.001$], of word length [$F(1, 23) = 15.91$, $p < 0.001$] and of sequence length [$F(2, 46) = 237.56$, $p < 0.001$]. There was a significant interaction between suppression and word

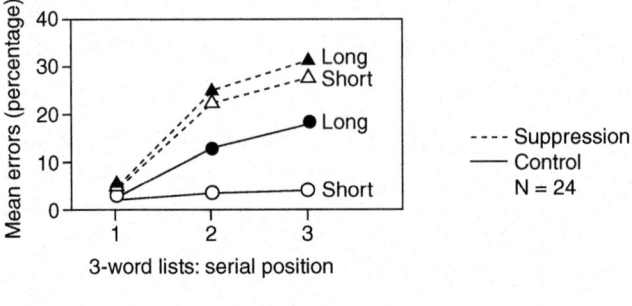

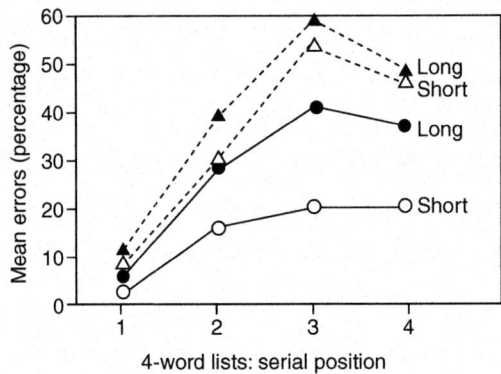

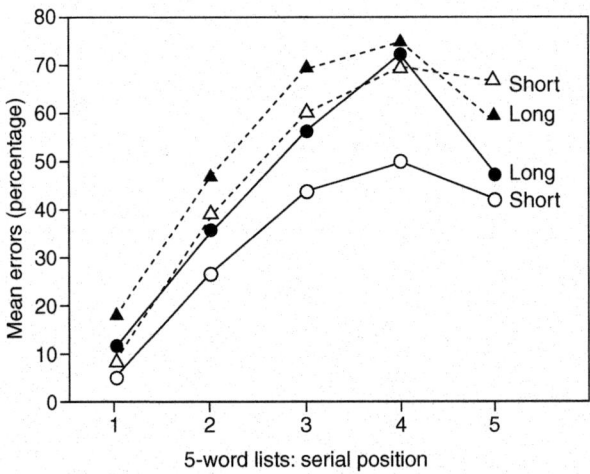

Figure 2 The influence of word length and articulatory suppression on immediate memory for sequences of three, four and five auditorily presented words.

length [$F(1, 23) = 7.78$, $p < 0.05$], while no other interactions reached statistical significance. When the suppression and control conditions were analysed separately, the word length effect failed to reach significance under articulatory suppression [$F(1, 23) = 2.62$, $p > 0.05$) but was very clearly significant under control conditions [$F(1, 23) = 39.45$, $p < 0.001$], while length of sequence was of course significant in both cases ($p < 0.001$). Examination of the serial position effects in Figure 2 indicates that any absence of a word length effect under suppression cannot be due to ceiling effects, since, if anything, the suggestion of a word length effect is greatest for the five-item sequences and least for sequences of three items. Finally, the tendency for the last item in the list to show no word length effect occurs again in the five-item sequence but is not present in either of the shorter sequences, again implying that performance on the last item may be particularly sensitive to strategy.

The results of Experiment 5 therefore reinforce those of Experiment 4 in indicating that articulatory suppression does substantially remove the word length effect, even with auditory presentation. Once again, however, there is a nonsignificant tendency for a small word length effect to remain, so that it would perhaps be unwise to conclude that the effect is completely abolished. This trace of an effect could, however, stem from the fact that suppression at the rates required might conceivably allow a subject to intersperse an occasional subvocal rehearsal, at least of the monosyllabic words. This could presumably be tested by increasing the required rate of suppression, although this does have the drawback that when subjects are required to suppress at the limit of their capacity, more general impairment occurs, possibly due to the involvement of the central executive when maximum articulation is required (Baddeley, Eldridge and Lewis, 1981; Besner, Davies and Daniels, 1981).

Experiments 4 and 5 therefore appear to indicate that the effects of articulatory suppression and word length do interact under conditions of auditory as well as visual presentation. Once again, the cautious reader might suggest that our results are limited to a single sample of words. It has, however, subsequently been replicated using a different set of words by Morris (1983) who investigated the role of both articulatory and phonological coding in elderly normal subjects and in patients suffering from senile dementia. His design was essentially similar to that used in the present study and examined the effect of articulatory suppression, phonological similarity and word length on immediate serial recall. At the time of carrying out the experiments, Morris was not familiar with the revised version of the articulatory loop, so that the replication was essentially done blind. Morris observed a virtually identical pattern of results, not only for his sample of normal elderly subjects but also for his demented patients, with the exception that overall level of performance was significantly lower in the case of the dements. Given the difficulty of collecting clean data from demented patients, this replication suggests that our results are relatively robust.

A final point that may be worrying the cautious reader is the question of whether the effects of articulatory suppression result from the disruption of phonological coding, or whether such effects might stem from a general disruption of information processing that would follow from the distracting effect of requiring the subject to perform any additional task, even one as apparently undemanding as constantly repeating the numbers 1 and 2. Some evidence for the potential disrupting effect of concurrent articulation is presented by Levy (1981) and by Margolin, Griebel and Wolford (1982).

This point has been discussed in greater detail elsewhere (Baddeley et al., 1981) where the effect of articulatory suppression on reading was compared with that of a secondary tapping task. Suppression had a marked effect on performance, whereas tapping led to no reliable change. Further evidence for a specific phonological effect of articulatory suppression comes from the study of a patient with impaired short-term memory who does not appear to use phonological coding with visually presented material (Vallar and Baddeley, in press). In her case, articulatory suppression had no effect on performance, as the articulatory loop interpretation but not the distraction hypothesis would predict. Finally, it is difficult to see how a simple attentional distraction hypothesis would predict the complex pattern of interaction between suppression, phonological similarity, word length and performance observed in the present series of experiments.

General discussion

A series of experiments exploring the role of articulatory suppression on the retention of material presented auditorily has produced the following results. First, we have have confirmed the previous observation (Murray, 1968), that with auditory presentation, the phonological similarity effect is not abolished by articulatory suppression; the effect remains whether suppression occurs during input or during both input and recall. In contrast to this result, our last two experiments show that the word length effect is abolished even with auditory presentation, provided suppression occurs during both input and recall.

Taken together, these results are entirely consistent with the modified version of the articulatory loop (Salame and Baddeley, 1982; Baddeley, 1983). This assumes a phonological input store supported by an articulatory control process. The persistence of the phonological similarity effect, despite articulatory suppression during both input and recall, is consistent with the assumption that auditory presentation guarantees access to a phonological store. Phonologically similar items will have similar and hence easily confused codes within this store, producing impaired performance. Further evidence for such a store comes from the demonstration by Colle and Welsh (1976) that unattended speech disrupts memory for visually presented items. As Salame and Baddeley (1982) showed, the amount of disruption is related to the phonological similarity between the remembered and the disrupting items. The meaning of the unattended items proved to be unimportant, while the disruptive effect was found not to occur when subjects were required to suppress articulation. Under these conditions, the visually presented material would presumably not gain access to the phonological store, so that the pollution of this store by unattended material would have no effect on performance.

In the case of auditory presentation, registration of material in the phonological store is guaranteed whether or not articulation is suppressed. Suppression will, however, prevent articulatory rehearsal and hence lead to poorer performance.

Finally, the observation in Experiments 4 and 5 that even with auditory presentation the word length effect can be abolished by articulatory suppression again supports the modified articulatory loop model. Since the word length effect is assumed to reflect the process of articulation per se, then preventing articulation should abolish the effect, regardless of modality of presentation. This occurs, though only when suppression is prevented during both input and recall. It is interesting to note in this connection that when the long and short words are presented visually, suppression during input is sufficient to remove the word length effect. This difference between visual and auditory presentation probably reflects the greater compatibility of an articulatory response to auditory material than to visual. It seems likely that part of the language learning process involves an in-built capacity for repeating heard stimuli. This is reflected both in the ease of such responses in adults (Davis et al., 1961; McLeod and Posner, 1984) and the much earlier age at which children are found to rehearse auditorily presented words as opposed to the names of visually presented pictures (Hitch and Halliday, 1983).

In conclusion, the modified articulatory loop model does appear to be capable of handling a very extensive range of results concerning immediate memory for verbal materials. While we do not dispute the ability of alternative models to account for selected parts of this corpus of results, we know of no other model that can adequately account for the whole range of phenomena.

References

Atkinson, R. C. and Shiffrin, R. M. (1968). Human memory: A proposed system and its control processes. In K. W. Spence (Ed.), *The psychology of learning and motivation: Advances in research and theory, Vol. 2*. Pp. 89–195. New York: Academic Press.

Baddeley, A. D. (1966). Short-term memory for word sequences as a function of acoustic, semantic and formal similarity. *Quarterly Journal of Experimental Psychology*, **18**, 362–365.

Baddeley, A. D. (1968). How does acoustic similarity influence short-term memory? *Quarterly Journal of Experimental Psychology*, **20**, 249–264.

Baddeley, A. D. (1979). Working memory and reading. In P. A. Kolers, M. E. Wrolstad and H. Bouma (Eds.), *Processing of visible language*. Pp. 355–370. New York: Plenum Press.

Baddeley, A. D. (1983). Working memory. *Philosophical Transactions of the Royal Society London B*, **302**, 311–324.

Baddeley, A. D., Eldridge, M. and Lewis, V. J. (1981). The role of subvocalization in reading. *Quarterly Journal of Experimental Psychology: Human Experimental Psychology*, **33**, 439–454.

Baddeley, A. D. and Hitch, G. J. (1974). Working memory. In G. Bower (Ed.), *Recent advances in learning and motivation Vol. VIII*. Pp. 47–90. New York: Academic Press.

Baddeley, A. D. and Hull, A. (1979). Prefix and suffix effects: Do they have a common basis? *Journal of Verbal Learning and Verbal Behavior*, **18**, 129–140.

Baddeley, A. D., Thomson, N. and Buchanan, M. (1975). Word length and the structure of short-term memory. *Journal of Verbal Learning and Verbal Behavior*, **14**, 575–589.

Besner, D., Davies J. and Daniels, S. (1981). Phonological processes in reading: The effects of concurrent articulation. *Quarterly Journal of Experimental Psychology*, **33**, 415–438.

Clarke, H. H. (1973). The language-as-a-fixed-effect-fallacy: A critique of language statistics in psychological research. *Journal of Verbal Learning and Verbal Behavior*, **12**, 335–359.

Colle, H. A. (1980). Auditory encoding in visual short-term recall: Effects of noise intensity and spatial location. *Journal of Verbal Learning and Verbal Behavior*, **19**, 722–735.

Colle, H. A. and Welsh, A. (1976). Acoustic masking in primary memory. *Journal of Verbal Learning and Verbal Behavior*, **15**, 17–32.

Conrad, R. (1960). Very brief delay of immediate recall. *Quarterly Journal of Experimental Psychology*, **12**(1), 45–47.

Conrad, R. (1964). Acoustic confusion in immediate memory. *British Journal of Psychology*, **55**, 75–84.

Conrad, R. (1970). Short-term memory processes in the deaf. *British Journal of Psychology*, **61**, 179–195.

Conrad, R. and Hull, A. J. (1964). Information, acoustic confusion and memory span. *British Journal of Psychology*, **55**, 429–432.

Craik, F. I. M. and Lockhart, R. S. (1972). Levels of processing: A framework for memory research. *Journal of Verbal Learning and Verbal Behavior*, **11**, 671–684.

Crowder, R. G. and Morton, J. (1969). Precategorical Acoustic Storage (PAS). *Perception and Psychophysics*, **5**, 365–373.

Davis, R., Moray, N. and Treisman, A. (1961). Imitative responses and rate of gain of information. *Quarterly Journal of Experimental Psychology*, **13**, 78–90.

Ellis, N. C. and Hennelley, R. A. (1980) A bilingual word-length effect: Implications for intelligence testing and the relative ease of mental calculation in Welsh and English. *British Journal of Psychology*, **71**, 43–52.

Hitch, G. and Halliday, S. (1983). The development of working memory. *Philosophical Transactions of the Royal Society London B*, **302**, 325–340.

Hull, A. J. (1973). A letter-digit matrix of auditory confusions. *British Journal of Psychology*, **64**, 579–585.

Laughery, K. R. and Pinkus, A. L. (1966). Short-term memory: Effects of acoustic similarity presentation rate and presentation mode. *Psychonomic Science*, **6**, 285–286.

Levy, B. A. (1981). Interactive processes during reading. In A. M. Lesgold and C. Perfetti (Eds.), *Interactive processes in reading*. Hillsdale, N.J.: Lawrence Erlbaum Associates.

Margolin, C. M., Griebel, B. and Wolford, G. (1982). Effect of distraction on reading vs. listening. *Journal of Experimental Psychology: Learning, Memory and Cognition*, **8**, 613–618.

McLeod, P. and Posner, M. I. (1984). Privileged loops from percept to act. In H. Bouma and E. Bouwhuis (Eds.), *Attention and performance X*. Hillsdale, N.J.: Lawrence Erlbaum Associates.

Morris, R. (1983). *Dementia and the functioning of the articulatory loop system*.

Murray, D. J. (1968). Articulation and acoustic confusability in short-term memory. *Journal of Experimental Psychology*, **78**, 679–684.

Nicholson, R. (1981). The relationship between memory span and processing speed. In M. Friedman, J. P. Das and N. O'Connor (Eds.), *Intelligence and learning*. Pp. 179–184. New York, Plenum Press.

Rabbitt, P. M. A. (1968). Channel-capacity, intelligibility and immediate memory. *Quarterly Journal of Experimental Psychology*, **20**, 241–248.

Salame, P. and Baddeley, A. D. (1982). Disruption of short-term memory by unattended speech: Implications for the structure of working memory. *Journal of Verbal Learning and Verbal Behavior*, **21**, 150–164.

Sperling, G. (1967). Successive approximations to a model for short-term memory. *Acta Psychologica*, **27**, 285–292.

Vallar, G. and Baddeley, A. D. (In press). Fractionation of working memory: Neuropsychological evidence for a phonological short-term store. *Journal of Verbal Learning and Verbal Behavior*.

Waugh, N. C. and Norman, D. A. (1965). Primary memory. *Psychological Review*, **72**, 89–104.

9 When long-term learning depends on short-term storage

A. D. Baddeley, C. Papagno and G. Vallar

Since the 1960s, there has been controversy as to whether long-term learning might depend on some form of temporary short-term storage. Evidence that patients with grossly impaired memory span might show normal learning was, however, particularly problematic for such views. We re-examine the question by studying the learning capacity of a patient, P.V., with a very pure deficit in short-term memory. A series of experiments compare her learning capacity with that of matched controls. The first experiment shows that her capacity to learn pairs of meaningful words is within the normal range. A second experiment examines her capacity to learn to associate a familiar word with an unfamiliar item from another language. With auditory presentation she is completely unable to perform this task. Further studies show that when visual presentation is used she shows evidence of learning, but is clearly impaired. It is suggested that short-term phonological storage is important for learning unfamiliar verbal material, but is not essential for forming associations between meaningful items that are already known. Implications for the possible role of a phonological short-term store in the acquisition of vocabulary by children are discussed.

During the 1960s, evidence for separate long- and short-term memory stores began to accumulate. This evidence led to a number of models, the most influential of which was the "modal model" of Atkinson and Shiffrin (1968). This assumed that information was fed from a series of sensory buffers into a limited capacity short-term store, which in turn fed information into and out of a much larger capacity long-term store. Learning was assumed to involve transfer of information from the short-term to the long-term store, with the probability of long-term learning being a function of the time spent by the relevant item in the short-term store. By the early 1970s evidence that did not fit into the modal model was beginning to accumulate. For example, Craik and Watkins (1973) found no relationship between the time an item was held in short-term storage and its probability of long-term learning.

One of the most telling sources of evidence against the modal model, however, came from the study of neuropsychological patients who had a specific impairment in short-term memory performance. Shallice and Warrington

(1970) reported data from a patient with a digit span of only two items who nevertheless showed normal long-term learning ability. As the modal model assumes that long-term learning depends crucially on the capacity of the short-term store, it is not clear how it can account for such results.

In the light of this and other data, Baddeley and Hitch (1974) proposed that the concept of a unitary short-term memory system should be replaced by the concept of a multicomponent working memory. The short-term memory patients studied by Shallice and Warrington might be assumed to have a deficit in one component of this, the short-term phonological store that comprises part of the articulatory loop sub-system of working memory. Subsequent research by Vallar and Baddeley (1984a) showed that the memory performance of P.V., a very pure short-term memory patient could be explained using the working memory framework by assuming that she had an impairment of the phonological store, but was otherwise intellectually unimpaired.

The concept of a phonological short-term store that forms part of the articulatory loop slave system is capable of explaining a wide range of short-term memory phenomena. These include the effects of acoustic similarity, word length, unattended speech, and articulatory suppression (Baddeley, 1986). The assumption that short-term memory patients have a deficit in this system is also a plausible one. It does, however, raise a rather important further question. The short-term memory patients studied by Shallice and Warrington and by Vallar and Baddeley show clear evidence of marked impairment on standard memory span tasks, but show remarkably few problems in coping with everyday life. This raises the somewhat depressing thought that the phonological short-term store may be essential for performing the kinds of short-term memory task studied by psychologists, but of little practical significance. We therefore decided to explore in more detail just what our short-term memory patient P.V. can and cannot achieve.

We began by studying her capacity to comprehend spoken and written prose. In casual conversation, she did not appear to show any obvious comprehension deficits, except where numbers and prices were involved, when she would supplement her mnemonic capacity by using notes. We found, however, that when her comprehension was tested using relatively demanding material in which it was necessary to maintain surface structure across several intervening words, her performance was significantly impaired (Vallar & Baddeley, 1984b).

Another possible role of the phonological short-term store was suggested by the study of children with developmental dyslexia. One of the most common characteristics of such children is their impaired memory span (Miles & Ellis, 1981). A more detailed analysis of their memory performance suggests that they may well have impaired phonological storage capacity (Baddeley, 1986; Jorm, 1983). This in turn raises the question of how the phonological store might be involved in learning to read.

One possibility is that short-term phonological storage may be used during the process of decoding an unfamiliar word. It is certainly the case that some children appear to read novel words by proceeding through the word decoding each individual letter and storing it as a speech sound. When they reach the end of the word, they attempt to blend the constituent sounds into a spoken response. A child who has impaired phonological storage capacity is likely to be able to hold fewer sounds and to be less capable of using this strategy effectively (Baddeley, 1979).

Another common feature of dyslexic children is their reported problems in rote verbal memory. For instance, Miles and Ellis (1981) report that such children typically have great difficulty learning multiplication tables or indeed learning the order in which the months of the year occur. They also report that such children tend to be particularly poor at learning foreign languages, often despite a high level of intelligence. Kail and Leonard (1986) report that children with specific reading difficulties tend to have a level of vocabulary that is lower than would be expected from children of equivalent intelligence. The deficit is present at an early stage of education, suggesting that it cannot readily be explained in terms of a failure to acquire vocabulary through reading. Again then, there appears to be some evidence that the phonological store may be important for certain aspects of verbal learning. How might such a vocabulary deficit arise? Would our patient P.V. show a deficit in new phonologically based nonword learning?

The most extreme deficit might be expected if the items to be learned were beyond the subject's memory span. If that were the case, then it is possible that at any given time the subject would only be able to retain part of the stimulus item, presumably making it difficult to interrelate the components and integrate them into a new phonological unit.

A second possibility is that even if an item is within the memory span, a subject might need additional storage capacity if the item is to be manipulated and related to other material in a relatively active way. For this reason, we determined the memory span for nonverbal material of our short-term memory patient P.V., and arranged that the new material to be learned included both items that were well within her span and also items on which she would be expected to make some mistakes if asked to repeat them back immediately. We then tested her capacity to learn to associate nonword items with meaningful words, as if learning the vocabulary of a foreign language. If the short-term phonological store is necessary for the long-term learning of such material, then we would expect her performance to be impaired when compared with that of a matched sample of normal subjects.

Case report

P.V., born in Italy in 1951, was a right-handed woman with 11 years of schooling who in February 1977 suffered a left-hemisphere stroke. This case has been described elsewhere (Basso, Spinnler, Vallar, & Zanobio, 1982) and only data relevant to the present experiments are summarized here. P.V. has a selective impairment of auditory memory span, being unable to repeat sequences longer than two or three digits, letters, or words. Performance was better when the stimuli were presented visually as is often the case with such patients. Her pattern of impairment was interpreted in terms of the reduced capacity of a phonological short-term input store. Vallar and Baddeley (1984a) suggested a fractionation of verbal working memory into a phonological short-term store, selectively defective in P.V.'s case, and an articulatory rehearsal process. Auditory stimuli are assumed to have obligatory access to the phonological store, while visual material may enter such a system through the rehearsal process or may be retained through an alternative nonphonological system.

P.V. was shown to perform normally on a number of long-term memory tasks involving meaningful words. These included verbal paired-associate learning, learning a list of 10 words, and delayed recall of a short story (Basso et al., 1982). The patient level of performance on nonverbal learning tasks such as a visuo–spatial maze was also well within the normal range (Basso et al., 1982).

Experiment 1: memory span for nonwords

We began by assessing P.V.'s span for nonwords as a function of length.

Material

This comprised 10 nonword items at each length ranging from two to five syllables, and comprising 4 to 11 letters. The nonwords were constructed so as to resemble Italian, P.V.'s native language, and hence were all readily pronounceable. Items were spoken at a rate of one per second for immediate spoken recall. We began by testing the 10 two-syllable items, first presenting each item individually, then testing sequences of two and so forth up to a point at which P.V. failed to recall in the correct order, the items comprising any of the 10 sequences. We then moved up to nonwords of three syllables, followed by four and five syllables. In each case, testing of a given set of items ceased when P.V. failed to recall correctly any of the 10 sequences.

Results

Table 1 shows P.V.'s performance on this task. It is clear that the only condition on which she performs perfectly is the repetition of single disyllabic items. Increasing the number of syllables to three leads to a 20% error rate, while she never succeeds in repeating back nonwords of four or five syllables. We therefore decided to use nonwords of two and three syllables for the subsequent studies of nonword learning.

Experiment 2: paired-associate word learning

The purpose of this experiment was to ensure that P.V. did indeed show normal long-term learning when meaningful material was employed. She was

Table 1 Immediate recall of sequences of nonwords by P.V.

Sequence length	Nonword length (syllables)			
	2	3	4	5
1	100 (100)	80 (80)	0 (0)	0 (0)
2	50 (70)	50 (65)	— —	— —
3	20 (60)	0 (27)	— —	— —
4	0 (30)	— —	— —	— —

Note
Percentage of sequences reported entirely correctly and in the correct order. Percentage items correct regardless of order is given in parentheses.

presented with eight pairs of words at a rate of 2 s per pair. Immediately after presentation, P.V. was tested by being given the first word of each pair and required to recall the second. This procedure was repeated until P.V. recalled all eight words for two successive trials, or for a maximum of 10 trials. A different random order was used on each presentation trial, and the order of testing differed from that of the presentation. The presentation was auditory, and the responses were spoken.

The material comprised pairs of four 2-syllable and four 3-syllable concrete Italian words of four to nine letters in length, with a frequence greater than 25 per 500,000 (Bortolini, Tagliavini, & Zampolli, 1972). Control data were collected from 14 subjects, matched with P.V. for age and educational level.

Results

P.V. required five trials to reach the learning criterion. Nine of the controls achieved criterion in three trials, 3 in four, and 2 in five. Hence P.V.'s performance is broadly within the normal range on this task. P.V.'s recall on individual trials, together with the average level for control subjects, is shown in Figure 1. Her recall is fully comparable to that of the control group on all learning trials with the possible exception of Trial 3. Here P.V. recalled six items, while the mean was 7.7. Controls ranged from six to eight items correct, so even if one selects the point of maximum difference, P.V. is still within the normal range. We therefore felt it appropriate to proceed to study her nonword learning.

Experiment 3: paired-associate nonword learning

This experiment differed from the previous one in only one respect; instead of pairs of Italian words, the stimuli comprised Italian words and the responses nonwords (e.g., *Rosa–Svieti*). The nonwords were based on Russian words, which where necessary were transliterated so as to be

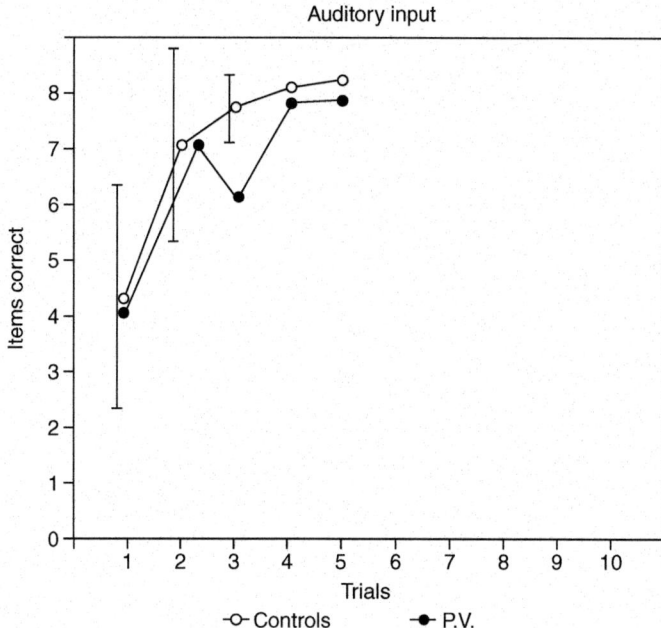

Figure 1 Associate learning of auditorily presented word–word pairs by P.V. and by matched control subjects.

readily pronounced in Italian. Five disyllabic items were used, and three trisyllables. The stimuli comprised two- to three-syllable concrete Italian words with a frequency greater than 25 per 500,000 (Bortolini et al., 1972), different from those employed in the previous experiment. Control data were collected from 14 subjects matched with P.V. for age and educational level. Neither P.V. nor the control subjects had any knowledge of Russian.

Results

Figure 2 shows the learning curve for the control subjects and for P.V., who shows a dramatic learning impairment on this task. She systematically failed to recall a single item on any of the 10 learning trials, failing even to come up with intrusions or erroneous responses. Of the 14 control subjects, 6 reached a criterion of two successive correct recalls, while a further 6 had managed perfect recall on the final trial.

There is no doubt that P.V. is dramatically impaired on this task. Indeed, her failure to learn is total for both disyllables that would clearly be within her span and trisyllabic items which might be just beyond it. It could,

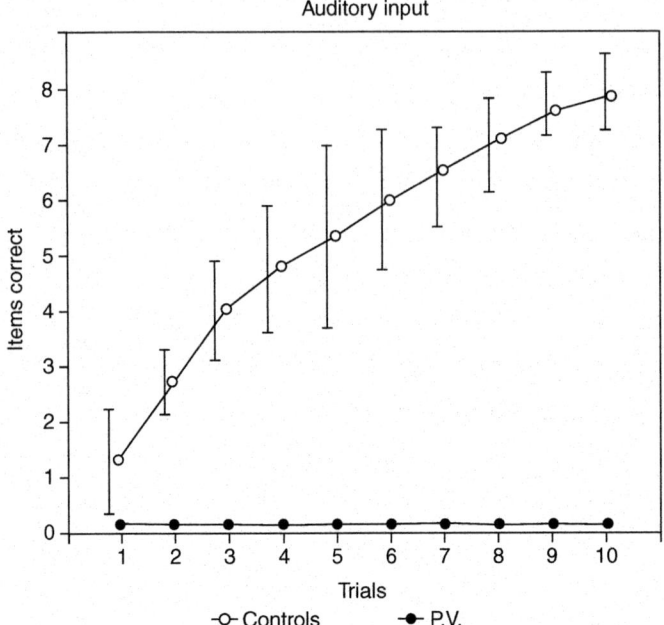

Figure 2 Performance on word–nonword paired associate learning with auditory presentation; data from P.V. and matched control subjects.

however, be argued that the rate of presentation used was sufficiently high to make conditions particularly difficult, placing a premium on the use of a temporary buffering system such as the phonological store might offer, and perhaps minimizing the possibility of utilizing long-term learning. We therefore decided to repeat the experiment using a parallel but different set of material and a slower rate of presentation.

Experiment 4: paired-associate nonword learning—slow presentation

This experiment differed from the previous one only in that rate of presentation was slowed down to one pair per 5 s. A fresh set of material was used, with the responses comprising five trisyllabic and three disyllabic nonwords.

Results

P.V. again showed a total lack of learning, failing to recall a single pair on any of the 10 learning trials.

Experiment 5: paired-associate learning—visual presentation

As mentioned previously, P.V. has a visual memory span that is consistently greater than her auditory span (Basso et al., 1982). Furthermore, visual presentation at least guarantees that the various components of the item are available simultaneously, provided P.V. can read the item. We tested her capacity for reading nonwords using the material that was previously used to test her memory span. She performed perfectly, reading all 10 items of two, three, four, and five syllables.

Procedure

We tested P.V. and the 14 control subjects under conditions equivalent to those used in Experiments 2 and 3, with the exception that presentation was visual rather than auditory. Items were written in black capital letters on white index cards and presented at a rate of 2 s per pair. The material again comprised word–word pairs and word–nonword pairs, and was made up from items of two or three syllables that were either high-frequency concrete Italian words or Russian words transliterated to be pronounceable in Italian. None of the items had been used previously. P.V.'s performance was compared with that of the 14 control subjects described previously.

Results

Figure 3 shows the performance of P.V. and the control subjects on learning the word–word and word–nonword pairs.

In the case of learning meaningful words, P.V. is performing within the normal range. She reaches the criterion of two successive correct recalls in 4 trials, while one control achieved criterion in 2 trials, eight in 3, and five in 4. Her learning curve also suggests that performance was well within the normal range, suggesting that her capacity for learning meaningful verbal material is not seriously impaired.

Performance on the word–nonword learning condition with rapid visual presentation is shown in Figure 3b. It is clear from this that P.V. benefits from visual presentation since she is able to learn six of the eight pairs within the 10 trials. She is still performing at a consistently poorer level than the controls, 12 of whom have reached criterion within the 10 trials. This difference is echoed in the learning curves which show that P.V. is more than 2 SD below the average score for the control group on all except the first learning trial.

It is clear that P.V. learns much more effectively with visual than with auditory presentation. What of the control subjects? In the case of learning pairs of words, no difference was found, with 9 subjects performing better on visual than auditory, and 5 the reverse. In the case of nonword learning,

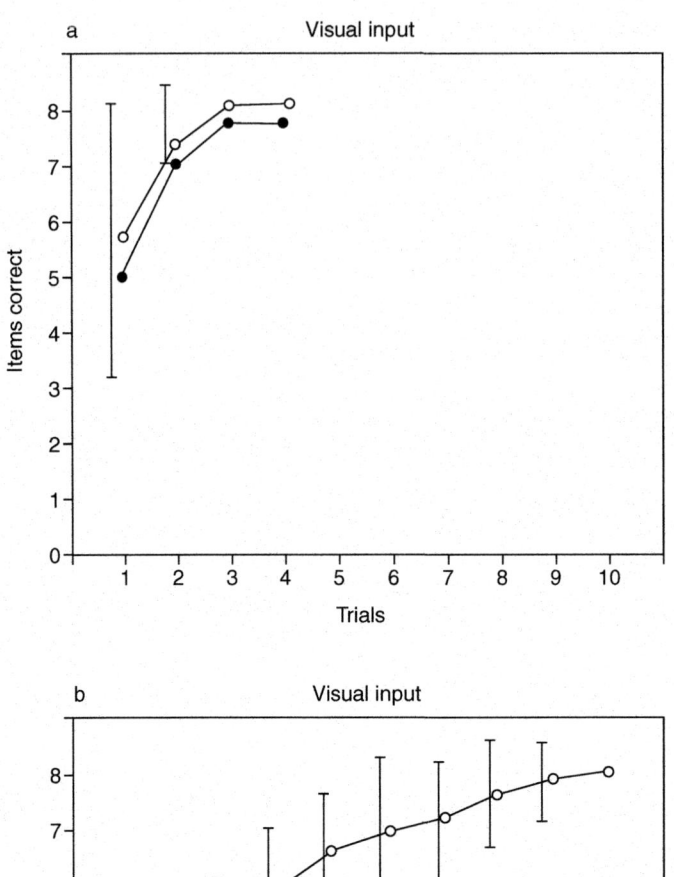
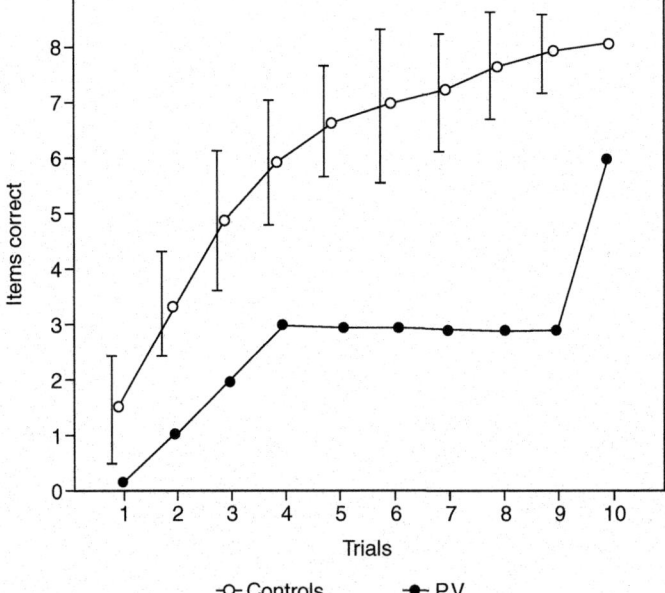

Figure 3 Learning of visually presented word–word pairs (a) and word–nonword pairs (b) by P.V. and matched control subjects.

however, controls do show an advantage for visual presentation, with 13 subjects requiring fewer trials to reach criterion in the rapid visual condition than the rapid auditory, and one showing no difference. Controls were not, of course, run under slow auditory conditions.

Since the order of the experimental conditions was not counterbalanced and subjects were tested with auditory presentation first, this apparent visual advantage should be treated with caution. However, it is consistent with earlier observations of visual modality advantage in the long-term or prerecency component of free recall (Vallar & Papagno, 1986; Watkins & Watkins, 1973). It is possible that memory for nonwords in our study might be enhanced by visual presentation because this allows both a phonological and a visual encoding of the unfamiliar material. This could be particularly useful in the case of nonwords which do not have an existing lexical representation. This point might be worth exploring further since it does have practical implications for the learning of foreign language vocabulary.

Discussion

Our results have shown that despite having a normal capacity to learn to associate pairs of meaningful words, P.V. shows gross impairment in her capacity to learn to associate pronounceable nonword responses with word stimuli. The deficit applies to both disyllabic words that are within her immediate memory span and to trisyllables, which are at span length. The deficit appears to be complete when auditory presentation is used, whether presentation is slow or fast. In the case of visual presentation she shows a clear capacity to learn, but still performs at a level well below that of matched control subjects. We begin by discussing the broad general implications of these results for models of human learning, before considering their implications for the possible function of the short-term phonological storage component of working memory. We then discuss possible practical implications.

As mentioned earlier, during the early 1970s, there was considerable controversy about the possible role of short-term phonological coding in long-term learning. Three different viewpoints were proposed: that short-term storage was an essential stage in long-term learning (Atkinson & Shiffrin, 1968), that long-term learning and short-term storage involved parallel and separate processes (Shallice & Warrington, 1970), and that long-term learning involved a shift from shallow visual and phonological codes to a deeper and richer semantic code (Craik & Lockhart, 1972). We consider our results in the light of each of these view-points.

The modal model of Atkinson and Shiffrin predicts that a deficit in short-term storage should lead to impaired long-term learning. This is exactly what we observe in the case of the nonword items. It is, however, far from clear how the modal model would account for the enhanced learning with visual presentation, let alone the normal rate of learning when both stimuli and responses comprise meaningful words. The data here and elsewhere

(Baddeley, 1986) are simply far too rich to be explained in terms of a single monolithic short-term store. The modal model could of course be adapted by abandoning the assumption of a unitary short-term store; this line of argument is, however, exactly what led Baddeley and Hitch (1974) to postulate a multi-component working memory in place of the unitary short-term store assumed by the modal model.

A second possible assumption is that access to long-term memory occurs in parallel with, and independent of, the operation of a short-term phonological store, a position espoused by Shallice and Warrington (1970). This assumption is certainly consistent with P.V.'s combination of impaired performance on digit span, together with normal long-term learning of meaningful material. From this viewpoint, however, it is not clear why P.V. should have shown the observed catastrophic deficit in nonword learning. Shallice and Warrington could certainly explain the present results by making a further assumption, namely, that the learning of nonword material depends on both short- and long-term storage systems. Such a position then becomes broadly equivalent to that assumed by the working memory framework (Baddeley, 1986).

How could our results be explained within a levels of processing framework? In its simplest form, this approach would suggest that items are first encoded in a superficial sensory way, followed by progressively deeper subsequent encoding. In the case of auditory presentation, those items that are beyond P.V.'s memory span presumably fail to be adequately encoded at a superficial level, hence leading to an absence of long-term learning. It is less clear why P.V. should have difficulty in the long-term learning of disyllabic nonwords, since these are within her memory span. In the visual presentation condition, material presumably will be initially encoded visually, followed by a deeper phonological encoding, which in turn will lead to a yet deeper encoding, perhaps in terms of semantic associations. Presumably in both the visual and the auditory cases the difficulty arises at the phonological level, possibly as a result of some kind of phonological processing difficulty. Presumably, the shallow visual code in the visual presentation condition supports performance by holding information while it gradually passes through the bottleneck created by impaired phonological processing, hence leading to better performance under visual than under auditory presentation.

A major difficulty with this view, however, is that P.V. shows no evidence that her memory deficit reflects impaired phonological *processing* (Vallar & Baddeley, 1984b). She was able to perform perfectly on a minimal pairs phonemic discrimination task in which she had to judge pairs of consonant–vowel sounds as the same (e.g., *ba–ba*) or different (e.g., *ba–da*). With visually presented words, she was capable of assessing whether the stress should come on the first or second syllable (e.g., *macchina* vs. *spaghetti*), and to judge whether the names of the pictured items rhymed or not. Finally, her speech production appears to be completely normal as assessed both by the distribution of pauses in spontaneous speech (Vallar, Vagges, Magno Caldognetto, & Botlini, unpublished data), and by her capacity to count and recite the alphabet, a task she performs just as rapidly and accurately as control subjects (Vallar & Baddeley, 1984a). In short, her problem appears to be with storage, not processing, and as such it is not clear to us how a levels of processing approach would handle our results.

How well can our results be fitted into a working memory framework? We have argued elsewhere that P.V. suffers from an impairment of the phonological storage component of the articulatory loop (Vallar & Baddeley, 1984a). Her unimpaired capacity to learn to associate meaningful words is explicable on the assumption that this task relies principally on semantic coding, hence placing minimal demands on her impaired phonological store. P.V.'s impaired capacity to repeat back polysyllabic nonwords is entirely consistent with the assumption that this task directly reflects the operation of the phonological short-term store. Impaired capacity to learn to associate nonwords with words when these are presented auditorily suggests that the process of long-term learning requires some form of maintenance of the incoming material. If the items are at or beyond her span, then she will have little or no spare storage capacity for manipulating the material or for relating it to material that she already knows, for example, by noting similarities between the nonwords and familiar Italian words.

With visual presentation, P.V. is able to take advantage of an alternative short-term code. Her span for visually presented items is consistently better than that for auditory (Basso et al., 1982), and hence short-term visual storage is likely to enhance her learning by allowing the maintenance of the unfamiliar nonwords. The fact that she does show clear evidence of learning under these conditions suggests that the problem of an impaired phonological store is not insurmountable.

Considered overall, our results suggest that an important function of the short-term phonological input store is that of learning new words. Such a function is likely to be particularly important during the early years of life when a child is first learning to speak, and when a severe impairment of this system might be expected to lead to an impoverished vocabulary.

While it is not uncommon to find children who have a somewhat reduced digit span, very few are likely to have a span as short as that of P.V. This raises the question of whether problems in long-term rote learning are found only with very severe impairments in span or whether they are likely to occur in the presence of much milder deficits. Some evidence that such impairments may occur are suggested by the previously described observations of Miles and Ellis (1981) on dyslexic children and of Kail and Leonard (1986) on children with impaired vocabulary.

Gathercole and Baddeley (1987) have found a similar deficit in vocabulary in a group of children designated as having normal intelligence but delayed language development. The feature that distinguished these children most clearly from controls proved to be their capacity to repeat spoken nonwords, suggesting the possibility that an impairment in the phonological short-term storage component of working memory may lie at the root of their reduced vocabulary.

Our evidence clearly suggests that short-term phonological storage is important for long-term phonological learning, but leaves open the question of the nature of this influence. One possibility is that the short-term phonological store is necessary as an intermediate stage, holding information and allowing it to be gradually transferred to some form of long-term phonological store. Such an assumption is similar to that made by the original Atkinson and Shiffrin (1968) model. The notion that time in store is necessarily related to amount of learning was challenged by the data of Craik and Watkins (1973) who found no relationship between the time an item was maintained in short-term storage and its probability of long-term learning. They argue that amount of maintenance rehearsal is not related to amount of learning. However, it is important to note that Craik and Watkins used meaningful words for which the phonological representation would already exist and where learning was likely to depend on semantic rather than phonological coding. In the case of nonsense material, Mechanic (1964) showed clear evidence that rote repetition of nonsense syllables in an incidental learning task led to enhanced learning, with longer rehearsal leading to better retention. The inability of P.V. to hold items in a short-term phonological store, resulting in an inability to transfer the information into long-term phonological storage, therefore remains a possible interpretation of our results.

An alternative way of conceptualizing such a deficit is suggested by some recent developments in connectionist models of memory and, in particular, by Hinton and Plaut's (in press) suggestion that learning may be based on the modification of weights between units and that such weights are of two kinds, fast and slow. Slow weights represent the gradual increment of learning over many successive trials, whereas the fast weights offer a way of setting up changes easily and rapidly, at the expense, however, of rapid dissipation. By analogy with this approach, one might argue that short-term phonological coding represents the distribution of fast weights within the phonological system, while long-term phonological learning is based on a modification of the underlying slow weights.

Within such a model, our results could reflect a process whereby the change in slow weights is dependent on the setting up of earlier fast weights. On this interpretation, our patient is defective in her capacity to set up fast weights, and these in turn cause an impairment in the slow weight change within her long-term learning system. A clear interpretation will obviously depend on formulating a more precise model; the development of connectionist modeling techniques would seem to have considerable promise in this particular area.

References

Atkinson, R. C., & Shiffrin, R. M. (1968). Human memory: A proposed system and its control processes. In K. W. Spence (Ed.), *The psychology of learning and motivation: Advances in research and theory* (Vol. 2). New York: Academic Press.

Baddeley, A. D. (1979). Working memory and reading. In P. A. Kolers, M. E. Wrolstad, and H. Bouma (Eds.), *Processing of visible language*. New York: Plenum.

Baddeley, A. D. (1986). *Working memory*. London: Oxford University Press.

Baddeley, A. D., & Hitch, G. J. (1974). Working memory. In G. Bower (Ed.), *Recent advances in learning and motivation* (Vol. VIII). New York: Academic Press.

Baddeley, A. D., Vallar, G., & Wilson, B. (1987). Sentence comprehension and phonological memory: Some neuropsychological evidence. In M. Coltheart (Ed.), *Attention and performance XII*. London: Erlbaum.

Basso, A., Spinnler, H., Vallar, G., & Zanobio, M. E. (1982). Left hemisphere damage and selective impairment of auditory verbal short-term memory. *Neuropsychologia*, **20**, 263–274.

Bortolini, U., Tagliavini, U., & Zampolli, A. (1972). *Lessico di frequenza della lingua Italiana*. Milano: Garzanti.

Craik, F. I. M., & Lockhart, R. S. (1972). Levels of processing: A framework for memory research. *Journal of Verbal Learning and Verbal Behavior*, **11**, 671–684.

Craik, F. I. M., & Watkins, M. J. (1973). The role of rehearsal in short-term memory. *Journal of Verbal Learning and Verbal Behavior*, **12**, 599–607.

Gathercole, S., & Baddeley, A. D. (1987). The processes underlying segmental analysis. *CPC: Cahiers de Psychologie Cognitive, European Bulletin of Cognitive Psychology*, **7**, 462–464.

Hinton, G. E., & Plaut, D. C. (in press). Using fast weights to deblur old memories. In *Proceedings of the Ninth Annual Conference of the Cognitive Science Society, Seattle, WA, 1987*.

Jorm, A. F. (1983). Specific reading retardation and working memory: A review. *British Journal of Psychology*, **74**, 311–342.

Kail, R., & Leonard, L. B. (1986). Word-finding abilities in language-impaired children (ASHA Monographs No. 25). Rockville, MD: American Speech–Language–Hearing Association.

Mechanic, A. (1964). The responses involved in the rote learning of verbal materials. *Journal of Verbal Learning and Verbal Behavior*, **3**, 30–36.

Miles, T. R., & Ellis, N. C. (1981). A lexical encoding difficulty II: Clinical observations. In G. Th. Pavlidis and T. R. Miles (Eds.), *Dyslexia research and its applications to education*. Chichester: Wiley.

Shallice, T., & Warrington, E. K. (1970). Independent functioning of verbal memory stores: A neuropsychological study. *Quarterly Journal of Experimental Psychology*, **22**, 261–273.

Vallar, G., & Baddeley, A. D. (1984a). Fractionation of working memory. Neuropsychological evidence for a phonological short-term store. *Journal of Verbal Learning and Verbal Behavior*, **23**, 151–161.

Vallar, G., & Baddeley, A. D. (1984b). Phonological short-term store, phonological processing and sentence comprehension: A neuropsychological case study. *Cognitive Neuropsychology*, **1**, 121–141.

Vallar, G., & Papagno, C. (1986). Phonological short-term store and the nature of the recency effect: Evidence from neuropsychology. *Brain and Cognition*, **5**, 428–442.

Watkins, M. J., & Watkins, O. C. (1973). The post-categorical status of the modality effect in serial recall. *Journal of Experimental Psychology*, **99**, 226–230.

10 The phonological loop as a language learning device

A. D. Baddeley, S. E. Gathercole and C. Papagno

A relatively simple model of the phonological loop (A. D. Baddeley, 1986), a component of working memory, has proved capable of accommodating a great deal of experimental evidence from normal adult participants, children, and neuropsychological patients. Until recently, however, the role of this subsystem in everyday cognitive activities was unclear. In this article the authors review studies of word learning by normal adults and children, neuropsychological patients, and special developmental populations, which provide evidence that the phonological loop plays a crucial role in learning the novel phonological forms of new words. The authors propose that the primary purpose for which the phonological loop evolved is to store unfamiliar sound patterns while more permanent memory records are being constructed. Its use in retaining sequences of familiar words is, it is argued, secondary.

Baddeley and Hitch (1974) considered the possibility that short-term memory (STM) may serve as a general working memory designed to support complex cognitive activities. This suggestion led to the development of a specific multicomponent model of working memory and has subsequently contributed to an enduring interest in the specific cognitive functions that are fulfilled by the separate subcomponents of working memory. The aspect of working memory for which the fullest theoretical account is now available is the phonological loop (Baddeley, 1986). The loop is specialized for the retention of verbal information over short periods of time; it comprises both a phonological store, which holds information in phonological form, and a rehearsal process, which serves to maintain decaying representations in the phonological store. This relatively simple model has proved capable of accommodating a great deal of experimental evidence from normal adult participants, children, and neuro-psychological patients (see Baddeley, 1997, and Gathercole & Baddeley, 1993, for reviews).

Although the evidence for the existence of such a short-term system is strong, it is not obvious why the phonological loop should be a feature of human cognition at all. People have a remarkable capacity to repeat what they hear, a capacity that has extensively been investigated by using lists of digits or unrelated words. When looking for a function that this capacity serves, Baddeley and Hitch (1974) concentrated on asking why it should be useful for people to remember sequences of words, and this led them to study comprehension and verbal reasoning. However, the evidence of a major role for the phonological loop was far from compelling (see Baddeley, 1986, for review). Indeed, much of the neuropsychological evidence that has led to the development of the current model of the phonological loop (e.g., Vallar & Baddeley, 1984) itself raises questions about its function. Many individuals with specific deficits in short-term phonological memory appear to have few problems in coping with everyday cognition: Despite dramatic reductions in the capacity of the phonological loop, such individuals typically have normal abilities to produce spontaneous speech (Shallice & Butterworth, 1977) and encounter few significant difficulties in language comprehension (Vallar & Shallice, 1990). Does this mean that the loop is of little practical significance and that at least this aspect of STM does not serve as a working memory? Some authors have argued that this is indeed the case (Butterworth, Campbell, & Howard, 1986).

The purpose of the present article is to propose that the phonological loop does indeed have a very important function to fulfill, but that it is one that is not readily uncovered by experimental studies of adult participants. We suggest that the function of the phonological loop is not to remember familiar words but to help learn new words. According to this view, the ability to repeat a string of digits is simply a beneficiary of a more fundamental human capacity to generate a longer lasting representation of a brief and novel speech event—a new word. For an experimental psychologist working exclusively with adults, this might at first seem a singularly arcane and useless skill for humans to possess. For a developmentalist, though, the point of such a skill is all too evident because the task of forming long-term representations of novel phonological material is a key component of language development. At a conservative estimate, the average 5-year-old child will have learned more than 2,000 words (Smith, 1926) and will learn up to 3,000 more per year in the coming school years (Nagy & Herman, 1987). Indeed, successful vocabulary acquisition has been claimed to be the single most important determinant of a child's eventual intellectual and educational attainments (Sternberg, 1987). Learning new words is clearly an important task facing the child's developing cognitive system.

Studies of language acquisition have highlighted numerous domains of skill that the young child has to command in order to become a competent speaker of the native language. There are well-established research traditions concerned with identifying, for example, the ways in which the phonological system develops during infancy and early childhood so that the child can start producing language that is comprehensible to others (e.g., Fowler, 1991; Ingram, 1974): how a child learns the concepts associated with words and their usage (Clark, 1983; Keil, 1979; Markman, 1994), and how the syntactic structure of a language is acquired (e.g., Brown, 1973; Gleitman, 1993; Pinker, 1984). However, little systematic attention has been directed at the processes and mechanisms by which the sound patterns of the words of the native language are learned by the child. This, we propose, is the function for which the phonological loop has evolved.

In the next section, we review evidence that the function of the phonological loop is to provide temporary storage of unfamiliar phonological forms while more permanent memory representations are being constructed (Gathercole & Baddeley, 1993). The contribution of this system to the short-term retention of familiar verbal material in conventional memory-span-type tasks is, we argue, merely an incidental by-product of the primary function of the phonological loop, which is to mediate language learning.

Vocabulary acquisition and the phonological loop in children

Childhood represents the most intensive period of new-word learning for most people, and it is during this period that a natural relationship between the phonological loop and word learning has proved easiest to observe. During childhood, large individual differences are found in STM capacity, even for samples of unselected children of the same age. For example, in a 3-year longitudinal study of children, we found that 10% of children aged between 2 years 10 months and 3 years 1 month could already achieve a digit span of four, whereas 36% of the same cohort did not reach this level until 2 years later (Gathercole & Adams, 1993).

Similar degrees of individual variation are found in children's knowledge of native vocabulary; furthermore individual differences in children's STM performance prove to be related to their vocabulary knowledge, with children who perform well on tests of verbal STM typically also having good vocabulary knowledge. Table 1 summarizes the correlation coefficients between two measures of verbal STM (auditory digit span and nonword repetition) and of vocabulary knowledge found in a range of studies (Gathercole & Adams, 1993, 1994; Gathercole & Baddeley, 1989; Gathercole, Hitch, Service, & Martin, 1997; Gathercole, Willis, & Baddeley, 1991; Gathercole, Willis, Emslie, & Baddeley, 1992; Michas & Henry, 1994).

Before considering the results in detail, it is worth commenting on these two tasks. Digit span (a measure of the maximum length of sequence of digits that an individual can correctly recall) is the most widely used measure of verbal STM ability and is present as a subtest in most major standardized ability test batteries such as the Wechsler Intelligence Scale for Children (Wechsler, 1974), Wechsler Adult Intelligence Scale (Wechsler, 1981), and the British Abilities Scales (Elliott, 1983). The digit span measure provides a useful indication of the capacity of an individual's phonological loop. Nonword repetition provides a measure of the accuracy with which a child can accurately repeat unfamiliar spoken forms such as *woogalamic* or *loddernaypish*. We originally became interested in this task because it appeared to provide a relatively pure measure of phonological loop capacity. Our

Table 1 Simple and partial correlations between phonological memory and vocabulary measures across different studies

Age	n	Nonword repetition		Digit span	
		Simple	Partial	Simple	Partial
3.00[a]	54	**.34**	**.31**	.15	.16
4.01[b,c]	70	**.49**	**.47**	**.28**	**.22**
4.07[d]	80	**.56**	**.46**	—	—
4.09[e]	57	**.41**	**.41**	**.28**	**.29**
5.03[b]	70	**.34**	**.36**	.20	.18
5.06[f]	48	**.48**	—	—	—
5.07[d]	80	**.52**	**.50**	—	—
5.09[e]	51	**.41**	**.31**	**.38**	**.28**
5.09[g,h]	65	**.61**	**.53**	**.44**	**.38**
6.07[d]	80	**.56**	**.48**	**.44**	**.33**
8.07[d]	80	**.28**	.22	**.36**	.23
13.10[g]	60	—	—	**.49**	**.46**

Note
Dashes indicate partial correlations are not available. Coefficients printed in bold are significant at the 5% level. In the partial correlations, the nonverbal ability measure (Raven's Coloured Progressive Matrices; Raven, 1986) was partialed out. Unless indicated otherwise (by Footnote c or h), vocabulary composite scores are based on the British Vocabulary Scale (Dunn & Dunn, 1982).

a Data are from Gathercole and Adams (1993).
b Data are from Gathercole and Adams (1994).
c Composite vocabulary score is based on British Picture Vocabulary Scales (Long Form; Dunn & Dunn, 1982) and the Oral Vocabulary subtest of the McCarthy Scales for Children (McCarthy, 1970).
d Data are from Gathercole, Willis, Emslie, and Baddeley (1992).
e Data are from Gathercole, Willis, and Baddeley (1991).
f Data are from Michas and Henry (1994).
g Data are from Gathercole, Hitch, Service, and Martin (1997).
h Composite vocabulary score is based on the measures indicated in Footnote c and the Expressive One-Word Picture Vocabulary Scale (Gardner, 1990).

reasoning was that owing to the absence of lexical support for these by unfamiliar sound patterns, the child would have to rely very heavily on the representation of the nonword in the phonological loop as a means of supporting its repetition (Gathercole & Baddeley, 1989, 1990a). Thus, nonword repetition may in fact turn out to be more sensitive to phonological loop function than the more conventional digit span measure.

Table 1 shows that across the early and middle childhood years, vocabulary knowledge is strongly associated with both digit span and nonword repetition scores. The significant values of the partial correlation coefficients shown in Table 1 (where available), in which the variance in vocabulary knowledge associated with general nonverbal ability was partialed out, establish that this relationship does not simply reflect a shared contribution to both STM performance and vocabulary knowledge of a general intelligence factor. It should be noted that the STM – vocabulary association is particularly high for measures of nonword repetition, in which coefficients typically fall in the range .4–.6; for digit span, the coefficients are consistently lower, in the range .25–.45. Possible reasons for this especially close link between vocabulary and nonword repetition are discussed later.

Of course, correlation does not imply causation. It is in principle as plausible that good vocabulary knowledge supports accurate nonword repetition as the reverse. One way of collecting further evidence on the direction of causality is to carry out a cross-lagged correlational analysis of longitudinal data. Such an analysis compares the correlation between two measures across a particular time period in the two possible causal directions (i.e., the correlations are calculated and compared between early x and later y and early y and later x). According to the logic of cross-lagged correlations (e.g., Crano & Mellon, 1978), the correlation should be stronger in the causal than in the noncausal direction. In other words, if verbal STM ability is the causal factor in the developmental relationship between nonword repetition and vocabulary, one would expect a stronger prediction from nonword repetition at the first assessment to vocabulary 1 year later than the reverse pattern. Gathercole, Willis, Emslie, and Baddeley (1992) applied a cross-lagged correlational analysis to data obtained in a longitudinal study of 80 children tested on three occasions between 4 and 8 years of age (see also Gathercole & Baddeley, 1989) and yielded results that were consistent with the phonological loop hypothesis. Nonword repetition at age 4 was found to be significantly associated with vocabulary test scores 1 year later (partial $r=.38$, $p<.001$, with variance associated with age and nonverbal ability controlled), whereas the vocabulary measure at age 4 was not a significant predictor of nonword repetition scores at age 5 (partial $r=.14$, $p>.05$). Although such a pattern does not provide watertight evidence for causation, it certainly lends further support to the view that ability to repeat nonwords influences learning of new words.

Research carried out by Service and colleagues on Finnish children learning English at school has extended the link between ability to repeat nonwords and word learning to the acquisition of foreign language vocabulary. The original group of children studied by Service (1992) started to learn English at school at 9 or 10 years. Before commencing the English course, the children were given a series of cognitive tests, one of which involved repeating pseudo-English nonwords. The measure of nonword repetition accuracy proved to a very strong predictor of English language learning when it was tested 2 years later. Further longitudinal analysis of Finnish children learning English as a foreign language has provided more direct evidence that the children's later success at acquiring English is principally mediated by a direct link between repetition ability and vocabulary acquisition (Service & Kohonen, 1995). Similar results have recently been reported in a study of 12-year-old children learning English as a second language (Cheung, 1996).

The data reviewed so far have established a close, natural association between children's phonological loop abilities and their knowledge of native vocabulary. Correlational studies of this kind are inevitably prey, however, to a number of important limitations. One such limitation is that studies of natural vocabulary learning do not permit close control of the word-learning opportunities of individual children. Could it therefore be the case that individuals with both good phonological loop skills and vocabulary knowledge are simply exposed to richer linguistic environments at home and that the greater variety of linguistic forms experienced will boost any language-related ability?

Another limitation concerns the specificity of the hypothesized causal relationship between the phonological loop and word learning. Although we have assumed that the loop serves to support the immediate retention and eventual learning of the novel phonological form of new words, the data from studies of children reviewed so far merely establish a link between loop function and ability to demonstrate knowledge of the meaning of a spoken word. Is it really the case that learning the *sound* of a new word taxes the phonological loop, or is it linked with all aspects of word learning, phonological and nonphonological? If so, the theoretical account of the relationship, which is that the temporary representation of the novel phonological form provided by the phonological loop provides the basis for the construction of a more enduring phonological specification, would clearly require substantial modification.

At least some of these concerns have been laid to rest by experimental studies of word learning in children. In an initial study, Gathercole and Baddeley (1990a) tested the abilities of 5-year-old children of either high or low nonword repetition ability (matched on a measure of nonverbal ability) to learn new names of toy animals. Across 15 trials, the experimenter named four toys and tested the children's memory for these names. The toys were either given familiar names such as Peter and Michael or phonologically unfamiliar names such as Pyemass and Meeton (constructed from the same phonological pool as the familiar names). The findings were clear: The children with the low nonword repetition scores were significantly poorer at learning the phonologically unfamiliar names than the high-repetition children. In contrast, there was no reliable difference in the rates at which the two groups of children learned the familiar names.

These results provide some reassurance that new-word learning is indeed linked to phonological memory skills, even when environmental exposure to new words is controlled across subjects. Furthermore, the specific pattern of findings, in which phonological loop function is significantly related to children's abilities to learn nonwords but not words, has turned out to be highly characteristic of studies of STM and long-term learning and is the signature of many of the studies discussed elsewhere in this article. Similar findings were obtained in a recent study of phonological memory and word learning in 65 5-year-old children (Gathercole, Hitch, Service, & Martin, 1997). The principal concern in this study was to investigate the specificity of the association between phonological loop function and the learning of the phonological form of new words. The children were tested, in separate sessions, on their abilities to learn either pairs of familiar words such as *table–rabbit* or word–nonword pairs such as *fairy–bleximus*. The main finding was that phonological loop ability in this sample of children, as indexed by their scores on nonword repetition and digit span tasks, was highly associated with rate of learning the word–nonword pairs ($r=.63$, $p<.001$) but not with word pair learning ($r=.23$, $p>.05$). Even after variance attributable to differences in age, nonverbal ability, and vocabulary knowledge were taken into account, the partial correlation between phonological memory and word–nonword learning remained strong (partial $r=.49$, $p<.001$); the corresponding partial correlation between memory and word pair learning diminished further to .07, $p>.05$. Thus, ability to learn to associate pairs of familiar words was quite independent of phonological loop function. In contrast, the children's ease of learning new words was strongly constrained by their phonological loop capacity.

A similar finding emerged from a recent study of experimental word learning by Michas and Henry (1994), in which young children were taught the names of three new words, such as *gondola, platypus*, and *minstrel*. An important degree of specificity to the memory–vocabulary association established by Michas and Henry was that it was independent of spatial memory skill.

Further explorations of the developmental relationship between nonword repetition in particular and vocabulary acquisition indicate that it is oversimplistic to claim that the phonological loop mediates long-term phonological learning in a unidirectional manner. Instead, vocabulary knowledge, phonological loop capacity, and nonword learning share a highly interactive relationship. There is accumulating evidence that, for at least some nonwords, the task of nonword repetition taps both the phonological loop and knowledge about the structure of the native language. This fact is demonstrated most simply by the finding that children are reliably more accurate at repeating nonwords that are high in degree of rated wordlikeness (Gathercole, 1995a; Gathercole, Willis, Emslie, & Baddeley, 1991). So even though the stimuli are by definition nonlexical, it appears that children are drawing on their knowledge of either specific familiar words in the language or generalized knowledge of the statistical properties of the language to support the repetition of the novel sound pattern.

The sensitivity of nonword repetition to word likeness provides an important clue as to the relationship between phonological loop function and vocabulary knowledge. It explains why nonword repetition is more highly correlated with vocabulary knowledge than digit span (see Table 1); the reason is that the repetition task itself draws to some degree on the child's vocabulary knowledge and on reflecting phonological loop constraints. For wordlike nonwords, the contribution of long-term knowledge will probably reduce the phonological loop contribution to repetition and hence the sensitivity of the task to phonological loop constraints.

Other evidence also points to a highly interactive relationship between the phonological loop, language knowledge, and long-term learning of the sounds of new words. In the study discussed earlier by Gathercole, Hitch, Service, and Martin (1997), in which experimental word-learning tasks were used as a means of assessing the cognitive components in vocabulary acquisition, speed of learning was correlated to a highly significant extent with children's vocabulary knowledge, even after shared variance with phonological STM had been partialed out. Thus, learning the sounds of new words appears to be mediated by both the phonological loop and long-term knowledge of the native language. The combination of these two types of learning support yields a highly flexible word-learning system in which, where possible, the capacity constraints of the phonological loop are offset by the use of stored knowledge about the language (Gathercole & Martin, 1996).

In summary, evidence from studies of children indicates that the phonological loop mediates the long-term phonological learning involved in acquiring new vocabulary items. This role appears to be particularly significant when the novel phonological forms to be learned have highly unfamiliar sound structures.

Experimental word learning and the phonological loop in adults

If the phonological loop is important for acquiring new vocabulary, it should be possible to hinder such acquisition by interfering with the operation of the loop. Any given manipulation may of course be regarded as affecting several underlying variables, and it could be one of these rather than phonological storage that plays a crucial role in vocabulary acquisition. It is at this point that a coherent model of the phonological loop, which is tied to well-explored experimental phenomena, becomes particularly valuable. Several quite distinct variables share a known impact on the phonological loop; if each of these has a corresponding influence on the learning of unfamiliar phonologically novel vocabulary items, it becomes much harder to provide an alternative account of the results. Accordingly, the three sets of experiments described later study the influence on vocabulary acquisition of variables that are known to influence the operation of the phonological loop in clearly specified ways. Detailed accounts of these empirical phenomena in terms of the phonological loop model are provided elsewhere (e.g., Baddeley, 1986, 1997). Briefly, the effective capacity of the phonological loop is diminished when list items have long names rather than short names, the *word length effect* (Baddeley, Thomson, & Buchanan, 1975), have names that are phonologically similar to one another, the *phonological similarity effect* (Conrad & Hull, 1964), and when participants are required to engage in irrelevant articulation during presentation of the memory list, the *articulatory suppression effect* (Murray, 1967). Although the word length and articulatory suppression effects appear to be located in the rehearsal process, the source of the phonological similarity effect is believed to be the phonological store (e.g., Baddeley, 1986). The important question for the hypothesis that the phonological loop mediates the long-term learning of the sounds of new words is the following: Do these three variables also influence phonological learning?

In an initial series of experiments, Papagno, Valentine, and Baddeley (1991) studied the effect of articulatory suppression on the acquisition rate of pairs of familiar words and items of a foreign language vocabulary by normal participants. It is important to note here that articulatory suppression places minimal demands on executive processes but has a precise effect on the capacity for phonologically encoding visually presented material and for actively maintaining it by rehearsal. It thus has no effect on performance for patients such as P.V. who do not use this mode of encoding (Baddeley, Papagno, & Vallar, 1988), but it does remove the phonological similarity effect with visual though not auditory presentation, in line with the phonological loop model (Murray, 1967). The participants in the initial experiments reported by Papagno et al. were native Italian speakers, and they were asked to learn two types of material. The first comprised pairs of unrelated Italian words such as *cavallo–libro*, whereas the second involved learning Italian–Russian pairs (e.g., *rosa–svieti*). Both visual and auditory presentation were used because articulatory suppression is likely to interfere with the two in slightly different ways; with visual presentation, the phonological recording of the material should be prevented (Baddeley et al., 1975), but on the other hand a visual code will be provided, and this may be helpful in vocabulary learning. With auditory presentation, the supplementary visual code will be absent, but the obligatory auditory access to the phonological store will allow a phonological representation to be set up, though not rehearsed (Vallar & Baddeley, 1984). Hence, suppression should impair performance under either mode of presentation. This is in fact what occurred, with suppression having a clearly deleterious effect on the acquisition of foreign language vocabulary. Suppression had little effect, though, on meaningful paired-associate learning in the participants' native language.

This pattern of findings has consistently emerged in experimental studies of paired-associate learning: Although the learning of phonological unfamiliar material is highly sensitive to variables known to influence the phonological loop, the learning of associations between already familiar phonological lexical forms proceeds more or less independently of these variables. The pattern is also notably similar to the one found in studies of children's word learning, in which learning of unfamiliar phonological forms is constrained by phonological memory skills, whereas learning of familiar names is not (Gathercole & Baddeley, 1990a; Gathercole, Hitch, Service, & Martin, 1997). The clear implication is that, when possible, people use existing language knowledge to mediate their attempts at verbal learning. When unfamiliar phonological forms are presented so that no such knowledge is available to support learning, participants are forced to rely solely on the more fragile phonological loop system to provide the necessary temporary storage of the phonological material while more stable long-term phonological representations are being constructed.

In line with this interpretation, an initial attempt to replicate the Papagno et al. (1991) findings described earlier by using English participants ran into difficulties because the participants found it too easy to form semantic associations to the forms of the Russian words. However, once the association values of the material were reduced (initially by using nonsense material and subsequently by using the more unfamiliar phonological structures provided by Finnish vocabulary), the initial results obtained with Italian participants were replicated (Papagno et al., 1991). A second way of exploiting the phonological loop model is to vary the degree of acoustic similarity among the items to be learned. There is abundant evidence to suggest that the long-term acquisition of pairs of items in one's native language depends on semantic rather than on acoustic coding (Baddeley, 1966). However if the unfamiliar new vocabulary items depend on the phonological loop for their initial acquisition, then one would expect an acoustic similarity effect to occur.

In a further series of experiments, Papagno and Vallar (1992) therefore manipulated the degree of phonological similarity among the items to be learned. The predictions are again straightforward. When learning meaningful paired associates, the principal mode of coding should be lexical–semantic, with the result that phonological similarity will have little impact. On the other hand, when participants are learning unfamiliar vocabulary from a foreign language, the phonological loop system should be crucial, hence phonologically similar items should be confusable and lead to slower learning. In a series of experiments using both auditory and visual presentation, Papagno and Vallar demonstrated that this was indeed the case.

A final variable that would be expected to specifically impair the operation of the phonological loop is item length, with immediate memory for long nonwords impairing immediate recall because of the impact of length both on rehearsal and on output delay. Papagno and Vallar (1992) therefore manipulated the number of syllables in the native and foreign response items to be learned in their paired-associate tasks. Once again, the prediction for lists containing familiar words is that length should not be an important variable because items will be acquired principally on the basis of semantic coding. In contrast, learning foreign language vocabulary (i.e., word–nonword pairs) should be impaired if the participant uses subvocal rehearsal as a crucial part of the phonological loop-based learning process. The data were consistent with this prediction: Word length had no influence on the participants' acquisition of pairs of items in their native language, but it had a substantial effect on the acquisition of unfamiliar Russian vocabulary.

In interpreting the influence of articulatory suppression, phonological similarity, and word length on the acquisition of foreign language vocabulary, three points are of particular importance. First, in each case the observed effect was predicted on the basis of specific well-established characteristics of the operation of the phonological loop. Second, the specific nature of the interaction between the type of material and each of the variables is important; in all cases, the relevant variable has no effect on the acquisition of words, coupled with a very clear effect on nonword learning. Finally, any possibility that we are simply picking up effects of added difficulty is ruled out by the absence of an influence of each of these variables, not only on word learning in normal participants but also in the absence of any effect on the performance of patients who do not utilize the phonological loop in the normal way (Baddeley et al., 1988).

Further evidence in favor of the phonological loop hypothesis comes from a series of studies concerned with optimizing foreign language learning. Ellis and Beaton (1993) investigated the role of visual imagery and rote verbal rehearsal in the acquisition of German vocabulary by English speakers. Imagery proved to be the most effective strategy for learning to produce the English equivalent of German words, but when the requirement was to generate the German translation of an English word, rote rehearsal proved more effective, again implicating the phonological loop in that aspect of the task that involves learning to produce novel phonological forms.

Indeed, it is an interesting possibility that imitating the sounds of new words may be a natural strategy that serves to boost vocabulary acquisition by enhancing phonological loop representations of the novel phonological structures. There is certainly considerable evidence that imitation does play a significant role in vocabulary learning, with many observations that some infants spontaneously imitate the language of others (Bloom, Hood, & Lightbown, 1974; Coggins & Morrison, 1981). Masur (1995) has recently provided a detailed quantitative evaluation of the links between imitation of words and later vocabulary development on the basis of longitudinal laboratory observations of 20 children between the ages of 10 months and 2 years. Children with larger vocabularies were found to imitate words spoken by the caregiver more than children with more restricted early vocabularies. Furthermore, Masur found that spontaneous imitations of words that were not in the children's current vocabulary significantly predicted their later vocabulary growth during the second year, even after the size of the children's vocabulary at the time of imitation had been taken into account. Whatever lies at the root of these differences in spontaneous imitation, these findings suggest that imitation of novel phonological forms may indeed serve to promote the long-term phonological learning of new words, possibly by increasing the period over which they are held in the phonological loop.

These various experimental studies converge on a simple model of new-word learning. According to this model, the long-term learning of the sound structures of novel, phonologically unfamiliar words depends on the availability of adequate representations of the sound pattern in the phonological loop. Thus, the phonological loop appears to provide a critical input to the construction of the more permanent phonological structures that are stored in the mental lexicon. Learning of associations that require the production of familiar lexical items, on the other hand, is achieved typically either without any reliance on the phonological loop or with reduced loop support and is presumably mediated instead by the use of existing knowledge of the native language.

In the sections that follow we summarize further evidence from a variety of participant populations and research laboratories that is consistent with the model of the function of the phonological loop as a word-learning device outlined earlier. In these sections, we chart the consistently close relationships between phonological loop capacity and abilities to learn new words, either in natural vocabulary acquisition or in experimental simulations of vocabulary learning in individuals with STM deficits arising from brain damage, developmental disorders, and specific mental handicaps. The weight of this converging evidence lends considerable force to the view that the primary function of the phonological loop is to support the long-term learning of the phonological forms of words in one's own language.

Cases of cognitive deficit

Following early studies by Shallice and Warrington (1970), the accepted view was that STM patients have a normal capacity for long-term learning. It is notable, though, that most long-term memory (LTM) tests give ample scope for semantic coding. Participants are usually required to learn arbitrary sequences of familiar words, not phonologically novel material. Baddeley et al. (1988) therefore decided to test the capacity of the STM patient P.V. for learning the vocabulary of an unfamiliar language, Russian.

P.V. and a group of 14 matched control participants were asked to learn the two types of paired associates used subsequently in the Papagno et al. (1991) study with normal adult participants. The pairs consisted of either unrelated Italian words (P.V.'s native language was Italian) or Italian–Russian equivalents. Because P.V. had difficulty repeating back polysyllabic Russian words, we restricted our list to comparatively short items. In each case, lists of eight pairs were presented by using either the auditory mode, which should place the maximum load on her phonological store, or the visual presentation of the transliterated stimuli. The results were clear. P.V. was perfectly normal at learning to associate pairs or words in her native language, but her capacity for learning Russian vocabulary was severely impaired. With auditory presentation, the control participants had learned the whole list before P.V. had mastered a single item, despite the fact that they were short enough for her to be able to hear and repeat back accurately. With visual presentation her performance was somewhat better, but it was still markedly worse than that of the control participants.

It appears then that P.V.'s short-term phonological deficit was indeed associated with a specific impairment in long-term learning of phonologically unfamiliar material. She showed a dissociation between her normal general long-term and learning capacity and her very marked deficit in long-term phonological learning. A similar pattern of immediate memory and long-term learning deficit was also reported by Trojano and Grossi (1995), in a study of a patient, S.C., with very poor phonological function who showed no evidence of rehearsal. Characteristically, S.C. was completely unable to learn auditorily presented word–nonword pairs, despite showing evidence of adequate learning ability in other tasks that did not share such a heavy phonological learning component.

Evidence that a long-term phonological learning deficit arises from impairments in the phonological loop has also been provided in a study of an individual who appears to have a developmental impairment of the loop. In attempting to collect control participants for an experiment, Baddeley (1993) identified a graduate student, S.R., with an unusual STM profile. Although highly intelligent and sophisticated in cognitive psychology, S.R. was not reliably able to repeat sequences of more than four digits and performed very poorly on a task involving the immediate repetition of multisyllabic nonwords. When compared with a group of six fellow students on a wide range of short-term phonological memory tests, S.R. invariably performed more poorly. On the other hand, his capacity for short-term visual memory was normal while his long-term visual recognition score on the Doors and People Test (Baddeley, Emslie, & Nimmo-Smith, 1994) was excellent.

The crucial question was how he would perform on phonological LTM. On two companion tests to the Doors Test, which involved the recognition and recall of names, S.R. performed more poorly than any of the control participants. Finally, he was tested by using a paradigm based on that developed with P.V., in which he learned pairs of meaningful English words and English–Finnish foreign language vocabulary. S.R. showed excellent use of mnemonics and was quite normal in his learning of meaningful paired associates. His capacity to learn Finnish vocabulary, though, was grossly impaired when compared with control participants. It is perhaps worth noting that S.R. had previously tried unsuccessfully to learn two languages, being eventually excused on a language qualification for admission to university on the grounds of his incapacity for such learning. His vocabulary was excellent as was his reading, but his spelling performance was very poor, despite the considerable energy and ingenuity he had invested in developing spelling mnemonics. In short, S.R.'s low nonword repetition and digit span were associated with very poor performance on name and foreign language learning and on English spelling.

The profiles presented by these three individuals with severely limited phonological loop function, due in two cases to acquired neurological damage (Baddeley et al., 1988; Trojano & Grossi, 1995) and in the other case to an unidentified developmental deficit (Baddeley, 1993), are very similar. Despite their STM limitations, both individuals were able to function adequately and indeed at a high level across a range of intellectual tasks. They did share, though, a highly specific deficit in learning verbal material that was phonologically unfamiliar, despite their normal long-term verbal learning of arbitrary pairs of familiar words.

Notably, neither individual had poor vocabulary knowledge in their native tongue. For P.V. this is unsurprising, as the vast majority of natural vocabulary acquisition takes place before adulthood, at a time before she suffered the neurological insult that resulted in her STM deficit. On the other hand, S.R.'s memory problems do seem likely to be part of a developmental disorder that extended back to early childhood. Although we did not have access to school or clinic records, he reported having been referred to a remedial program in connection with his spelling and language learning problems. Nevertheless, his vocabulary acquisition problems appear to have been restricted to foreign language learning.

There is no doubt that S.R. represents an important paradox for our hypothesis; if he has an impaired phonological loop, how has he acquired a good vocabulary? To resolve this paradox, the process of vocabulary acquisition needs to be considered. During the early years, the words that are first acquired are likely to be highly frequent and often relate to concrete objects. Vocabulary in children is typically assessed by requiring them to either name or point to pictures, and in the early years these are likely to represent objects that most children would be likely to encounter. Under these circumstances, the rate of learning is likely to be set by the child's capacity to master the new phonological forms and to attach them to their referents rather than to the likelihood that the word has been encountered. The probable importance of phonological factors is indicated by the data on age of acquisition, in which for an equivalent level of word frequency, long words tend to be acquired later than short words (Brown & Hulme, 1996). As vocabulary develops, it is likely to depend increasingly on acquiring low-frequency words. These in turn will often be abstract in nature and relatively unlikely to be encountered with any frequency in day-to-day conversation. Testing tends to be by synonym matching, allowing the participant to use sophisticated guessing strategies to rule out at least some of the alternatives. General intelligence is likely to be important in this context and to be even more important in determining whether the listener or reader is able to gain some idea as to the meaning of a novel word when it is encountered in context. Hence, a phrase such as "the lawyer was searching sedulously through his papers" may give some idea as to the meaning of "sedulous" even though the word is never specifically looked up in a dictionary or defined.

We now return to S.R. who is highly intelligent, well motivated, and well educated but with poor phonological loop capacity: His rate of acquisition of new words is initially likely to be relatively slow, but over 20 years he is likely to have plenty of opportunity to acquire the type of word that occurs frequently within the language. His performance on relatively frequent words is thus likely to approach a similar plateau to other participants, although more slowly. In the case of those rarer words that it is necessary to know in order to score more highly on vocabulary tests, he is favored by his general intelligence, his education, and his motivation. Hence, while he might not have as high a vocabulary as he would have done had he been phonologically well endowed, his cognitive and educational advantages are likely in the long run to substantially outweigh the limitations set by the slower rate of acquisition of new phonological forms.

To uncover a direct relationship between verbal STM and natural vocabulary acquisition, it is therefore necessary to study either children still in the process of acquiring their first language or individuals without exceptional cognitive abilities to compensate for specific memory problems in vocabulary learning. It is to these participant populations that we turn to in the next two sections.

Learning disabilities

There is increasing awareness of the diverse patterns of cognitive ability that may be seen in genetic syndromes associated with mental handicap. Bellugi and her colleagues have conducted a series of studies contrasting the phenotypic profiles of individuals with Williams syndrome and those with Down's syndrome, and they noted that while both are associated with mental handicap, the profile of abilities is very different (e.g., Bellugi, Marks, Bihrle, & Sabo, 1988). Individuals with Williams syndrome demonstrate relatively good language skills in relation to their mental ages and are more likely to produce unusual and low-frequency words both in spontaneous speech and in a verbal fluency task, whereas Down's syndrome is usually associated with poor communicative skills. Wang and Bellugi (1994) explicitly compared memory span in individuals with Williams and Down's syndromes, using groups who were matched on overall IQ. Those with Williams syndrome had a mean digit span of 4.6, whereas the Down's group had a significantly lower mean span of 2.9. Wang and Bellugi found a contrasting pattern of differences between the two groups on a measure of nonverbal span, the Corsi Blocks Test (DeRenzi & Nichelli, 1975), with superior performance by those with Down's syndrome.

Although the innovative research by Bellugi and her colleagues (Bellugi et al., 1988) clearly demonstrated different patterns of performance in Down's and Williams syndromes, the absence of appropriate control groups makes it difficult to be sure whether the pattern represents a particular weakness in Down's syndrome or a particular strength in Williams. A tendency for Down's syndrome to be associated with hearing problems presents a further complication. Recent studies by Jarrold and Baddeley (1997) and Jarrold, Baddeley, and Hewes (in press) suggested that Down's syndrome is indeed associated with impaired digit span when compared not only with Williams syndrome but also with younger mainstream children and participants with minimal learning difficulties when the groups were matched for verbal mental age. Furthermore, hearing problems were ruled out as a possible explanation of the deficit. It is of interest to note that even when the Down's group was matched with the comparison groups on current vocabulary, their digit span was significantly lower. When matched for vocabulary, however, the Down's group tended to be significantly older, suggesting that their impaired phonological loop performance may have resulted in their taking longer to acquire the same amount of vocabulary as the comparison groups.

Recent work on Williams syndrome has made it clear that although language development in this group is better preserved than nonverbal skills, nonetheless, the verbal IQ scores are typically in the delayed range (Arnold, Yule, & Martin, 1985; Karmiloff-Smith, Grant, Berthoud, Davies, Howlin, & Udwin, in press). A recent study by Grant et al. (1997) has specifically looked at nonword repetition in Williams syndrome, finding that repetition performance was not correlated with chronological age, presumably because the degree of learning disability was varied across participants but finding that it was associated with digit span ($r=.59$, $p=.12$), Raven's Matrices (Raven, 1986; $r=.50$, $p=.039$), Bishop's Test for the Reception of Grammar ($r=.68$, $p=.003$), and vocabulary ($r=.77$, $p<.001$). The mean test age of the group was 107 months on the vocabulary measure, compared with test ages in the region of 80 months on the other tests. This presumably reflects the fact that vocabulary represents a "crystallized" measure that accumulates over time, whereas the other measures are based on current capacity. Finally, Grant et al. found a higher correlation between repetition of low wordlike items and vocabulary than occurred with high wordlike items—a pattern that resembled Gathercole's (1995a) observations on 4-year-old normally developing children rather than on her results for 5-year-olds. They interpreted this pattern as indicating that repetition performance in this group relies principally on phonological memory and is less influenced by existing vocabulary, concluding that "the good vocabulary scores of older children and adults with WS [Williams syndrome] may be simply due to their relatively good phonological short-term memory" (Grant et al., 1997, p. 82).

Although these authors did not directly explore the relationship between the phonological loop and vocabulary knowledge in the groups of individuals with Down's and Williams syndromes, unusually good vocabulary knowledge is a characteristic of Williams syndrome (Bellugi et al., 1988). The profile of superior phonological STM skills and precocious vocabulary knowledge in this syndrome, accompanied by very depressed levels of more general cognitive function, is therefore entirely consistent with the notion that the phonological loop serves a word-learning function.

A notable feature of Down's syndrome is the wide degree of variability in the cognitive abilities of different cases. In fact, a small proportion of individuals with Down's syndrome has achieved near normal language abilities by adulthood (Rondal, 1994). Vallar and Papagno (1993) investigated the word-learning capabilities of one such case, a young woman with Down's syndrome who had a full scale IQ of 71 but a digit span of 5.7, well within the normal range. She was Italian but had lived abroad with her parents and could speak good English and reasonable French. When given the same paired-associate tests as P.V., she proved to be normal in her capacity for learning Russian vocabulary but impaired when compared with control participants in learning pairs of words in her native language, showing the converse deficit to P.V. Thus, as has been reported with Williams syndrome, this woman's intact phonological memory skills appear to have been sufficient to mediate normal levels of vocabulary learning, despite her substantial cognitive deficits in other areas.

Children with specific language impairment (SLI)

The developmental studies reviewed earlier in this article focused on the consequences of individual variation in phonological loop function in normal children for their capacity to acquire new vocabulary. In the present section, a brief overview is provided of the memory and word-learning profiles of children with specific language impairment (SLI). This condition is diagnosed when a child fails to develop language at a normal rate for no obvious reason and despite adequate progress in other areas. The particular profile of language problems varies from child to child, but problems with syntax and morphology are particularly common, with expressive language usually more severely impaired than receptive language (Bishop, 1992). In line with the general profile of impaired language skills, SLI children typically lag behind their peers in terms of vocabulary development (Stark & Tallal, 1981). Could a phonological loop deficit lie at the root of their word-learning problems?

Phonological memory problems have certainly been implicated in SLI (Kirchner & Klatzsky, 1985; Menyuk & Looney, 1976). When compared with age-matched controls, children with SLI perform poorly on both conventional verbal memory span tests (Locke & Scott, 1979, Raine, Hulme, Chadderton, & Bailey, 1991) and on tests of nonword repetition (Kamhi & Catts, 1986; Taylor, Lean, & Schwartz, 1989). Studies of incidental word learning have shown that SLI children are poorer at recalling phonologically novel names for new concepts than age-matched controls, although nonphonological aspects of their acquisition of new words are unimpaired (Dollaghan, 1987; Rice, Buhr, & Nemeth, 1990).

Although abundant evidence exists for patterns of association between cognitive processes in SLI children, it is difficult to tease apart cause and effect. Poor memory in children with SLI is consistent with the hypothesis that memory limitations are the root cause of the language impairment. However, the opposite is also possible; poor verbal memory could result from weak language-impaired skills. One approach to this dilemma is to compare language-impaired children with younger control children matched on language level. If their memory is poor even in relation to this language-matched younger control group, then it can be assured that the problem is not just a secondary consequence of the language limitations. However, a finding of no difference is difficult to interpret; it does not rule out the possibility that memory deficits are holding the child's language development back, but it is also compatible with the view that the memory deficits are secondary (Bishop, 1992).

Studies using the language-matched control design have obtained mixed results. Leonard and Schwartz (1985) found that young SLI children at the one-word stage of language development tended to imitate adult's speech, just like normal young children at this language level, and in an experimental task they showed facility at learning nonsense names for novel items equal to that of younger language-matched controls. However, Haynes (1982) conducted an intentional word-learning study and found that her group of SLI children was far poorer than younger language-matched controls at identifying the target nonwords to which they had earlier been exposed. It is possible that the different results reflect the fact that the children studied by Haynes were older with more severe problems.

In a study of a group of SLI children, Gathercole and Baddeley (1990b) found evidence of a phonological loop deficit that was even more severe than the generalized language delay of the children. In terms of vocabulary knowledge, sentence comprehension and reading achievement, the SLI participants (who had a mean age of 8 years) were lagging, on average, between 18 and 24 months behind their chronological age peers. The SLI group showed normal evidence of rehearsal in their immediate serial recall. However, on tests of immediate nonword repetition, the SLI children performed significantly more poorly than even their 6-year-old language controls; on a 40-item nonword repetition test, all of the SLI children performed more poorly than any of the control children. The nonword repetition abilities of the SLI children were equivalent to those of the average 4-year-old, a full 4 years behind the mean chronological age of this group. Van der Lely and Howard (1993) reported no difference in nonword recall between an SLI group and age-matched controls. Their conclusions are, however, open to dispute (Gathercole & Baddeley, 1995), and subsequent studies from other laboratories report substantial phonological memory deficits in SLI children (Jones, von Stienbrugge, & Chicralis, 1994; Montgomery, 1995).

Intriguingly, Bishop, North, and Donlan (1996) found severe limitations of nonword repetition in children with a history of SLI, whose language problems had been resolved as well as in those who still had measurable language deficits. Because the resolved group no longer had significant impairments on standardized language tests, their memory failures could not be regarded as a secondary consequence of generally weak verbal skills. However, if poor phonological memory causes language delay, how can a child have a major impairment of nonword repetition without showing major limitations of language skills? Bishop et al. argued that children can use good general ability to compensate for early language deficits. As previously suggested in our discussion of the graduate student S.R., weak phonological loop function will delay language development, but it does not necessarily result in lasting deficits, especially if the child is bright and can adopt compensatory strategies.

Gifted language learners

So far, the principal focus of this review has been on the poor vocabulary acquisition abilities that are associated with below-average phonological loop function, in both children and adults. Intriguing evidence has recently been presented that the converse is also true and that the source of the "natural talent" that some individuals have for acquiring foreign languages may be the result of exceptional phonological loop skills.

Papagno and Vallar (1995) compared the performance of groups of polyglot and nonpolyglot university students on a range of memory and long-term learning tasks. The polyglots were able to speak at least three languages fluently and were enrolled at the Language Faculty of Milan University. The non-polyglot students were not studying any languages at an advanced level and had only studied one foreign language at school. The polyglot and non-polyglot participants performed indistinguishably on tests of nonverbal ability, visuospatial STM span, and visuospatial learning and were equivalent in general intellectual skills.

Interestingly, though, the polyglots performed significantly better on the two phonological memory tests: auditory digit span and nonword repetition. On the span measure, the memory advantage to the polyglot group corresponded to an extra 1.6 digits, a substantial gain. Performance on the two phonological memory measures correlated highly with participants' abilities to learn new word–nonword pairs by using the stimuli and methods developed by Baddeley et al. (1988). Memory scores were, however, independent of word–word paired-associated learning.

Once again, good phonological memory performance shares a highly specific link with fast and efficient learning of unfamiliar phonological material, but it is independent of both nonverbal STM skills and the ability to learn combinations of familiar lexical items. This profile, recurring as it does across children, adults, and several special developmental populations, provides the substantive basis for our claim that the primary function of the phonological loop is to provide a mechanism for the temporary storage of new words while more stable long-term phonological representations are being constructed. The case of gifted language learners suggests that a natural talent for language learning may arise directly as a consequence of excellent phonological loop function.

A device for the acquisition of syntax?

Learning the vocabulary of one's native language is one of the most important aspects of language acquisition. Words represent the basic building blocks of language, and vocabulary knowledge limits both the speaker's production of spoken language and the comprehension of language produced by others. The role we ascribe here to the phonological loop, of supporting the learning of new words, is therefore by no means trivial.

This view may nonetheless represent an underestimation of the contribution of the phonological loop in language acquisition. A further possibility is that the loop system mediates the acquisition of syntactic knowledge, as well as the learning of individual words. The preschool years are characterized by children's rapid learning of syntactic rules, and this syntactic knowledge is itself the source of very considerable individual variation. Many researchers in the area of child language have argued that one of the ways in which this syntactic development is achieved is by the child learning a storehouse of multiword language patterns that are used both as models for his or her own utterances and for the abstraction of the rules governing connected language (e.g., Brown & Fraser, 1963; Nelson, 1987; Pinker, 1984; Plunkett & Marchman, 1993). Speidel (1993) has proposed that the multiword utterances to be learned must first be held in phonological working memory. By this account, the integrity of the temporary phonological representations of the utterances will constrain the speed and accuracy with which more permanent LTM representations will be constructed.

Speidel (1989, 1993) based this suggestion on a detailed longitudinal analysis of the developing language abilities of two bilingual siblings, Mark and Sally. Both children had excellent and comparable general intellectual abilities and had no problems in understanding either language. However, although Sally's language production was as good as her comprehension, Mark had difficulties in speaking both languages. His parents reported that he was slow to start producing single words, and articulatory and word order problems were apparent by the time he started producing multiword utterances. By 5 years of age, his speech was intelligible but marked by syntactic errors and difficulties in retrieving the precise phonological forms of familiar words. Notably, Mark performed much more poorly than Sally on phonological memory tests such as auditory digit span and serial recall. Speidel (1989, 1993) suggested that as a consequence of Mark's relatively poor abilities to hold phonological material temporarily, he failed to develop adequate long-term representations of the words and phrases that are used to build syntactic patterns in speech. Thus, he had a much more limited repertoire of templates to guide the construction of his own utterances and also to provide the basis for his abstraction of the syntactic rules governing the two languages.

Correlational studies of normally developing children also support a link between phonological memory ability in young children and speech output. A study by Daneman and Case (1981) showed that word span was better than chronological age at predicting performance of 2- to 6-year-old children in an artificial grammar-learning task, and Blake, Austin, Cannon, Lisus, and Vaughan (1994) reported that word span predicted mean length of utterance in 2- to 3-year-olds better than either chronological age or mental age. Similar findings have emerged from a study by Adams and Gathercole (1995) who contrasted the spontaneous speech of two groups of 3-year-old children: one group with good scores on the phonological memory tests of digit span and nonword repetition and one group with relatively poor performance on these measures. Clear differences emerged in both the quantity and quality of their utterances. The high-memory-performance children produced lengthier utterances (see also Adams & Gathercole, 1996) and used a wider vocabulary than the low-memory children. Furthermore, a significantly wider range of syntactic structures was also present in the speech of the high- than the low-memory-performance group. Again, there is evidence of a relationship between phonological loop function during language acquisition and syntactic as well vocabulary development.

In summary, the phonological loop may play a crucial role in syntactic learning and in the acquisition of the phonological form of lexical items. This line of inquiry is very much in its early stages at present. The results so far, though, are certainly consistent with recent views developed on the basis of computational models of language acquisition that a single mechanism underpins the learning of single words and of the morphological properties of the language (e.g., Plunkett & Marchman, 1993). According to this position, there is no functional distinction between the way that words and, for example, inflectional morphology are learned. Our suggestion here is that the operation of this single system is significantly constrained by the phonological loop.

What part of the phonological loop supports language learning?

We have as yet said little about the more detailed aspects of how the loop system mediates learning. A full account of the mechanisms by which the sounds of new words are learned has still to be developed, but some advances in the direction of a fuller understanding have been made. Some of the major issues concerning the microstructure of the phonological loop and how it supports language learning are considered in the next two sections.

Phonological storage or rehearsal?

The current model of the phonological loop consists of two components: the phonological short-term store and a subvocal rehearsal process that serves to preserve decaying representations in the phonological store (Baddeley, 1986). The body of evidence reviewed earlier indicates a close association between the phonological loop and long-term phonological learning. An obvious question to ask is the following: What aspect of the phonological loop is critical to this learning function?

The answer seems to be that the fundamental mechanism linking phonological memory and vocabulary acquisition is the phonological store. It is now widely believed that although the phonological store is in place in children as soon as language abilities begin developing, their use of subvocal rehearsal as a means of silently maintaining the contents of the phonological store does not emerge until around 7 years of age (see Cowan & Kail, 1996, and Gathercole & Hitch, 1993, for reviews). There is a variety of evidence supporting this claim, including the emergence of overt signs of articulatory activity in immediate memory tasks at around this age (Flavell, Beach, & Chinsky, 1966) and in the absence of significant correlations between articulation rate (which appears to provide an index of rate of subvocal and overt articulation) and memory span in children below this age (Gathercole & Adams, 1994).

However, as Table 1 illustrates, there is ample evidence of close links between phonological memory performance and vocabulary learning under 7 years of age and, indeed, from children as young as 3 years (Gathercole & Adams, 1993). This makes it unlikely that it is the subvocal rehearsal process that mediates long-term phonological learning in young children, although rehearsal does appear to play a role in second-language learning in adults, as evidenced by both the negative effects of articulatory suppression (Papagno et al., 1991) and the positive effects of a rehearsal strategy (Ellis & Beaton, 1993).

The importance of storage rather than rehearsal processes is reinforced by consideration of the demands of the nonword repetition task. The task involves the child attempting the immediate repetition of a single unfamiliar item. Repetition attempts typically commence within 1 s of the end of the nonword, and the nonwords in the test we have developed typically have spoken durations of considerably less than 1 s (Gathercole, Willis, Baddeley, & Emslie, 1994). Given the usual estimate of the temporal capacity of the phonological store of about 2 s (Baddeley et al., 1975), and the fact that the child is allowed to repeat the item aloud as soon as he or she wishes, the likelihood that the rehearsal process significantly contributes to individual differences in nonword repetition ability seems remote. Rather, we suggest, performance on the task is constrained by the quality of the phonological representation of the just-spoken unfamiliar item. In other words, nonword repetition provides a measure of the phonological store, not phonological rehearsal.

There is, however, at least one other possible account of the memory–vocabulary association that merits consideration. It has been suggested by Snowling, Chiat, and Hulme (1991) that the association is mediated by articulatory output skills (see also Wells, 1995). The argument is that children with poor articulatory function will perform at a low level on verbal tasks that require speech output and that these output difficulties will be particularly manifest in tasks such as nonword repetition in which output of the spoken form is unpracticed by the child, and absolute phonological accuracy is required.

At one level, this account must be true: Children with impaired articulation such as J.B. (Snowling & Hulme, 1989) will necessarily be poor at repeating nonwords, and it would clearly be inappropriate to interpret low nonword repetition scores in individuals with articulatory deficits as reflecting a phonological loop impairment. A more important issue for the present purposes is whether the link between phonological memory function and word learning in the populations that have been studied (normal children and adults, gifted language users, neuropsychological cases, and cases of developmental disorder) is entirely mediated by individual differences in speed of articulation (Gathercole, 1995b). This seems unlikely as such an interpretation was explicitly ruled out by the observation of impaired phonological long-term learning despite normal articulation rate in studies involving both SLI children (Gathercole & Baddeley, 1989) and the STM deficit patient P.V. (Baddeley et al., 1988).

In the case of normal 4-year-old children, it has been shown that a speech output requirement in immediate memory performance is not crucial to the link between memory and vocabulary (Gathercole, Hitch, Service, Adams, & Martin, 1997). A non-word matching span task was used in which the children heard sequences of nonwords repeated in either the same or transposed order. The sequences increased in length over successive trials, and the child's task was simply to identify the two sequences as either the same (e.g., *guk, dar, lus* ... *guk, dar, lus*) or different (e.g., *pes, vip, mel* ... *pes, mel, vip*), thereby removing any significant output component to the task. Nonword matching span measured in this way was found to be significantly related to vocabulary knowledge ($r=.56$, $p<.05$), as was both digit span ($r=.59$, $p<.05$) and nonword repetition ($r=.39$, $p<.05$). There was therefore no evidence that the link between immediate memory performance and vocabulary was critically mediated by differences in the abilities of these children to accurately output nonwords. Rather, the data suggest that a common phonological loop constraint underpins the relationship between all three memory measures and vocabulary knowledge.

Finally, the relationship between phonological memory and long-term phonological learning of new words is not simply restricted to cases in which the individual has to recall the new word form. Gathercole, Hitch, Service, and Martin (1997) found significant links between young children's performance on tests of phonological memory and experimental word learning even in a task in which the child was required not to recall the phonological form of a new word but simply to recognize it and to supply its associated semantic attributes.

In summary, the broad sweep of available evidence indicates that it is the phonological store that plays a critical role in the learning of the phonological forms of new words. Although rehearsal may be important for maintaining the quality of its representations, it is the store that is the primary language learning device.

What is the phonological store?

We have so far depended on an extremely simple model of the phonological loop as some form of store supplemented by an articulatory rehearsal process. It is far from obvious how such a store might operate or why indeed it would be helpful in the acquisition of novel long-term phonological representations. In particular, the relationships between short- and long-term aspects of the phonological loop are clearly close, but quite unspecified.

One popular way of conceptualizing working memory is in terms of the activation of some aspect of LTM (for a discussion see Gathercole & Martin, 1996). This simple generalization does indeed capture one important feature of the operation of the phonological loop; namely its capacity to exploit prior learning. The evidence for this capacity is now extensive. (a) Non-words that closely resemble the phonological structure of English are more readily repeated than less wordlike items (Gathercole, 1995a) and are better recalled in a serial recall paradigm (Gathercole, Frankish, Pickering, & Peaker, 1997). (b) We have found that English children's abilities to repeat unfamiliar nonwords constructed to conform to the phonotactic rules of the French language are directly related to their knowledge of French vocabulary (Gathercole & Thorn, 1997; Thorn & Gathercole, in press), (c) Memory span for sequences of nonwords increases when the items are familiarized through a training procedure, whereas the use of already familiar words further enhances performance (Hulme, Maughan, & Brown, 1991). (d) Span increases dramatically from about 5 words to 16 or more when sequences of unrelated words are replaced by text. Although this clearly reflects the importance of syntactic and semantic factors in text recall, performance still appears to have a phonological basis, as a similar sentential advantage is also found in patients with a phonological loop deficit who might have a span of 1 or possibly 2 unrelated words, but who will have a sentence span of 6 or 7 (Baddeley, Vallar, & Wilson, 1987).

Under these circumstances, it is tempting to argue that the phonological loop simply represents the activation of those systems necessary for the perception of language (Allport, 1984; Brown & Hulme, 1996; Gathercole & Martin, 1996). There are, however, three problems with this interpretation. First of all, it predicts that patients with marked phonological loop deficits should have equivalent problems in speech perception, production, or both. Although Allport has argued for the presence of subtle deficits in one of the classic cases of a STM patient, Shallice (1988) has argued that the evidence for such a view is extremely weak, particularly because patient P.V. (Vallar & Baddeley, 1984) has shown no evidence of speech perception or production problems.

The second reason for doubting a simple association between phonological processing and memory comes from the study of articulatory suppression. Suppression has a very marked effect on memory span performance while having little or no effect on the capacity to perform a phonological judgment such as assessing whether two items are homophonous (Baddeley & Lewis, 1981; Besner, Davies, & Daniels, 1981).

A third reason for dissatisfaction with interpretations of STM processes, purely in terms of structures within the LTM, comes from considering the development of vocabulary. It is clear that in the case of an adult or older child, vocabulary growth is associated with a range of variables that probably involves both phonological and lexical development. A relevant model is that proposed by Brown and Hulme (1996), shown in Figure 1. The problem with this model is that it provides no explanation for why some children develop vocabulary more rapidly than others. Furthermore, each component of the model implicitly depends on prior learning, but no mechanism for such learning is provided. Our own evidence suggests that while such a multiply

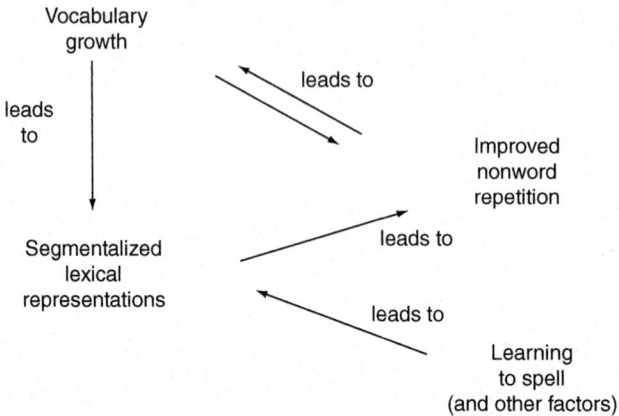

Figure 1 Causal relationships between vocabulary size, nonword repetition, and other factors as proposed by Brown and Hulme (1996).

interactive model may well apply to older children, the earlier stages of development are better characterized by a model in which differences in the capacity to repeat back unfamiliar items will lead to differences in vocabulary, which only subsequently begin to have a reciprocal influence on nonword repetition performance (Gathercole et al., 1992).

Despite the absence of a primary function for phonological STM in their own model, Brown and Hulme (1996) have proposed an important role for a separate STM system in vocabulary acquisition. They emphasized the computational advantage of having such a temporary system by using as an analogy the case made by McClelland, McNaughton, and O'Reilly (1994) for the role of the hippocampus. McClelland et al. argued that efficient LTM involves representing the underlying structure of the environment in the neocortex, which in turn requires the assignment of similar patterns to similar representations by a slow incremental process. However, to allow the rapid encoding and representation of novel experiences, a more temporary registration is necessary, a process that is dependent on the operation of the hippocampus. Brown and Hulme suggested that a phonological short-term store, although operating over a much briefer time scale, could serve a similar function to that of the hippocampus, in allowing the precise registration of phonological sequences while they are recoded into a more durable form in phonological and lexical LTM.

Note that it is important for such a system to make use of prior phonological and lexical knowledge but not to allow that knowledge to override the short-term representation of novel stimuli. A system that simply reflects existing knowledge will be inherently conservative and insensitive to novel inputs and as such will represent a poor learning system. Conversely, a system that attempts to learn all novel events will run the risk of unnecessarily committing valuable storage resources. Consider the case of spoken language. Many of the utterances people hear are spoken in a range of different voices, accents, or are partially masked by ambient noise. Permanent storage of the novel tokens in each of these cases would be premature and of little value. However, a system in which learning occurs incrementally over time, on the basis of the detection of repeated features of temporary memory representations, would allow a long-term record of new words to be based on abstractions of sound patterns consistent over several exposures. In this way learning of mispronounced stimuli or strangely accented forms is minimized, allowing effective use of limited learning resources to be focused on real new words. What is required therefore is some form of temporary representation that can both provide an accurate if brief record of specific potentially novel input while relating that input to the long-term system that represents the prior knowledge of the structure of language. The system thus needs some form of temporary activation, which might in connectionist terms be represented as "fast weights," which in turn may gradually influence some more durable representation, "slow weights" (Hinton & Plaut, 1987).

The association between the phonological loop and long-term phonological learning does not, however, appear to run only in one direction. Earlier, we reviewed a range of evidence showing that more permanent knowledge about the structure of individual words and of language more generally influences immediate memory performance (Gathercole, Frankish, et al., 1997; Hulme et al., 1991; Gathercole & Thorn, 1997). The implication of such findings is that long-term knowledge is used to "fill in" representations in the phonological loop that are incomplete, as a consequence of either decay or interference, by using a process of "redintegration" (Brown & Hulme, 1995), whereby partial traces resulting from familiar words or highly wordlike novel words have a greater likelihood of being correctly reinstated than nonwords with unusual sound patterns such as words in a foreign language. This application of stored knowledge about the sound structure of the language to the phonological loop will, of course, result in a temporary memory system, which is effectively tailored for storing words in one's native language. Given the role played by the phonological loop in long-term learning, this will also mean that such words are relatively easy to learn.

Our broad overview of the phonological loop is summarized in Figure 2. Auditory information is analyzed and fed through to a phonological store, where the input is represented by means of a STM trace. The trace involves the temporary activation of a structure or network that reflects the influence, though not dominance, of a phonological long-term system. The activation is short lived but has the capacity, in turn, to influence the long-term representation. However, although the short-term system depends on fast weights, the capacity to modify the long-term system depends on slow weights and is likely to require substantial learning, particularly in the case of the acquisition of very novel material by an already mature phonological system. It is important to note that the phonological LTM is not an episodic memory

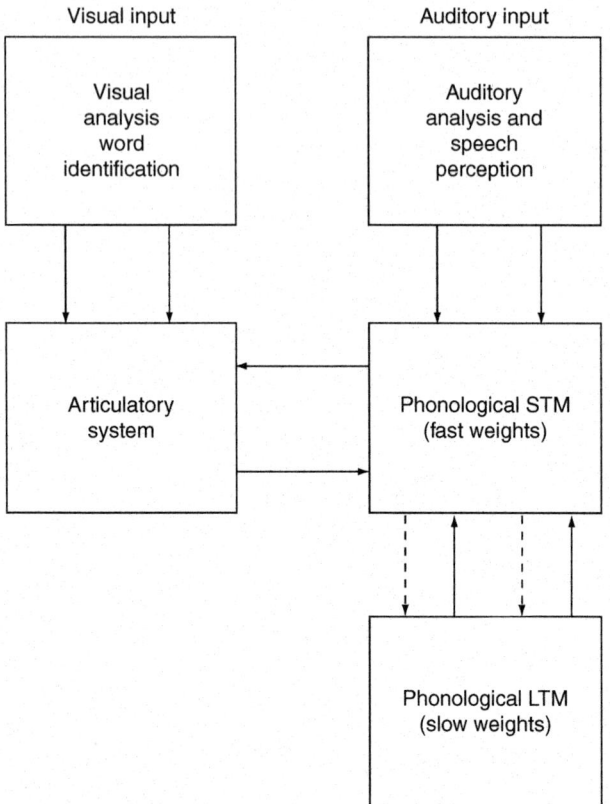

Figure 2 A simple model of the proposed components involved in short-term storage and long-term phonological learning. STM = short-term memory; LTM = long-term memory.

system but rather represents the residue of accumulated long-term phonological knowledge. Immediate serial recall can of course be influenced by both semantic and episodic memory. However, these influences are beyond the scope of the present discussion.

In the case of written input of verbal material, the visual analysis will be fed into the phonological store by means of subvocal speech by using the articulatory system. This system is also used for verbal output, which in the case of overt speech will lead to an auditory input that in turn will enter the phonological store. This process can be operated in the absence of overt output, as in the case of silent rehearsal, in which case the phonological store is activated in the absence of auditory input.

Neuropsychological evidence is broadly supportive of this structure, with recent neuroradiological imaging studies supporting earlier findings based on the location of brain lesions. In particular, the phonological store appears to be located in the perisylvian region of the left hemisphere, whereas the articulatory rehearsal component appears to be associated with Broca's area (Paulesu, Frith, & Frackowiak, 1993). A review of this and subsequent evidence is provided by Smith and Jonides (1995).

In recent years, much more detailed and specific models of the phonological store have begun to be developed. Examples include that of Burgess and Hitch (1992, 1996); Hartley and Houghton (1994); and by Page, Norris, and collaborators (e.g., Henson, Norris, Page, & Baddeley, 1996). Although these models differ in detail, they have many features in common, including a tendency to separate the storage of order and item information. In all three cases, however, the models so far concentrate on short-term retention and have not yet tackled the question of how short-term storage leads to long-term phonological learning.

Conclusions

We have suggested in this article that the phonological loop component of working memory has evolved as a system for supporting language learning. The evidence reviewed points to direct links between phonological loop function and word learning in a variety of participant populations and also identifies significant contributions of existing knowledge of the structure of the spoken language to both immediate memory performance and to vocabulary acquisition. The general model we advocate is of a highly flexible language learning system in which the valuable but limited-capacity resource of the phonological loop is available to support the construction of more permanent representations of the phonological structure of new words, but in which established knowledge of the language is used to offset this fragile temporary storage component whenever possible. Many of the details of this model have yet to be fully fleshed out, and this process of theoretical development represents a major goal of our current program. The general structure of the model characterized in this article, however, appears to us to be grounded securely in empirical evidence.

Given the overwhelming importance of language learning to cognitive development, this position attributes considerable significance to a component of memory, the phonological loop, in which its practical significance has in the past been attributed principally to dealing with telephone numbers. It seems likely that one reason for underestimating the developmental significance of the phonological loop is precisely because of the traditional emphasis on indexes of verbal STM based on recall of unrelated words, and especially digit span. We propose that the primary function of the phonological loop is the processing of novel speech input. Participants who are asked to memorize familiar words will make use of the phonological loop, but in so doing they exploit a supplementary function of a device that evolved for other, more important, purposes.

References

Adams, A.-M., & Gathercole, S. E. (1995). Phonological working memory and speech production in preschool children. *Journal of Speech and Hearing Research, 38*, 403–414.

Adams, A.-M., & Gathercole, S. E. (1996). Phonological working memory and spoken language development in young children. *Quarterly Journal of Experimental Psychology, 49A*, 216–233.

Allport, D. A. (1984). Auditory–verbal short-term memory and conduction aphasia. In H. Bouma & D. G. Bouwhuis (Eds.), *Attention and performance X: Control of language processes* (pp. 313–325). Hillsdale, NJ: Erlbaum.

Arnold, R., Yule, W., & Martin, N. (1985). The psychological characteristics of infantile hypercalcaemia: A preliminary investigation. *Developmental Medicine and Child Neurology, 27*, 49–59.

Baddeley, A. D. (1966). The influence of acoustic and semantic similarity on long-term memory for word sequences. *Quarterly Journal of Experimental Psychology, 18*, 302–309.

Baddeley, A. D. (1986). *Working memory.* Oxford, England: Oxford University Press.

Baddeley, A. D. (1993). Short-term phonological memory and long-term learning: A single case study. *European Journal of Cognitive Psychology, 5*, 129–148.

Baddeley, A. D. (1997). *Human memory: Theory and practice* (2nd ed.). Boston: Allyn & Bacon.

Baddeley, A. D., Emslie, H., & Nimmo-Smith, I. (1994). *Doors and People: A test of visual and verbal recall and recognition.* Flempton, Bury St. Edmunds, England: Thames Valley Test Co.

Baddeley, A. D., & Hitch, G. (1974). Working memory. In G. A. Bower (Ed.), *Recent advances in learning and motivation* (Vol. 8, pp. 47–90). New York: Academic Press.

Baddeley, A. D., & Lewis, V. J. (1981). Inner active processes in reading: The inner voice, the inner ear and the inner eye. In A. M. Lesgold & C. A. Perfetti (Eds.), *Interactive processes in reading* (pp. 107–129). Hillsdale, NJ: Erlbaum.

Baddeley, A. D., Papagno, C., & Vallar, G. (1988). When long term learning depends on short-term storage. *Journal of Memory and Language, 27*, 586–595.

Baddeley, A. D., Thomson, N., & Buchanan, M. (1975). Word length and the structure of short-term memory. *Journal of Verbal Learning and Verbal Behavior, 14*, 575–589.

Baddeley, A. D., Vallar, G., & Wilson, B. A. (1987). Sentence comprehension and phonological memory: Some neuropsychological evidence. In M. Coltheart (Ed.), *Attention and performance XII: The psychology of reading* (pp. 509–529). Hillsdale, NJ: Erlbaum.

Bellugi, U., Marks, S., Bihrle, A., & Sabo, H. (1988). Dissociation between language and cognitive functions in Williams syndrome. In D. V. M. Bishop & K. Mogford (Eds.), *Language development in exceptional circumstances* (pp. 177–189). Edinburgh, Scotland: Churchill Livingstone.

Besner, D., Davies, J., & Daniels, S. (1981). Phonological processes in reading: The effects of concurrent articulation. *Quarterly Journal of Experimental Psychology, 33*, 415–438.

Bishop, D. V. M. (1992). The underlying nature of specific language impairment. *Journal of Child Psychology and Child Psychiatry, 33*, 1–64.

Bishop, D. V. M., North, T., & Donlan, C. (1996). Nonword repetition as a phenotypic marker for inherited language impairment: Evidence from a twin study. *Journal of Child Psychology and Child Psychiatry, 37*, 391–404.

Blake, J., Austin, W., Cannon, M., Lisus, A., & Vaughan, A. (1994). The relationship between memory span and measures of imitative and spontaneous language complexity in preschool children. *International Journal of Behavioral Development, 17*, 91–107.

Bloom, L., Hood, L., & Lightbown, P. (1974). Imitation in language development: If, when and why. *Cognitive Psychology, 6*, 380–420.

Brown, R. A. (1973). *A first language: The early stages.* Cambridge, MA: Harvard University Press.

Brown, R., & Fraser, C. (1963). The acquisition of syntax. In C. N. Cofer & A. Musgrave (Eds.), *Verbal behavior and learning: Problems and processes* (pp. 158–210). New York: McGraw-Hill.

Brown, G. D. A., & Hulme, C. (1995). Modelling item length effects in memory span: No rehearsal needed? *Journal of Memory and Language, 34*, 594–621.

Brown, G. D. A., & Hulme, C. (1996). Non-word repetition, STM, and word age-of-acquisition: A computational model. In S. E. Gathercole (Ed.), *Models of short-term memory* (pp. 129–148). Hove, England: Psychology Press.

Burgess, N., & Hitch, G. J. (1992). Toward a network model of the articulatory loop. *Journal of Memory and Language, 31*, 429–460.

Burgess, N., & Hitch, G. J. (1996). A connectionist model of STM for serial order. In S. E. Gathercole (Ed.), *Models of short-term memory* (pp. 51–72). Hove, England: Psychology Press.

Butterworth, B., Campbell, R., & Howard, D. (1986). The uses of short-term memory: A case study. *Quarterly Journal of Experimental Psychology, 38*, 705–737.

Cheung, H. (1996). Nonword span as a unique predictor of second-language vocabulary learning. *Developmental Psychology, 32*, 867–873.

Clark, E. V. (1983). Meanings and concepts. In J. H. Flavell & E. M. Markman (Eds.), *Handbook of child psychology. Vol. 3: Cognitive development* (pp. 787–840). New York: Wiley.

Coggins, T. E., & Morrison, J. A. (1981). Spontaneous imitations of Down's syndrome children: A lexical analysis. *Journal of Speech and Hearing Research, 24*, 303–307.

Conrad, R., & Hull, A. J. (1964). Information, acoustic confusion and memory span. *British Journal of Psychology, 55*, 429–432.

Cowan, N., & Kail, R. (1996). Covert processes and their development in short-term memory. In S. E. Gathercole (Ed.), *Models of short-term memory* (pp. 29–50). Hove, England: Psychology Press.

Crano, W. D., & Mellon, P. M. (1978). Causal influence of teachers' expectations of children's academic performance: A cross-lagged panel analysis. *Journal of Educational Psychology, 70*, 39–49.

Daneman, M., & Case, R. (1981). Syntactic form, semantic complexity, and short-term memory: Influences on children's acquisition of new linguistic structures. *Developmental Psychology, 17*, 367–378.

DeRenzi, E., & Nichelli, P. (1975). Verbal and nonverbal short-term memory impairment following hemispheric damage. *Cortex, 11*, 341–353.

Dollaghan, C. (1987). Fast mapping in normal and language-impaired children. *Journal of Speech and Hearing Disorders, 52*, 218–222.

Dunn, L. M., & Dunn, L. M. (1982). *The British Picture Vocabulary Scale.* Windsor, England: NFER-Nelson.

Elliott, C. D. (1983). *British Abilities Scales.* Windsor, England: NFER-Nelson.

Ellis, N., & Beaton, A. (1993). Factors affecting the learning of foreign-language vocabulary: Imagery keyword mediators and phonological short-term memory. *Quarterly Journal of Experimental Psychology, 46A,* 533–558.

Flavell, J. H., Beach, D. R., & Chinsky, J. M. (1966). Spontaneous verbal rehearsal in a memory task as a function of age. *Child Development, 37,* 283–299.

Fowler, A. E. (1991). How early phonological development might set the stage for phoneme awareness. In S. Brady & D. Shankweiler (Eds.), *Phonological processes in literacy.* Hillsdale, NJ: Erlbaum.

Gardner, F. M. (1990). *The Expressive One-Word Picture Vocabulary Scale* (Rev. ed.). Novato, CA: Academic Therapy Publications.

Gathercole, S. E. (1995a). Is nonword repetition a test of phonological memory or long-term knowledge? It all depends on the nonwords. *Memory & Cognition, 23,* 83–94.

Gathercole, S. E. (1995b). Nonword repetition: More than just a phonological output task. *Cognitive Neuropsychology, 12,* 857–861.

Gathercole, S. E., & Adams, A. (1993). Phonological working memory in very young children. *Developmental Psychology, 29,* 770–778.

Gathercole, S. E., & Adams, A. (1994). Children's phonological working memory: Contributions of long-term knowledge and rehearsal. *Journal of Memory and Language, 33,* 672–688.

Gathercole, S. E., & Baddeley, A. D. (1989). Evaluation of the role of phonological STM in the development of vocabulary in children: A longitudinal study. *Journal of Memory and Language, 28,* 200–213.

Gathercole, S. E., & Baddeley, A. D. (1990a). The role of phonological memory in vocabulary acquisition: A study of young children learning new names. *British Journal of Psychology, 81,* 439–454.

Gathercole, S. E., & Baddeley, A. D. (1990b). Phonological memory deficits in language-disordered children: Is there a causal connection? *Journal of Memory and Language, 29,* 336–360.

Gathercole, S. E., & Baddeley, A. D. (1993). *Working memory and language.* Hillsdale, NJ: Erlbaum.

Gathercole, S. E., & Baddeley, A. D. (1995). Short-term memory may yet be deficient in language-impaired children: A comment on van der Lely & Howard. *Journal of Speech and Hearing Research, 38,* 463–472.

Gathercole, S. E., Frankish, C., Pickering, S. J., & Peaker, S. (1997). *Influences of nonword frequency and lexicality on children's serial recall.* Manuscript submitted for publication.

Gathercole, S. E., & Hitch, G. J. (1993). Developmental changes in short-term memory: A revised working memory perspective. In A. Collins, S. E. Gathercole, M. A. Conway, & P. E. Morris (Eds.), *Theories of memory* (pp. 189–210). Hillsdale, NJ: Erlbaum.

Gathercole, S. E., Hitch, G. J., Service, E., Adams, A. -M., & Martin, A. J. (1997). *Phonological short-term memory and vocabulary acquisition: Further evidence on the nature of the relationship.* Manuscript submitted for publication.

Gathercole, S. E., Hitch, G. J., Service, E., & Martin, A. J. (1997), Phonological short-term memory and new word learning in children. *Developmental Psychology, 33,* 966–979.

Gathercole, S. E., & Martin, A. J. (1996). Interactive processes in phonological memory. In S. E. Gathercole (Ed.), *Models of short-term memory* (pp. 73–100). Hove, England: Psychology Press.

Gathercole, S. E., & Thorn, A. S. C. (1997). Phonological short-term memory and foreign language learning. In A. F. Healy & L. E. Bourne (Eds.), *Foreign language and learning: Psycholinguistic experiments on training and retention.* Hillsdale, NJ: Erlbaum.

Gathercole, S. E., Willis, C., & Baddeley, A. D. (1991). Differentiating phonological memory and awareness of rhyme: Reading and vocabulary development in children. *British Journal of Psychology, 82*, 387–406.

Gathercole, S. E., Willis, C., Baddeley, A. D., & Emslie, H. (1994). The Children's Test of Nonword Repetition: A test of phonological working memory. *Memory, 2*, 103–127.

Gathercole, S. E., Willis, C., Emslie, H., & Baddeley, A. D. (1991). The influences of syllables and wordlikeness on children's repetition of nonwords. *Applied Psycholinguistics, 12*, 349–367.

Gathercole, S. E., Willis, C., Emslie, H., & Baddeley, A. D. (1992). Phonological memory and vocabulary development during the early school years: A longitudinal study. *Developmental Psychology, 28*, 887–898.

Gleitman, L. (1993). The structural sources of verb meanings. In P. Bloom (Ed.), *Core readings in language acquisition* (pp. 174–221). Cambridge, England: Harvester Wheatsheaf.

Grant, J., Karmiloff-Smith, A., Gathercole, S. E., Paterson, S., Howlin, P., Davies, M., & Udwin, O. (1997). Phonological short-term memory and its relationship to language in Williams Syndrome. *Cognitive Neuropsychiatry, 2*, 81–99.

Hartley, T., & Houghton, G. (1994). A linguistically-constrained model of short-term memory for nonwords. *Journal of Memory and Language, 35*, 1–31.

Haynes, C. (1982). *Vocabulary acquisition problems in language-impaired children.* Unpublished master's thesis, Guy's Hospital Medical School, University of London, London, England.

Henson, R. N. A., Norris, D. G., Page, M. P. A., & Baddeley, A. D. (1996). Unchained memory: Error patterns rule out chaining models of immediate serial recall. *Quarterly Journal of Experimental Psychology, 49A*, 80–115.

Hinton, G. E., & Plaut, D. C. (1987). Using fast weights to deblur old memories. In *Proceedings of the Ninth Annual Conference of the Cognitive Science Society* (pp. 177–186). Hillsdale, NJ: Erlbaum.

Hulme, C., Maughan, S., & Brown, G. D. A. (1991). Memory for familiar and unfamiliar words: Evidence for a longer term memory contribution to short-term memory span. *Journal of Memory and Language, 30*, 685–701.

Ingram, D. (1974). Phonological rules in young children. *Journal of Child Language, 1*, 97–106.

Jarrold, C., & Baddeley, A. D. (1997). Short-term memory for verbal and visuospatial information in Down's Syndrome. *Cognitive Neuropsychiatry, 2*, 102–122.

Jarrold, C., Baddeley, A. D., & Hewes, A. K. (in press). Verbal and non-verbal abilities in the Williams Syndrome phenotype: Evidence for diverging developmental trajectories. *Journal of Child Psychology and Psychiatry.*

Jones, D., von Stienbrugge, W., & Chicralis, K. (1994). Underlying deficits in language-disordered children with central auditory processing difficulties. *Applied Psycholinguistics, 15*, 311–328.

Kamhi, A. G., & Catts, H. W. (1986). Toward an understanding of developmental language and reading disorders. *Journal of Speech and Hearing Research, 51*, 337–347.

Karmiloff-Smith, A., Grant, J., Berthoud, I., Davies, M., Howlin, P., & Udwin, D. (in press). Language and Williams Syndrome: How intact is "intact"? *Child Development.*

Keil, F. C. (1979). *Semantic and conceptual development.* Cambridge, MA: Harvard University Press.

Kirchner, D., & Klatzky, R. L. (1985). Verbal memory and rehearsal in language-disordered children. *Journal of Speech and Hearing Disorders, 28*, 556–565.

Leonard, L. B., & Schwartz, R. G. (1985). Early linguistic development of children with specific language impairment. In K. E. Nelson (Ed.), *Children's language* (Vol. 5, pp. 291–318). Hillsdale, NJ: Erlbaum.

Locke, J. L., & Scott, K. K. (1979). Phonetically mediated recall in the phonetically disordered child. *Journal of Communication Disorders, 12*, 125–131.

Markman, E. M. (1994). Constraints on word meaning in early language acquisition. *Lingua*, 199–227.

Masur, E. F. (1995). Infants' early verbal imitation and their later lexical development. *Merrill-Palmer Quarterly, 41*, 286–306.

McCarthy, D. (1970). *McCarthy Scales of Childrens' Abilities.* New York: Psychological Corporation.

McClelland, J. L., McNaughton, B. L., & O'Reilly, R. C. (1994). *Why are there complementary learning systems in the hippocampus and neocortex?* (Tech. Rep. No. PDP. CNS.94.1). Pittsburgh, PA: Carnegie Mellon University.

Menyuk, P., & Looney, P. L. (1976). A problem of language disorder: Length versus structure. In D. M. Morehead & A. E. Morehead (Eds.), *Normal and deficient child language* (pp. 259–279). Baltimore, MD: University Park Press.

Michas, I. C., & Henry, L. A. (1994). The link between phonological memory and vocabulary acquisition. *British Journal of Developmental Psychology, 12*, 147–163.

Montgomery, J. (1995). Examination of phonological working memory in specifically language-impaired children. *Applied Psycholinguistics, 16*, 355–378.

Murray, D. J. (1967). The role of speech responses in short-term memory. *Canadian Journal of Psychology, 21*, 263–276.

Nagy, W. E., & Herman, P. A. (1987). Breadth and depth of vocabulary knowledge: Implications for acquisition and instruction. In M. G. McKeown & M. E. Curtis (Eds.), *The nature of vocabulary acquisition* (pp. 19–35). Hillsdale, NJ: Erlbaum.

Nelson, K. E. (1987). Some observations from the perspective of the rare event cognitive comparison theory of language acquisition. In K. E. Nelson & A. van Kleeck (Eds.), *Children's language* (Vol. 6, pp. 229–331). Hillsdale, NJ: Erlbaum.

Papagno, C., Valentine, T., & Baddeley, A. D. (1991). Phonological short-term memory and foreign-language vocabulary learning. *Journal of Memory and Language, 30*, 331–347.

Papagno, C., & Vallar, G. (1992). Phonological short-term memory and the learning of novel words: The effect of phonological similarity and item length. *Quarterly Journal of Experimental Psychology, 44A*, 47–67.

Papagno, C., & Vallar, G. (1995). Verbal short-term memory and vocabulary learning in polyglots. *Quarterly Journal of Experimental Psychology, 38A*, 98–107.

Paulesu, E., Frith, C. D., & Frackowiak, R. S. J. (1993). The neural correlates of the verbal component of working memory. *Nature, 362*, 342–345.

Pinker, S. (1984). *Language learnability and language development.* Cambridge, MA: Harvard University Press.

Plunkett, K., & Marchman, V. (1993). From rote learning to system-building—Acquiring verbal morphology in children and connectionist models. *Cognition, 48,* 21–69.

Raine, A., Hulme, C., Chadderton, H., & Bailey, P. (1991). Verbal short-term memory span in speech-disordered children. *Child Development, 62,* 415–423.

Raven, J. (1986). *Coloured progressive matrices.* Oxford, England: Oxford Psycholinguistic Press.

Rice, M. L., Buhr, J. C., & Nemeth, M. (1990). Fast mapping word learning abilities of language-delayed preschoolers. *Journal of Speech and Hearing Disorders, 55,* 33–42.

Rondal, J. A. (1994). *Exceptional language development in Down's syndrome: Implications for the cognition–language relationship.* New York: Cambridge University Press.

Service, E. (1992). Phonology, working memory, and foreign-language learning. *Quarterly Journal of Experimental Psychology, 45A,* 21–50.

Service, E., & Kohonen, V. (1995). Is the relationship between phonological memory and foreign language learning accounted for by vocabulary acquisition? *Applied Psycholinguistics, 16,* 155–172.

Shallice, T. (1988). *From neuropsychology to mental structure.* Cambridge, England: Cambridge University Press.

Shallice, T., & Butterworth, B. (1977). Short-term memory impairment and spontaneous speech. *Neuropsychologia, 15,* 729–735.

Shallice, T., & Warrington, E. K. (1970). Independent functioning of verbal memory stores: A neuropsychological study. *Quarterly Journal of Experimental Psychology, 22,* 261–273.

Smith, M. E. (1926). An investigation of the development of the sentence and the extent of vocabulary in young children. *University of Iowa Studies in Child Welfare, 3*(5).

Smith, E. E., & Jonides, J. (1995). Working memory in humans: Neuropsychological evidence. In M. Gazzaniga (Ed.), *The cognitive neurosciences* (pp. 1009–1020). Cambridge, MA: MIT Press.

Snowling, M., Chiat, S., & Hulme, C. (1991). Words, nonwords and phonological processes: Some comments on Gathercole, Willis, Emslie, & Baddeley (1991). *Applied Psycholinguistics, 12,* 369–373.

Snowling, M., & Hulme, C. (1989). A longitudinal case study of developmental phonological dyslexia. *Cognitive Neuropsychology, 6,* 379–401.

Speidel, G. E. (1989). A biological basis for individual differences in learning to speak. In G. E. Speidel & K. E. Nelson (Eds.), *The many faces of imitation in language learning* (pp. 199–229). New York: Springer-Verlag.

Speidel, G. E. (1993). Phonological short-term memory and individual differences in learning to speak: A bilingual case study. *First Language, 13,* 69–91.

Stark, R., & Tallal, P. (1981). Selection of children with specific language deficits. *Journal of Speech and Hearing Disorders, 46,* 114–122.

Sternberg, R. (1987). Most vocabulary is learned from context. In M. McKeown & M. Curtis (Eds.), *The nature of vocabulary acquisition* (pp. 89–106). Hillsdale, NJ: Erlbaum.

Taylor, H. G., Lean, D., & Schwartz, S. (1989). Pseudoword repetition ability in learning-disabled children. *Applied Psycholinguistics, 10,* 203–219.

Thorn, A. S. C., & Gathercole, S. E. (in press), Language-specific knowledge and short-term memory in bilingual and non-bilingual children. *Quarterly Journal of Experimental Psychology.*

Trojano, L., & Grossi, D. (1995). Phonological and lexical coding in verbal short-term memory and learning. *Brain & Cognition, 21,* 336–354.

Vallar, G., & Baddeley, A. D. (1984). Phonological short-term store, phonological processing and sentence comprehension: A neuropsychological case study. *Cognitive Neuropsychology, 1,* 121–141.

Vallar, G., & Papagno, C. (1993). Preserved vocabulary acquisition in Down's syndrome: The role of phonological short-term memory. *Cortex, 29,* 467–483.

Vallar, G., & Shallice, T. (Eds.). (1990). *Neuropsychological impairments of short-term memory.* Cambridge, England: Cambridge University Press.

van der Lely, H. K. J., & Howard, D. (1993). Children with specific language impairment: Linguistic impairment or short-term memory deficit? *Journal of Speech and Hearing Research, 36,* 1193–1207.

Wang, P. P., & Bellugi, U. (1994). Evidence from two genetic syndromes for a dissociation between verbal and visual–spatial short-term memory. *Journal of Clinical and Experimental Neuropsychology, 16,* 317–322.

Wechsler, D. (1974). *Wechsler Intelligence Scale for Children: Revised.* New York: Psychological Corporation.

Wechsler, D. (1981). *Wechsler Adult Intelligence Scale.* New York: Psychological Corporation.

Wells, B. (1995). Phonological considerations in repetition tests. *Cognitive Neuropsychology, 12,* 847–855.

Part IV
The visuo-spatial sketchpad[1]

My interest in visual memory went back to attempts to test some of the early Gestalt predictions on visual memory, followed by the application of the Peterson technique to memory for location in both healthy and amnesic patients (Warrington & Baddeley, 1974), but became more theoretically focused in the first chapter in this section. This was prompted by an ingenious and influential paper by Posner and Keele (1967) based on same–difference judgements on pairs of letters. This seemed to suggest separate contributions from both a verbal code and a very short-lived visual code. The brief time course of the visual component, however, appeared to be inconsistent with our work on visual location. My colleague, Bill Phillips, and I suggested that the apparent brevity of the visual trace may have resulted from its being displaced by the slower but ultimately dominant verbal code rather than its inherent brevity. We decided to test the durability of visual short-term memory using complex non-nameable matrix patterns demonstrating that relatively complex visual stimuli were indeed more durable than suggested by Posner and Keele. We published it as a brief report in *Psychonomic Science* which at the time did not require review and was largely ignored. Bill then went on to use our matrix change detection-paradigm in a classic series of studies in which he measured forgetting over intervals as a function of pattern complexity (Phillips, 1974). The change detection technique is now used widely in the study of visual working memory following a seminal paper by Luck and Vogel (1997).

Work on the sketchpad within the multicomponent working memory framework, however, focused initially on studies of visual imagery using relatively complex tasks. My own interest was prompted by the personal experience of attempting to drive along the freeway while listening to an American football game on the radio. I had a clear image of the field, but only at the expense of weaving from lane to lane! Maintaining and updating the visual image seemed to interfere with controlling the car. I was able to simulate this effect in the laboratory using an immediate memory task involving the use of visual imagery to recall verbal sequences while at the same time performing a tracking task, keeping a stylus in contact with a moving spot of light (Baddeley, Grant, Wight & Thomson, 1975). On describing this series of experiments at a meeting, I was asked whether I regarded the sketchpad system as visual or spatial. The second chapter in this section describes my attempt to answer this question. As you will see, we came to the conclusion that the system was spatial in nature, a conclusion that was later challenged by my colleague Robert Logie (1986) who produced convincing evidence that the system has both visual and spatial components. The final chapter in this section is a further demonstration with Logie of the separability of the visuo-spatial and phonological systems, together with a discussion of their role within the working memory framework.

Note

1 Again we abandoned our original term of *scratchpad* as this implies an aid that can be used for either written notes or drawings, whereas *sketchpad* suggests only the latter.

References

Baddeley, A. D., Grant, W., Wight, S., & Thomson, N. (1975). Imagery and visual working memory. In P. M. A. Rabbitt & S. Dornic (Eds), *Attention and performance V* (pp. 205–217). London: Academic Press.

Logie, R. H. (1986). Visuo-spatial processing in working memory. *Quarterly Journal of Experimental Psychology, 38A*, 229–247.

Luck, S. J., & Vogel, E. K. (1997). The capacity of visual working memory for features and conjunctions. *Nature, 390*, 279–281.

Phillips, W. A. (1974). On the distinction between sensory storage and short-term visual memory. *Perception and Psychophysics, 16*, 283–290.

Posner, M. I., & Keele, S. W. (1967). Decay of visual information from a single letter. *Science, 158*, 137–139.

Warrington, E. K., & Baddeley, A. D. (1974). Amnesia and memory for visual location. *Neuropsychologia, 12*, 257–263.

Papers

1 Phillips, W. A., & Baddeley, A. D. (1971). Reaction time and short-term visual memory. *Psychonomic Science, 22*, 73–74.
2 Baddeley, A. D., & Lieberman, K. (1980). Spatial working memory. In R. S. Nickerson (Ed.), *Attention and performance VIII*. Hillsdale, NJ: Lawrence Erlbaum Associates, pp. 521–539.
3 Logie, R. H., Zucco, G. M., & Baddeley, A. D. (1990). Interference with visual short-term memory. *Acta Psychologica, 75*, 55–74.

11 Reaction time and short-term visual memory*
W. A. Phillips and A. D. Baddeley

Posner's method of using differences in RT for physical and name matches to estimate the time constant of visual STM is criticized as confounding the decay of the visual trace with the development of a name code. When this confounding is avoided by using stimuli that are hard to name (a 5 by 5 matrix of randomly filled squares), the time constant shown by both RT and errors is consistently longer than that reported by Posner.

Posner and his co-workers have recently devised an ingenious technique by which reaction-time (RT) measures may be used to study visual encoding, and from which they have drawn interesting conclusions about short-term visual memory. The technique involves presenting S with a letter, followed after a brief interval by a second letter which may be either the same or a different letter and may be either upper- or lowercase. When the second letter immediately follows the first, RT is faster if the two are physically identical (i.e., same letter, same case) than if the letters have the same name but are in a different case. As the interval between the two increases, this difference decreases, till it disappears at delays of 1.5 sec or longer (Posner & Keele, 1967; Posner, Boies, Eichelman, & Taylor, 1969). They interpret this as evidence for a short-term visual trace with a decay time of about 1.5 sec.

The existence of such a trace is of considerable interest, since its time constant is clearly longer than that of iconic storage (Neisser, 1967). On the other hand, this time constant is considerably shorter than that for visual STM suggested by Posner (1967), Blick (1969), and Dale & Folarin (personal communication). Posner found little or no decay over a 20-sec interval; Blick found decay continuing beyond 5 sec; and Dale and Folarin found decay continuing beyond 10 sec. Closer examination of the rationale behind the Posner result, however, suggests that it may be based upon a logical error. It assumes that the discrepancy between the RT to physically identical terms and terms with the same name but different case simply reflects the strength of the physical trace, with the two RTs becoming equal when the trace becomes indistinguishable from the visual background noise. It is clear from Posner's own experiments, however, that at the same time as the physical trace is fading, the item is being translated into a name code. The point at which the difference between the physical and name match RT disappears, therefore, represents the combined effect of a fading visual trace and a developing name code. Once the name code has developed to a point at which it allows faster RTs than the visual code, S will presumably use it in preference to the visual trace, even though visual trace continues to be available. Since S need no longer use the visual trace, his RTs will no longer reflect its strength. In short, Posner's technique confounds the fading of the visual trace with the development of the name code, and as such cannot give a valid indication of the time course of visual STM.

This suggests that the method of Posner & Keele (1967) may give an underestimate of the duration of visual STM, and that the longer times suggested by the other experiments are more accurate. However, the discrepancy may be due to differences in procedure. All three experiments suggesting longer times measured accuracy rather than reaction time, and studied memory for position or length rather than memory for form. Furthermore, Posner (1967) and Dale and Folarin used recall rather than recognition. The following experiment, therefore, studies visual STM using a technique analogous to that employed by Posner and Keele, but material which cannot easily be named in order to avoid the confounding of visual trace decay with name-code development.

Method

Stimuli comprised a 5 by 5 matrix of squares in which each square had a 0.5 probability of being filled. A new pattern was used each trial. After a randomly selected delay of 0.3, 1.0, 3.0, or 9.0 sec, a second matrix was presented. This was either identical to the first or differed by having one square more or one square less filled. Ss were instructed to decide "as quickly as possible without error," whether the second pattern was the same or different, and press one of two display-console keys accordingly. The second pattern remained in view until the key was pressed. Stimuli were exposed for 0.5 sec on the visual display of an Elliot 4130 computer. The 5 by 5 matrix measured 0.6×0.6 in., and was viewed from a distance of about 12 in. In order to minimize iconic storage effects, a masking field comprising a 10 by 10 matrix of randomly filled squares was interpolated between the two 5 by 5 stimuli.

Ss were given 60 practice trials comprising 15 trials at each of the four delays in random order. The actual test involved a further six such blocks of 60 trials. Seven Ss comprising undergraduates and research workers were tested individually.

Results

Figure 1 shows the overall mean percentage of correct Rs as a function of delay. In contrast to the results of Posner et al. our data clearly suggest that

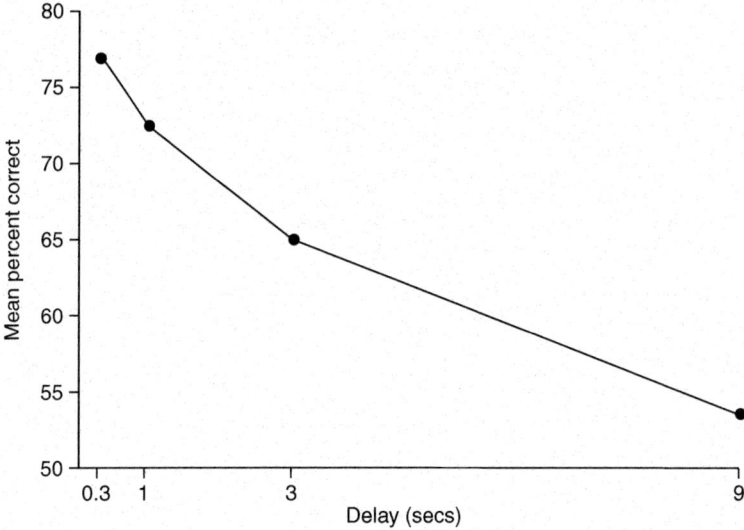

Figure 1 Percentage correct recognition responses as a function of delay (chance level = 50%).

forgetting continues after the first 3 sec. All Ss show overall forgetting and all show a drop in retention between 3 and 9 sec (T = 0, N = 7, p < .02 on the Wilcoxon test). Mean RT was plotted separately for correct and incorrect Rs and is shown in Figure 2. All Ss show an overall tendency for RT to increase with delay (T = 0, N = 7, p < .02). The increase in RT between 3 and 9 sec is also significant (T = 1, N = 7, p < .05). The trend is closely parallel for both correct Rs and errors, which are associated with slower RTs in all Ss (T = 0, N = 7, p < .02).

Discussion

These results are clearly at variance with the time constant for visual STM suggested by Posner et al. and, as such, they reinforce the claim that the physical vs name-match technique does not provide a valid indicator of short-term visual memory. They suggest a rate of forgetting which is consistent with the experiments suggesting a longer duration, despite considerable differences in procedure between this and the previous studies. Both accuracy and RTs give mutually consistent results, which suggests that RT may indeed provide a useful indicator of short-term forgetting.

Posner, Boies, Eichelman, & Taylor (1969) suggest that the decay of visual STM may be reduced by appropriate rehearsal strategies. They provide evidence for this by studying a pure-list condition in which, as in the present

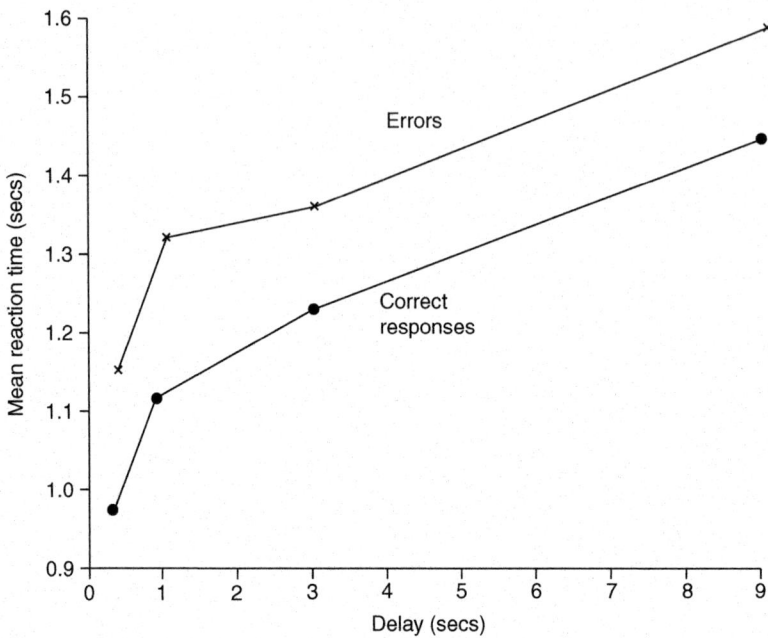

Figure 2 Mean reaction time for correct and incorrect responses as a function of delay.

experiment, a physical match is always required. RTs increased little over the 1-sec interval studied. They interpreted this as due to increased rehearsal of the visual aspect in the pure-list condition. In terms of this view, the results of the present experiment would suggest that rehearsal can maintain a visual STM for the kinds of patterns used beyond 3 sec, but not beyond 9 sec.

As Posner et al. note, their experiments do not show whether the decay observed in a matter of reducing ease of accessibility, or a matter of complete loss of availability. By showing a large increase in errors, as well as in RT, the present experiment suggests that the decay observed involves a complete loss of availability.

Note

* The authors are grateful for financial support to the Medical Research Council. The computer facilities were made available through a grant from the Science Research Council.

References

Blick, K. I. Decision and decay processes in the short-term memory for length. Journal of Experimental Psychology, 1969, 82, 224–230.

Neisser, U. *Cognitive psychology*. New York, Appleton-Century-Crofts, 1967.
Posner, M. I. Characteristics, of visual and kinesthetic codes. Journal of Experimental Psychology, 1967, 75, 103–107.
Posner, M. I., Boies, S. J., Eichelman, W. H., & Taylor, R. L. Retention of visual and name codes of single letters. Journal of Experimental Psychology Monograph, 1969, 79, No. 1, Part 2.
Posner, M. I., & Keele, S. W. Decay of visual information from a single letter. Science, 1967, 158, 137–139.

12 Spatial working memory

A. D. Baddeley and K. Lieberman

The role of imagery in verbal memory is explored using the framework of a working memory system comprising a central executive and two hypothetical "slave" systems, an articulatory loop and a visuo-spatial "scratch pad." Experiments 1 and 2 suggest that the scratch pad is sensitive to disruption by concurrent spatial rather than visual activity. It is further shown that a concurrent spatial task, pursuit tracking, interferes with the utilization of a visual imagery mnemonic for remembering word lists. Comparable disruption does not occur in the case of a mnemonic based on alphabetic order rather than imagery. Some broader implications of the assumption of a spatial working memory system are discussed.

Introduction

Following a series of experiments exploring the role of memory in reasoning, learning, and comprehension, Baddeley and Hitch (1974) proposed that the concept of short-term memory should be modified. The alternative proposed was termed *working memory* and comprised a central executive component that was assumed to be responsible for control processes and at least two supplementary "slave" systems, an articulatory loop and a visuo-spatial temporary store or "scratch pad." The articulatory loop was assumed to allow speech-based material to be maintained using subvocalization, a process making comparatively light demands on the central executive and being particularly appropriate for maintaining serial order. The articulatory loop was assumed to be responsible for the speechlike characteristics of performance in many short-term memory tasks, including the phonemic similarity effect (Conrad, 1964) and the word-length effect (Baddeley, Thomson, & Buchanan, 1975a), both of which can be eliminated if the material is presented visually and vocalization suppressed.

The present study is concerned with the second hypothetical slave system, the visuo-spatial scratch pad. This was discussed only briefly by Baddeley and Hitch (1974) but began to be explored in a series of studies by Baddeley, Grant, Wight, & Thomson (1975b). The experiments that follow are a continuation of this series. The experiments utilized the technique pioneered by

Lee Brooks (1967, 1968) whereby the process of imaging is disrupted by a concurrent visuo-spatial task. Our previous study demonstrated that a concurrent tracking task using a pursuit rotor dramatically impaired the subject's memory performance on a task that Brooks had previously shown to depend on imagery, whereas no such decrement occurred on a task relying on verbal encoding. We were also able to show that the memory task based on imagery impaired tracking performance, whereas the verbal memory task did not.

Having established that pursuit tracking was a suitable task for disrupting the use of visuo-spatial imagery, we then used tracking to study the encoding of verbal material of high- or low-rated concreteness/imageability (Baddeley et al, 1975b). Subjects performed a paired-associate task involving concrete nouns and highly imageable adjectives such as *strawberry—ripe* and *bullet—grey* or pairs involving abstract nouns and low-imageability adjectives such as *gratitude—infinite* and *mood—cheerful*. There is abundant evidence (Paivio, 1971) that concrete and imageable material is remembered very much better than abstract material of low imageability, an effect that is usually attributed to the subject's strategy of actively forming and using visual images in order to assist his learning. If this were the case we argued, it should be possible to disrupt the process by means of a concurrent visuo-spatial task. The effect of this should be to reduce the advantage enjoyed by concrete over abstract material. We therefore required our subjects to learn both types of material either unencumbered by a supplementary task or while pursuit tracking. Tracking caused a small overall impairment in performance; despite the absence of instructions to use imagery there was a massive difference between the abstract and concrete material, but there was no trace of the predicted interaction. On the basis of these results it was suggested that a distinction should be drawn between the abstract–concrete distinction and imagery as a control process whereby material is manipulated in a visuo-spatial working memory. The former, it was suggested, represents the manner in which a given type of material is registered in semantic memory. Its manner of registration will influence performance whether or not the control process of visuo-spatial imagery is employed in learning or recall.

The experiments to be described represent a continuation of this study. More specifically we try to answer three questions: (1) Does the system involved in the control process of imagery employ *visual* or *spatial* coding? (2) Is the system limited to the type of immediate memory task studied by both Brooks and ourselves, or is it also involved, as we predicted, in the use of imagery in mnemonics? (3) What is the relationship between this subsystem and the rest of working memory on the one hand and semantic memory on the other?

Visual or spatial working memory?

In presenting our previous results at the Attention and Performance meeting, I referred throughout to their implication for a *visual* working memory system. In the discussion, Daniel Kahneman raised the issue of whether the evidence necessarily implied a visual as opposed to a spatial system, and it became obvious that our data would not allow us to choose between these possibilities. Experiments 1 and 2 therefore aim to answer this question. In the first experiment, the influence of a spatial but nonvisual task on imagery is studied, whereas in the second the disrupting effect of a task involving visual but nonspatial processing is examined.

Experiment 1

In this study subjects were required to perform the immediate memory tasks that had previously been shown by Brooks (1967) and Baddeley et al (1975b) to emphasize either visuo-spatial or verbal coding. Subjects are required to perform both tasks alone and while performing an auditory tracking task that involves pointing to a moving sound source while blindfolded. The task has a clear spatial component, but it does not depend on visual input.

Apparatus

The subject was seated in front of a pendulum made from a 6-ft 4-in. light metal rod attached to the ceiling at a height of 9 ft 6 in. above the floor and 3 ft 11 in, from either wall. The bob of the pendulum contained a sound-emitting device, a 2½-in. 35-ohm loud speaker and a photocell. The sound emitter produced a continuous tone under ambient illumination, and a sequence of higher-pitched bleeps when light was shone on the photocell. The subject held a flashlight in his hand and attempted to keep the beam of the light on the photocell; when he did so, the emitted sound became intermittent, providing feedback.

Material

This comprised the memory-span material used by Brooks (1967) and Baddeley et al (1975, Experiment 2). Two types of sequence were used, one easily visualized and termed the *spatial material*; the other formally equivalent but not easily visualized termed the *nonsense material*. The subject is told to imagine a 4 × 4 matrix and is taught that one particular square (the second square in the second row) will be designated the *starting square*. Each message described the location of the digits 1 to 8 within the matrix, and in each case the digit 1 was in the starting square and successive digits appear in adjacent squares. Because the message is always presented in the sequence 1 to 8, it is possible to remember it in terms of a path through the matrix, with each successive digit being located as above, below, to the right, or to the left of the previous location (e.g., in the starting square, put a 1; in the next square to the right, put a 2; in the next square up, put a 3; etc.). The nonsense messages are formally equivalent except that the words *up* and *down* are replaced by *good* and *bad*, whereas the words *left* and *right* are replaced by *slow* and *quick* (e.g., in the starting square, put a 1; in the next square to the quick, put a 2; in the next square to the good, put a 3; etc.). Both we and Brooks find that the nonsense sequences are more difficult than the spatial but that reducing them from eight to six items gives an approximately equivalent probability of correct reproduction. Throughout therefore we use eight-digit messages for the spatial material and six-digit sequences for the nonsense material.

Design

All subjects were tested on four conditions comprising the two types of material tested with and without concurrent auditory tracking. Subjects received eight messages in each of the four conditions. Eight research students served as subjects and were paid £2.00 for participating in Experiments 1 and 2. Half began with Experiment 1 and half with Experiment 2. The order of conditions was determined by a Greco–Latin square within which the two spatial and the two nonsense conditions were blocked so as to minimize disruption of strategy by change of memory task.

Procedure

Subjects were first trained on the auditory tracking task up to a point at which they were able to achieve a score of at least 80% time on target over two successive 30-sec runs. Subjects were then given practice on the spatial memory task. Here they were given six practice trials, unless they reached a criterion of two successive perfect runs in less than six trials. They were then given practice on combining the two tasks until the subject had reached a level of skill whereby the tracking task was not substantially lowered by the memory task, and the memory task did not produce a large number of hesitations. The nonsense task was then practiced to the same criterion, and subjects were finally practiced at combining the nonsense task and tracking, again to the same criterion. They were then tested on the four conditions in the appropriate order. In the dual-task conditions, subjects tracked during both presentation and recall.

Results

Performance was scored in terms of number of sequences in which reproduction was perfect, for which the maximum is eight in each condition. Because the number of observations is small and scores are not normally distributed, the data were analyzed using nonparametric tests. There was a clear tendency for tracking to disrupt recall on the spatial task, with mean number of correct sequences dropping from 6.75 to 3.14, a decrement being shown by all subjects ($p < .005$, sign test). In the case of the nonsense task, mean memory performance dropped from 6.50 to 5.14 sequences correct during tracking. A decrement was present for five of the eight subjects but was not significant statistically. For seven of the eight subjects, tracking led to a greater recall decrement on the spatial task than on the nonsense task ($p < .05$, sign test).

Tracking performance was slightly poorer when doing the spatial task (80.1% time on target) than during the nonsense task (84.8%), an effect that was shown by seven of the subjects ($p < .05$, sign test). In general, however, the fact that not only was the spatial task more vulnerable to tracking but also that tracking itself was more vulnerable to the effect of the spatial task combine to rule out a speed error tradeoff interpretation of our results. Our data are clearly consistent with the assumption that the memory task relying on imagery has a spatial component that does not depend on direct visual input. It may of course have a purely visual component in addition to the spatial component; this possibility was explored in Experiment 2.

Experiment 2

In this study we used the same spatial and nonsense memory tasks, but this time we attempted to disrupt performance using a concurrent visual task, brightness judgment, for which the spatial demands were minimal.

Procedure

The material and experimental design were identical with Experiment 1, but, instead of the auditory tracking task, subjects were required to judge the brightness of a series of light patches on a screen. These were provided by a Carousel projector loaded with slides, each containing either two or three layers of tracing paper. The subjects were seated about 1 meter in front of a screen that was totally illuminated by the projector. The room was in darkness except for a low level of ambient illumination provided by a flashlight, so as to allow the experimenter to record the subject's responses. Stimuli were presented at a rate of 1 every 2.5 sec. In order to minimize any spatial component in the response, subjects were given only a single key that they were instructed to press whenever they saw a bright stimulus. Half the stimuli were bright, and half were dim.

Subjects commenced the session by practicing on the brightness discrimination task. They were given a minimum of 30 practice slides and continued to practice until they had reached a consistent level of performance of at least 90% correct responses. Subjects then received training on the memory tasks and on combining the memory and supplementary tasks, as in Experiment 1. Subjects were then tested on eight sequences in each of the four conditions, the order of conditions for any given subject being equivalent to that experienced in Experiment 1. In the dual-task conditions, subjects began the brightness judgment task before the presentation of the first sequence and continued performing the task until that condition was completed.

Results

Considering first the number of sequences correct, the pattern of Experiment 1 was reversed, with the retention of nonsense sequences being significantly disrupted by the concurrent task (from a control mean of 7.25 to an interference mean of 4.14 sequences), an effect that occurred with all eight subjects ($p < .005$, sign test). The disruption effect was not significant for the spatial memory task (from 5.75 to 4.72), where it was shown by only five of the eight subjects. Comparing the two conditions, seven of the eight subjects showed a greater degree of disruption on the nonsense than on the spatial tasks ($p < .05$, sign test). Subjects were successful in maintaining their discrimination performance at around the 90% mark, and there were no differences between the performance of subjects doing the two types of memory task, with three doing better while processing spatial messages, three doing the opposite, and two showing no difference.

Discussion

It is clear form Experiment 2 that the concurrent visual task of judging brightness was sufficient to cause a significant decrement in performance on the memory task, but that this was not specific to the spatial condition. Indeed, the evidence suggests that the disruption was rather less on the spatial than the nonsense conditions.

The combined results of the two experiments allow us to reject two rival interpretations to that of a spatial working memory. The first of these argues that the decrement shown in Experiment 1 and in Baddeley *et al* (1975) simply reflects the greater vulnerability of the spatial task, possibly stemming from its comprising longer sequences. The greater disruption of recall in the nonsense condition of Experiment 2 rules this out. The second is the hypothesis that a modality-specific visual memory system is involved. The fact that a spatial task disrupts performance in the imagery condition whereas a visual task does not implies spatial rather than visual coding.

Spatial memory and imagery mnemonics

In our previous paper (Baddeley et al., 1975b) we suggested distinguishing between the control process of imagery that represents the manipulation of visuo-spatial information in some form of working memory and the variable of rated imageability/concreteness, representing the way in which material is registered in semantic memory. It was suggested that it is this latter characteristic of material that accounted for the powerful effect of rated concreteness on memory performance. It was suggested on the other hand however that imagery mnemonics do rely on spatial working memory. As Bower (1972) has pointed out, for an imagery mnemonic to be successful, the two items visualized must be made to interact in some way; if the plausible assumption is made that the items are held and manipulated in a spatial working memory, then it follows that a concurrent tracking task should interfere with the use of an imagery mnemonic. Experiments 3 and 4 test this prediction.

Experiment 3

Subjects were instructed to use either rote memory or a pegword mnemonic employing visual imagery to remember lists of 10 words, each of which was paired with a digit from 1 to 10. On half the trials subjects were required to track on a pursuit rotor, whereas on the other half they were free to concentrate on the memory task. It was predicted that tracking would differentially disrupt performance based on the imagery mnemonic.

Materials

These comprised eight lists of 10 words of which half were high in rated concreteness (>6.5) and half were low (<2.0), as rated in the Paivio, Yuille, and Madigan (1968) norms. The two types of material were matched for Thorndike–Lorge frequency, with all words having a frequency of at least 20 per million.

Design and subjects

The design involved two types of learning instruction, visual imagery and rote, which were combined with the presence or absence of a concurrent tracking task to produce four conditions. Subjects were tested on two lists in each condition, with each list containing five abstract and five concrete words presented in random order. The two lists in any given condition were blocked, but the order of the four conditions was determined by a Latin square. Subjects were 28 Stirling University undergraduates, with seven subjects being assigned at random to each row of the Latin square.

Procedure

Subjects were first trained to perform the pursuit rotor task. This involved tracking a circular path with a stylus on a Lafayette pursuit rotor. During this initial phase the rotor was adjusted to a point at which subjects were able to achieve a tracking performance of 80–90% time on target over three successive 30-sec practice trials. This was typically within the range of 20–30 rpm.

They were then taught the one-is-a-bun pegword mnemonic (Paivio, 1971) to a criterion of two successive perfect recalls. They were then instructed to use the mnemonic by creating a visual image of each word and integrating it with the image of the item associated with the pegword. For example, if the pair was *one—ship*, the subject was to associate the number *one* with the word *bun* and form an interacting image of a ship and a bun, perhaps of a ship sailing into a huge floating bun. Subjects were then given a practice list of 10 number–word pairs in ascending order and were tested by being presented with the numbers 1–10 in random order and required to call out the word that had been associated with that particular number in that list. It was then pointed out that for half the lists this strategy should be used; for the other half the pairs would be presented much more rapidly and the subject was instructed to repeat each pair to himself and avoid attempting to form visual images. The test then began with pairs of items in the visual imagery condition being presented at a rate of 6 sec per pair, whereas the rote condition involved presenting each pair three times at a rate of 2 sec per pair. In both cases recall was at a rate of 6 sec per item. In half the conditions subjects were required to track during both presentation and recall, whereas in the remaining conditions no supplementary task was required. They were instructed to regard the tracking task as primary, using the analogy of driving a car while listening to the radio.

Results

Under control conditions the imagery mnemonic was effective giving a mean 67.3% correct pairs compared with 50.0% under rote learning instructions. Concurrent tracking reduced performance in the imagery condition to a level of 57.5%, whereas rote performance was unaffected at 49.5% correct. Analysis of variance showed significant effects of instructions ($F1, 27 = 21.63, p < .001$) indicating that the imagery mnemonic was proving helpful. There was also an overall effect of concurrent tracking ($F1, 27 = 4.76, p < .05$), but this was modified by an interaction with instructions ($F1, 27 = 5.33, p < .05$) indicating as predicted that the spatial concurrent task disrupted performance *only* when subjects were using the visual imagery mnemonic. There was a tendency for concrete words to be better recalled than abstract ($F1, 27 = 8.33, p < .01$). The concreteness/imageability effect interacted with learning instruction ($F1, 27 = 7.45, p < .05$) but not with tracking ($F1, 27 < 1.0$).

Subjects were successful in maintaining their tracking performance at the 80% level, with no differences appearing across conditions.

Discussion

The results of Experiment 3 are consistent with the hypothesis that the spatial working memory system is involved in the pegword imagery mnemonic, because the advantage provided by the imagery mnemonic is abolished by the spatial tracking task. At the same time, the lack of an interaction between suppression by tracking and rated concreteness supports our previous conclusion (Baddeley *et al.*, 1975b) that concreteness does not depend for its effect on the operation of the spatial working memory system. It does however appear to be sensitive to either presentation rate or encoding instructions, because when subjects were instructed to use a rote rehearsal strategy and material was presented rapidly, the effect of concreteness disappeared. It is tempting to suggest that under these conditions subjects were relying on a relatively shallow phonemically based code that was hence unaffected by the semantic characteristics of the remembered material (Baddeley, 1966).

Taken overall therefore, the results of Experiment 3 conform closely to predictions. The effects were however much weaker than those obtained in earlier experiments in the series. One possible reason for this is suggested by Experiments 1 and 2, which imply that the crucial factor about the tracking task is its *spatial* component. Although the pegword mnemonic involves manipulating visual images so as to make them interact spatially, it could be argued that the degree of spatial precision necessary is relatively slight. We therefore decided to study the influence of tracking on a mnemonic that we chose as having a much clearer and more important spatial component than the pegword mnemonic. For this we employed a version of the classical location mnemonic in which we instructed our subjects to imagine a walk through the University of Stirling campus. A sequence of 10 locations along this walk was selected, and subjects were instructed to remember the various items by imagining them located at the appropriate point along the walk.

Experiment 4

Procedure

The material and experimental design followed that used in Experiment 3, with the exception that only 12 subjects were tested, three being assigned at random to each row of the Latin square. Once again Stirling University undergraduates participated in the experiment for course credit. In place of the pegword mnemonic they were taught a location mnemonic based on a path through the University campus that contained 10 well-known and easily discriminable locations. They were told to imagine themselves following the route and the 10 locations were pointed out, the procedure being repeated until they successfully reported the correct locations themselves in the appropriate order on two successive trials. The second major difference from previous experiments was in requiring serial recall rather than associating each item with a digit. Hence in the mnemnic condition the subjects were presented with the 10 words at a rate of one every 6 sec and were instructed to associate an image of each successive item with the appropriate location along the walk. In the rote learning condition the items were presented at a rate of one every 2 sec with the subjects instructed to avoid the use of imagery. In order to keep total presentation time constant, the whole sequence was repeated three times in the rote conditions.

Results

Mean recall scores for the various conditions are shown in Figure 1. There was a significant overall effect of tracking ($F1, 11 = 18.87$, $p < .01$). The

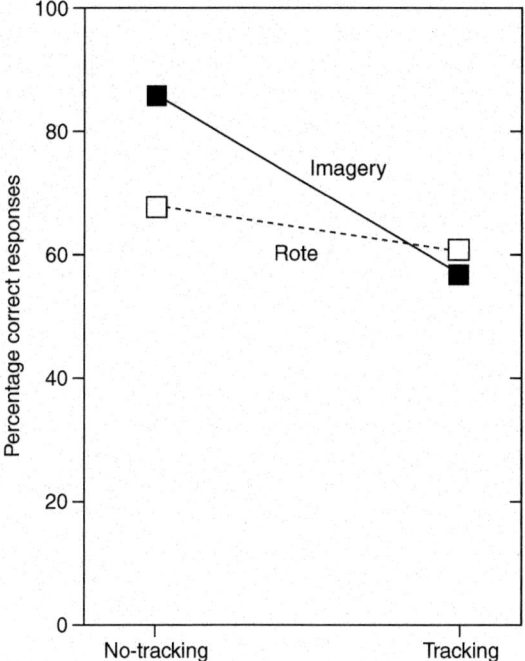

Figure 1 Influence of pursuit tracking on serial verbal recall based on either rote learning or a location mnemonic.

overall effect of the imagery mnemonic failed to reach significance, ($F1, 11 = 1.93$, $p > .05$) but the predicted interaction occurred between imagery mnemonic and tracking ($F1, 11 = 15.62$, $p < .01$). There was no effect of the abstractness of the material ($F1, 11 < 1$), and none of the further interactions achieved statistical significance. In the unloaded control condition the imagery strategy led to significantly better recall than the rote strategy ($t = 2.44$, $p < .05$). The difference between the two conditions was however abolished by the addition of the tracking task, with the imagery and tracking condition in fact being nonsignficantly worse than the rote condition while tracking ($t = 0.13$, $p > .05$). The effect of tracking on learning using imagery was very consistent, being shown by all 12 subjects tested ($t = 4.91$, $p < .001$), whereas the effect of pursuit tracking on rote learning failed to reach significance ($t = 1.99$, $p > .05$).

Once again subjects maintained their tracking performance at the recommended level, achieving 84.9% time on target for the imagery condition and 86.1% for the rote, with six subjects tracking better under imagery and six under rote conditions.

Discussion

Experiment 4 broadly replicates the effects shown in previous experiments. Once again, a visuo-spatial tracking task is found to interfere with the use of an imagery mnemonic, although having no comparable effect on the rote learning strategy. In contrast to the previous study, the effect is shown clearly by all 12 subjects tested, hence lending further support to the suggestion that tracking will have a particularly clear effect on a mnemonic making heavy demands on precise spatial coding.

Unlike the previous study where there was an effect of the concreteness of material, provided the subject was using the imagery mnemonic, in this experiment we found no effect of type of material on recall. This is consistent with Paivio and Csapo's (1969) suggestion that visuo-spatial coding is much less appropriate in serial recall than in free recall or paired-associate learning. The fact that an imagery mnemonic may nevertheless work very effectively in serial recall provides yet another example of the need to distinguish between the control process of imagery and its use in mnemonics on the one hand and the variable of rated concreteness or imageability on the other.

So far, our experiments appear to fit our predictions reasonably accurately. In discussing them, however Patrick Rabbitt pointed out that our results could equally well be explained by the simple generalization that organized learning is more easily disrupted by a secondary task than rote learning. In all cases, the task that is disrupted by tracking is that in which the subject uses the more complex organizational strategy, and it can plausibly be argued that anything that occupies processing capacity will be particularly detrimental to complex learning strategies such as those involved in the two mnemonics. Experiment 5 attempts to test this interpretation by studying the influence of pursuit tracking on a first-letter alphabetic mnemonic. If pursuit tracking has a general effect on the subject's organizational skills, we should observe a similar decrement.

Experiment 5

Design and material

The basic design was the same as that used in Experiment 4 with the exception that the location mnemonic was replaced by a mnemonic based on the initial letter of the words to be recalled. This necessitated changing the material. Two types of lists were created; both comprised items from a single semantic category such as *foreign cities* or *reading material*. The items were selected such that one item started with each of the first 10 letters of the alphabet. In the mnemonic condition, the order of the items was determined by their initial letter, *A* through to *J*, (e.g., *Amsterdam, Berlin, Chicago, Durban, ..., Jerusalem*). Half the lists were alphabetically ordered, whereas the remainder also comprised items beginning with each of the first 10 letters of the alphabet but were presented in random order. Finally, for each condition, half of the categories were selected as concrete items and half as abstract; this latter distinction was based on the experimenter's judgment, because standardized ratings were not available for many of the items used. The lists were balanced for word frequency. Lists were balanced across subjects such that any given category appeared equally frequently as an alphabetically ordered and unordered list. Subjects were tested on two abstract and two concrete lists in each of the four conditions produced by combining two types of list, alphabetic and random, and two load conditions, with and without tracking. A total of 16 women from the Applied Psychology Unit Subject Panel were tested, four being assigned at random to each row of a Latin square.

Procedure

Subjects were first trained on the pursuit rotor up to a level of 75–80% time on target. This took between five and ten 20-sec trials. They were then given instruction on the two memory tasks in which the subject was first given a cue as to whether an alphabetical or random list was involved, followed by the 10 items each preceded by the relevant number (e.g., *Foreign Cities, Alphabetical—One—Amsterdam, Two—Berlin, Three—Chicago,* etc). Recall was cued by presenting the digits *One, Two, Three,* etc. in ascending order. Rate of presentation and recall was 8 sec per word, a rate that was found to yield approximately the same overall recall score as was obtained in the previous study using a location mnemonic.

Results

The mnemonic worked, with subjects recalling a mean of 9.17 items as compared with 6.58 in the rote condition. With concurrent tracking these scores fell to 8.92 and 5.81, respectively. Analysis of variance indicated a highly significant effect of alphabetic ordering ($F1, 15 = 49.24, p < .001$), a significant effect of tracking ($F1, 15 = 5.77, p < .05$), but no significant effect of level of concreteness ($F1, 15 < 1$) and no statistically significant interactions. This result seems to rule out the simple generalization that tracking differentially interferes with memory organization. We obtained a mnemonic effect that is very close in magnitude to that obtained using the location mnemonic. There is however no evidence to suggest that the mnemonic condition here is particularly susceptible to interference, indeed the trend if anything is in the opposite direction. As in the previous study we find no evidence for an abstract versus concrete difference; although too much emphasis should not be placed on this because the materials were not standardized, it nevertheless reinforces the previous suggestion that serial learning does not appear to be very sensitive to the degree of concreteness of the material being learned.

Implications for the structure of the memory system

Is there a separate visuo-spatial memory?

In this and the previous study (Baddeley et al., 1975), we have consistently argued for a separate visuo-spatial working memory system. Such an interpretation was suggested by Brooks (1967, 1968), and it has been widely accepted since that time (e.g., Bower, 1972). Our results suggest that the system relies on spatial rather than visual coding because it can be disrupted by a spatial task that is free from visual input and is relatively resistant to disruption by a visual task that makes only a slight spatial processing demand. This conclusion is consistent with that of Byrne (1974) who observed that a spatially incompatible pointing task disrupted the use of imagery in recall.[1]

Further evidence for a spatial working memory system is provided by the mental rotation task first studied by Shepard and Metzler (1971). Although early studies with this have all used visual stimuli, a recent study by Carpenter and Eisenberg (1978) has shown similar effects with haptic stimuli explored tactually by blindfolded normal or congenitally blind subjects.

In a paper published since the completion of the present experimental series, Phillips and Christie (1977a) have argued that the evidence at present available forces the assumption of at least two components in tasks relying on imagery but does not force the conclusion that one of these is visuo-spatial in nature. Using a working memory framework, it is possible, for example, to explain all the data in terms of a general-purpose central executive and an articulatory loop. Although we would contend that it is difficult to give a plausible interpretation of all the available data in these terms, none of the available experiments logically forces the assumption of three separate processes. The problem of designing such a study presents an interesting technical challenge.

Neuropsychological evidence for separate systems

Further evidence for both long- and short-term visual and spatial memory systems comes from the clinical literature and, in particular, from the work of De Renzi and his co-workers. For example, De Renzi and Nichelli (1975) have contrasted verbal and spatial memory span. Verbal memory span is measured using either a conventional digit span or a modified version wherein the subject points to a series of digits so as to avoid any potential difficulties due to possible limitations in speech. The spatial memory span was measured using a task devised by Corsi (quoted by Milner, 1971) in which the subject has placed in front of him a board containing nine randomly spaced cubes. The experimenter taps a sequence of from two to eight cubes, and the subject is required to reproduce the sequence of taps. Verbal memory span was a clearly separable function that tended to be associated with left-hemisphere damage. Impaired spatial memory span on the other hand was associated with posterior lesions in either hemisphere. In a subsequent study, De Renzi,

Faglioni, and Preveidi (1977) drew a further distinction between spatial memory-span performance, which again they found to be impaired following damage to the posterior part of either hemisphere, and long-term visual memory, which was disrupted only by right-hemisphere damage. The distinction is further supported by a number of more detailed case histories of patients showing grossly impaired spatial learning while having a normal spatial memory span. De Renzi and Nichelli (1975) cite examples of the converse, namely patients who are grossly defective on spatial memory span but who show no difficulty in learning a visual maze and show no evidence for topographical disorientation in the real world. Such results are consistent with the concept of a spatial working memory system that is used for the Corsi block tapping test but is separate from the long-term visuo-spatial memory system necessary for geographical orientation.

The neuropsychological evidence we have discussed so far has been largely confined to spatial memory; imagery may of course be involved but the evidence for this is at present far from clear. A more direct involvement of imagery is however suggested by an intriguing recent observation by Bisiach and Luzzatti (1978) who describe the performance of two patients suffering from posterior right-hemisphere lesions following stroke. In both cases the patient had a hemianopia in the left visual field together with unilateral neglect, a tendency to ignore items presented to the left of the midline, although such items could be reported if requested. The patients were asked to imagine themselves in the Cathedral Square of their home city of Milan. They were asked to imagine themselves standing outside the cathedral and to describe the scene. In both cases subjects successfully described buildings that would have been in their right visual field but completely ignored buildings on the left. They were then asked to imagine that they had crossed the square and were pointing in the opposite direction and were asked to repeat the exercise. They then described the buildings that had previously been in their neglected hemifield perfectly adequately, presumably because performance was now dependent on the intact right hemifield. Those buildings that had previously been described perfectly adequately were not omitted, presumably because report relied on the operation of the neglected hemifield. Although only two cases are reported, it appears that this is by no means an uncommon phenomenon. It appears to imply a very close relationship between some of the processes involved in visual perception and those involved in imagery. Although it would be premature to suggest that such patients have a defect that is associated with spatial working memory, this clearly presents a hypothesis that is worth further exploration. For example, using the simple analogy of the visuo-spatial scratch pad as a screen on which spatial information may be represented, one might conclude either than part of the screen was inoperative or that the process of scanning the screen was defective, in each case allowing only part of the system to be used.

One central executive or two?

Bearing in mind the evidence from both normal and neuropsychological sources, there would appear to be a good case for assuming a separate spatial working memory system. Baddeley and Hitch (1974) tentatively suggested a system comprising a central executive served by two slave systems, an articulatory loop and a spatial scratch pad. The evidence we have discussed is so far consistent with such a notion. It is however also consistent with an alternative version assuming completely parallel and separate visual and spatial working memory systems. Fortunately, Phillips and Christie (1977a, b) have provided some very compelling evidence for a single central executive. Their task required the subject to remember a 4×4 or 5×5 matrix of calls, any one of which had a 0.5 probability of being filled. An initial study (Phillips and Christie, 1977a) showed that subjects are able to maintain one such stimulus pattern reasonably accurately over a matter of seconds. If a series of such patterns is presented, performance on all but the last item drops to a common asymptotic level. In their second study Phillips and Christie (1977b) studied the factors that disrupt performance. In a series of five experiments they showed very clearly that the most crucial factor is not the *visual* similarity between a subsequent event and the pattern being retained but rather the extent to which the interpolated task demands central processing capacity. Hence, retention of the matrix is far better following the reading of a series of visually presented digits than it is following the addition of visual or auditory digits. They argue convincingly that maintenance of such a spatial representation is crucially dependent on general processing capacity. Phillips and Christie's results suggest therefore that the maintenance of a precise and detailed visual representation may make very heavy general processing demands, implying a single central executive for both visuo-spatial and verbal tasks.

Is there a nonspatial visual memory?

We have so far argued strongly for a spatially based working memory system. The existence of such a system does not of course preclude the occurrence of a parallel system or component concerned with pictorial or nonspatial visual representation. Indeed, work on animals has suggested just such a distinction between a spatial component of visual perception concerned with *where* an object is seen and a feature system concerned with what is seen. Neurological evidence from goldfish (Ingle, 1967) and golden hamsters (Schneider, 1967, 1969) suggests that locational information is mediated by the superior colliculi of the optic tectum, whereas pattern discrimination involves a cortical system. Although exactly equivalent data are not available for man, Weiskrantz, Warrington, Sanders, and Marshall (1974) have shown that certain patients may be able to respond by pointing to the location of an object despite the fact that it is not consciously perceived, due to cortical damge. Such "blindsight" is again compatible with two separate visual systems, one cortically mediated and the other relying on neurologically separate pathways.

Probably the best evidence for a nonspatial component to visual imagery is presented by Janssen (1976a, b, c). He describes a series of experiments following up the report by Atwood (1971) that memory for high imagery phrases such as *nudist devouring bird* is disrupted if followed by a task requiring the subject to process a visually presented digit in contrast to abstract phrases such as *the intellect of Einstein was a miracle*, which are more disrupted by the processing of an auditory digit. We ourselves found difficulty in replicating Atwood's result, as have a number of other investigators (see Baddeley et al, 1975b). Janssen succeeds in producing the Atwood effect consistently over several experiments and argues that other failures to replicate, including our own, stem from using an interfering task that is spatial rather than visual. His results and those of Atwood therefore do seem to point quite strongly toward a visual component that differs from the spatial component studied in our own experiments in a number of respects. In particular, Janssen's effect interacts with type of material, with visual disruption present for imageable material but not for material of low imageability, even though the subject is attempting to use an imagery strategy. This contrasts with our own results that suggest that the nature of the material is of minimal importance, with spatial interference being present for both abstract and concrete material, provided the subject is attempting to use an imagery strategy. Understanding the relation between spatial working memory and the more visual system suggested by Janssen's work presents an important but difficult problem. A related and perhaps even more important problem concerns the nature of rated concreteness/imageability. How do these powerful variables have their effect, and why do their effects differ, with concreteness tending to be more important in learning (Richardson, 1975a, b) and imageability in perception (Marcel and Patterson, 1978)?

In conclusion, the evidence seems to suggest the existence first of a visuospatial working memory system that is used both in immediate memory and in long-term verbal memory when a spatial mnemonic is employed. The work of Phillips and Christie (1977b) suggests that such a system is probably also heavily dependent on a central processor of limited capacity. There appears in addition to be evidence for a nonspatial system or possibly a component of a more complex system that is used for maintaining highly imageable material and may be disrupted by a very simple visual processing task (Janssen 1975a, b, c). These systems are in turn dependent on a semantic memory system that is sensitive to separable effects of both concreteness and imageability.

Acknowledgments

We are grateful to the Medical Research Council and Social Science Research Council for financial support; to Graham Hitch, Karalyn Patterson, and Bill Phillips for discussion and comments; and to Mark Roberts, who ran Experiment 5.

Note

1 Subsequent to the performance of Experiment 1; a somewhat less Gothic means of disrupting spatial coding has been developed by Ian Moar at the Applied Psychology Unit (Moar, 1978). This involves a matrix of buttons that are hidden from the subject by a cover. He is required to press the buttons one after the other following a boustrophedal path through the array, up one row and back down the next. Although the path is entirely predictable, the subject must maintain a reasonably accurate representation of his location if he is to avoid missing buttons as he moves along, and this appears to be sufficient to cause a clear decrement in tasks involving visuo-spatial imagery. Because it is easier to instrument than our auditory tracking, it is a preferable means of disrupting imagery, particularly because we found our auditory tracking task extremely difficult for some undergraduate subjects tested in a pilot study.

References

Atwood, G. E. An experimental study of visual imagination and memory. *Cognitive Psychology*, 1971, *2*, 290–299.

Baddeley, A. D. The influence of acoustic and semantic similarity on long-term memory for word sequences. *Quarterly Journal of Experimental Psychology*, 1966, *18*, 302–309.

Baddeley, A. D., & Hitch, G. J. Working memory. In G. Bower (Ed.), *Recent advances in learning and motivation 8*, New York: Academic Press, 1974.

Baddeley, A. D., Thomson, N., & Buchanan, M. Word length and the structure of short-term memory. *Journal of Verbal Learning and Verbal Behavior*, 1975, *14*, 575–589. (a)

Baddeley, A. D., Grant, S., Wight, E., & Thomson, N. Imagery and visual working memory. In P. M. A. Rabbitt & S. Dornic (Eds.), *Attention and performance V*. London: Academic Press, 1975. (b)

Bisiach, E., & Luzzatti, C. Unilateral neglect of representational space. *Cortex*, 1978, *14*, 129–133.

Bower, G. H. Mental imagery and associative learning. In L. W. Gregg (Ed.), *Cognition in learning and memory*. New York: Wiley, 1972.

Brooks, L. R. The suppression of visualization by reading. *Quarterly Journal of Experimental Psychology*, 1967, *19*, 289–299.

Brooks, L. R. Spatial and verbal components in the act of recall. *Canadian Journal of Psychology*, 1968, *22*, 349–368.

Byrne, B. Item concreteness versus spatial organization as predictors of visual imagery. *Memory & Cognition*, 1974, *2*, 53–59.

Carpenter, P. A., & Eisenberg, P. Mental rotation and the frame of reference in blind and sighted individuals. *Perception & Psychophysics*, 1978, *23*, 117–124.

Conrad, R. Acoustic confusion in immediate memory. *British Journal of Psychology*, 1964, *55*, 75–84.

De Renzi, E., Faglioni, P., & Previdi, P. Spatial memory and hemispheric locus of lesion. *Cortex*, 1977, *13*, 424–433.

De Renzi, E., & Nichelli, P. Verbal and non-verbal short-term memory impairment following hemispheric damage. *Cortex*, 1975, *11*, 341–354.

Ingle, D. Two visual mechanisms underlying the behavior of fish. *Psychologische Forschung*, 1967, *31*, 44–51.

Janssen, W. H. Selective interference in paired-associate and free recall learning: Messing up the image. *Acta Psychologica*, 1976, *40*, 35–48. (a)

Janssen, W. H. Selective interference during the retrieval of visual images. *Quarterly Journal of Experimental Psychology*, 1976, *28*, 535–539. (b)

Janssen, W. H. *On the nature of the mental image*. Soesterberg, The Netherlands: Institute for Perception TNO, 1976. (c)

Marcel, A. J., & Patterson, K. E. Word recognition and production: reciprocity in clinical and normal studies. In J. Requin (Ed.), *Attention and performance VII*. Hillsdale, N.J.: Lawrence Erlbaum Associates, 1978.

Milner, B. Interhemispheric differences in the localization of psychological processes in man. *British Medical Bulletin*, 1971, *27*, 272–277.

Moar, I. T. *Mental triangulation and the nature of internal representations of space*. Unpublished PhD thesis, University of Cambridge, 1978.

Paivio, A. *Imagery and verbal processes*. New York: Holt, Rinehart & Winston, 1971.

Paivio, A., & Csapo, K. Concrete-image and verbal memory codes. *Journal of Experimental Psychology*, 1969, *80*, 279–285.

Paivio, A., Yuille, J. C., & Madigan, S. Concreteness imagery and meaningfulness values for 925 nouns. *Journal of Experimental Psychology Monograph Supplement*, 1968, *76*(1, Pt. 2).

Phillips, W. A., & Christie, D. F. M. Components of visual memory. *Quarterly Journal of Experimental Psychology*, 1977, *29*, 117–133. (a)

Phillips, W. A., & Christie, D. F. M. Interference with visualization. *Quarterly Journal of Experimental Psychology*, 1977, *29*, 637–650. (b)

Richardson, J. T. E. Imagery and deep structure in the recall of English nominalizations. *British Journal of Psychology*, 1975, *66*, 333–339. (a)

Richardson, J. T. E. Concreteness and imageability. *Quarterly Journal of Experimental Psychology*, 1975, *27*, 235–250. (b)

Schneider, G. E. Contrasting visuo-motor functions of tectum and cortex in the golden hamster. *Psychologische Forschung*, 1967, *31*, 52–62.

Schneider, G. E. Two visual systems. *Science*, 1969, *163*, 895–902.

Shepard, R. N., & Metzler, J. Mental rotation of three-dimensional objects. *Science*, 1971, *171*, 701–703.

Weiskrantz, L., Warrington, E. K., Sanders, M. D., & Marshall, J. Visual capacity in the hemianopic field following a restricted occipital ablation. *Brain*, 1974, *97*, 709–728.

13 Interference with visual short-term memory*

R. H. Logie, G. M. Zucco and A. D. Baddeley

Working memory (Baddeley and Hitch 1974) incorporates the notion of a visuo-spatial sketch pad; a mechanism thought to be specialized for short-term storage of visuo-spatial material. *However, the nature and characteristics of this hypothesized mechanism are as yet unclear.* Two experiments are reported which examined selective interference in short-term visual memory.

Experiment 1 contrasted recognition memory span for visual matrix patterns with that for visually presented letter sequences. These two span tasks were combined with concurrent arithmetic or a concurrent task which involved manipulation of visuo-spatial material. Results suggested that although there was a small, significant disruption by concurrent arithmetic of span for the matrix patterns, there was a substantially larger disruption of the letter span task. The converse was true for the secondary visuo-spatial task.

Experiment 2 combined the span tasks with two established tasks developed by Brooks (1967). Span for matrix patterns was disrupted by a visuo-spatial task but not by a secondary verbal task. The converse was true for letter span.

These results suggest that the impairment in short-term visual memory resulting from secondary arithmetic reflects a small general processing load, but that the selective interference due to mode of processing is by far the stronger effect. Results are interpreted as being entirely consistent with the notion of a specialized visuo-spatial mechanism in working memory.

General introduction

There is a variety of tasks that appear to involve the storage of visuo-spatial information in short-term memory; visualizing number-matrix patterns (Baddeley et al. 1975b; Baddeley and Lieberman 1980; Brooks 1967), remembering abstract, irregular patterns (Broadbent and Broadbent 1981), the use of visual imagery mnemonics (Baddeley and Lieberman 1980; Logie 1986), and random square matrices (Phillips and Christie 1977a, 1977b). However, the nature of the short-term storage function involved in these tasks has been a topic of some considerable debate.

Baddeley and his colleagues have proposed that short-term visuo-spatial storage is served by a specialized mechanism within the framework of a working memory system. Working memory (Baddeley and Hitch 1974) is thought to provide the functions necessary for tasks that involve short-term storage and processing of information. It is assumed to involve moment to moment updating and rehearsal of information to prolong storage. This notion of working memory comprises a central executive, thought to be involved in reasoning, decision making and supervising a number of specialist slave systems. One slave system, the articulatory loop, is thought to be involved in storage and processing of verbal material. A second slave system, the visuo-spatial sketch pad (VSSP), performs a similar function for visuo spatial material.

The model of the articulatory loop is based on a variety of converging evidence and the specification of its characteristics is now relatively sophisticated (see Baddeley 1986). In contrast, the characteristics of the VSSP are somewhat less clear, although some progress has been made (see Baddeley 1986; Logie and Baddeley 1990). Baddeley et al. (1975b) demonstrated that concurrent visuo spatial tracking disrupts a task that involves storage of the relative positions of numbers in a square matrix (Brooks 1967). Memory for equivalent verbally encoded material was largely unaffected by tracking. The authors interpret these results in terms of a specialized VSSP that is involved in retention of the Brooks (1967) matrix material, and is also involved in tracking.

Baddeley and Lieberman (1980) showed that the disruption of the matrix task occurs with a non-visual tracking task in which subjects were blindfolded and received auditory feedback about spatial location. Tracking was also shown to affect the superiority of an imagery mnemonic (Method of Loci) over rote rehearsal. An initial letter alphabetic mnemonic was unaffected. However, an alternative mnemonic involving a smaller spatial component but with a high visual component (peg-word) was significantly but much less drastically disrupted by tracking than was the spatial mnemonic. The authors suggested that visual imagery depends on a system that is primarily spatial rather than visual in nature.

More recently, Logie (1986) has reported that use of the less spatially demanding peg-word mnemonic is affected by a secondary visual task that involves matching successive abstract patterns. Disruption also occurs where the patterns are merely presented as unattended, irrelevant visual input. That is, subjects are required to passively watch the patterns. Rote learning is unaffected by such irrelevant patterns. In addition, where the verbal material for recall is presented visually, then rote learning is affected by concurrent, irrelevant speech. Use of the peg-word mnemonic is largely unaffected by the use of irrelevant speech.

The short-term storage and manipulation of information has been examined from a different perspective in a number of studies by Phillips and his colleagues. The paradigm they developed involves memory for sequences of matrix patterns, typically square matrices with half the cells randomly filled. Phillips and Christie (1977a) have shown a marked serial position curve for visual recognition memory. Typically, items were tested in reverse serial order, so that the last item in the presented series was tested first using a recognition procedure. Under these conditions, recognition performance for the last item is substantially greater than that for items earlier in the series. Performance on the earlier items is just above chance for all serial positions including the first. Phillips (e.g., 1983) has suggested that the single item recency effect for the last item may reflect the operation of short-term visual storage, while earlier items are contained in long-term memory.

The experiments by Phillips and his colleagues also studied the effect of secondary tasks interpolated between presentation of the series and the recognition test. The major finding is that the one-item recency effect is removed by a variety of secondary tasks, and in particular by mental arithmetic (Phillips and Christie 1977b). This they suggest makes it unclear as to whether visualization occurs in a self-contained visual short-term memory system, or requires the use of general purpose resources. Indeed, on the grounds that mental arithmetic would not appear to be a specifically visual task, it seems a fairly convincing case. However, the situation may not be quite so straightforward.

Broadbent and Broadbent (1981) have shown visual recency effects for the last three items in a series of irregular patterns. Following presentation of a series of abstract and irregular patterns, subjects were shown probe items, only half of which had appeared in the presented series. Target probes were selected from a variety of positions in the series and subjects were requested to indicate whether or not they recognized the presented item as having been presented in the series. On some of the trials, between presentation of the series and the forced-choice recognition procedure, subjects were given an interpolated task that involved counting the number of symbols on a single presented card. The secondary task disrupted overall performance, but the recency effect remained intact. Broadbent and Broadbent suggested that this was due to the higher overall level of performance in their task, thus making it more sensitive than that used in the Phillips and Christie studies. They also suggested that performance on the earlier (i.e. non-recency) items in the Phillips and Christie series may be aided by the use of verbal coding of easily labelled patterns, such as vertical and horizontal lines or letter shapes. They argue that without such verbal labelling, performance on the earlier items would be at chance level. Broadbent and Broadbent concluded that the recency effect appeared to reflect some form of specialized short-term store.

There is a further difficulty with the suggestion by Phillips and his colleagues, that retention of matrix patterns relies heavily on general purpose resources. Even if we accept that interpolated mental arithmetic results in a disruption of memory for patterns (Phillips 1983), it is crucial to test whether a similar interpolated task may interfere as much or more with memory for verbal sequences. It would then be possible to explore whether the Phillips finding suggested that general purpose resources were responsible primarily for visual short-term memory, or that VSTM function involved both a specialized visuo-spatial resource and a general purpose monitoring function. This is the major issue to be tackled by the present experiments.

Within the working memory framework both verbal and visuo-spatial STM may involve general purpose resources at a 'supervisory level' in addition to their specialized storage and processing function. Indeed, the central executive of working memory is thought to have a supervisory role in the activities of its slave systems (Baddeley 1986). The finding of a disruption by mental arithmetic, of verbal short-term memory would also support the conclusion by Broadbent and Broadbent that interpolated tasks were affecting overall performance rather than just the functioning of a specialized visuo-spatial store.

We attempted to test this suggestion directly in a series of two experiments. The primary visuo-spatial task involved the use of the Phillips type matrices, adapted to memory for single patterns of varying complexity. This was developed as a span procedure by Wilson et al. (1987). The use of a span procedure has two major advantages. First, it is less prone to problems of individual differences among the abilities of our subjects. Changes in performance due to secondary or interpolated tasks can be measured against an individual subject's optimal score. Second, the one-item recency effect reported by Phillips and Christie is all-or-none on any one trial. The span procedure gives a wider range of scores and therefore can give a clearer measure of the size of interference effects.

Experiment 1

Method

Subjects

Subjects were 19 female and 13 male members of the Applied Psychology Unit Subject Panel. Their mean age was 38 years.

Visual span

The visual span for each subject was measured using a procedure devised by Wilson et al. (1987). This involved the presentation of square matrix patterns of increasing complexity. The patterns comprised squares on a black background, which could be either white (filled) or black (unfilled). Each square measured 2 cm by 2 cm and the patterns were displayed on a Microvitec Cub, RGB monitor. The display was controlled by an Apple II computer.

The first pattern consisted of two squares, one above the other, displayed at the left of the screen, with one of the squares filled. The patterns were made more complex by adding squares, two at a time. The next two squares were added alongside the first two and so on until the pattern reached the right hand edge of the screen. Further pairs of squares were added along the top of the pattern to a maximum of 60 pairs of squares. For each pattern, half of the squares were filled at random.

Subjects viewed each pattern build up on the screen, a square at a time at a rate of about four squares per second. Once the pattern was complete, it remained on the screen for two seconds. The screen was then blank for two seconds and the pattern reappeared with one of the previously white squares changed to a blank square. The subject's task was to point to the position in the matrix pattern, of the changed square. The correct square was then filled in, the screen was cleared and a new trial began with a fresh pattern.

Subjects were given three trials with patterns at each level of complexity. The number of squares in the pattern increased until the subject selected an incorrect square on three successive occasions. Visual memory span was taken as the average of the three most complex patterns to which the subject responded correctly.

Letter span

The measure of letter span for each subject involved presenting a random sequence of consonants on the computer screen. The letters appeared one after the other in the centre of the screen and each letter remained on the screen for three seconds, before it was replaced by the next letter. This display time was used in an attempt to equate its duration as far as possible, with the overall presentation time per trial for the visual span task. It was also shown in pilot studies to result in performance which avoided floor and ceiling effects.

Subjects were asked to memorize the letter sequence in the order in which the letters appeared. There was then a two-second gap, followed by the same letter sequence presented in the same way as before, but with one of the letters replaced by another, different letter. The subjects' task was to point to the screen whenever they detected the new letter in the sequence. On the first three trials, only two letters were presented in the sequence. Sequence length was increased by one letter after each group of three trials. The procedure continued until subjects were incorrect for three consecutive sequences.

Secondary tasks

The letter span and visual span task described above, were combined with each of two secondary tasks. The first of these involved mental arithmetic. Subjects were presented aurally with five single digits, for mental addition. After presentation of the second, third, fourth and fifth digit, they were required to respond vocally with the relevant intermediate total. For example, if the experimenter said *5 – 7*, the subject would respond *12*, experimenter – *4*, subject – *16*, experimenter – *9*, subject – *25*, experimenter – *6* subject – *31*. This completed the sequence. The experimenter then announced a new sequence and the task commenced again from the beginning. Presentation rate was dependent on the speed of the subjects' response, but they were required to respond as rapidly as possible, and reminded of this if any responses were hesitant.

Subjects were given two problems for practice, and then given ten problems to measure control performance.

The second task involved asking subjects to imagine a three by five arrangement of squares as shown in Figure 1. The experimenter then read aloud a sequence of instructions regarding whether each square was to be filled in or left blank, starting with the top left square and working along each row in turn. The resulting pattern of filled and blank squares was in the form of a digit between zero and nine. Figure 1 shows an example for the digit three. The sequence of instructions for this example would be:

Filled, Filled, Filled,
Unfilled, Unfilled, Filled,
Filled, Filled, Filled,
Unfilled, Unfilled, Filled,
Filled, Filled, Filled.

The subject's task was to generate a visual image of the blank matrix, and then imaginally fill squares or not as instructed. On completion of a full set of instructions, the subject was required to respond vocally with the digit specified by the resulting pattern.

For practice, subjects were first given a sheet of squared paper and for each of the digits were required to fill in or leave blank squares in a three by five matrix as directed by the experimenter. Once it was clear that subjects understood this aspect of the task, and could accurately identify each of the digits, the squared paper was removed and they were required to carry out the task by visualizing

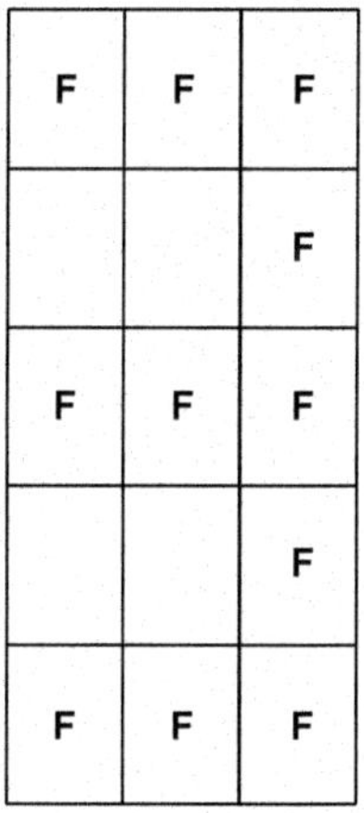

Figure 1 Example of the number matrix task used in experiment 1.

the matrix. The patterns for the digits one to nine were given in a random order and the subject was required to identify each digit correctly on two occasions before continuing with the next experimental condition.

Design and procedure

Subjects were tested individually. All subjects were given the visual span and letter span tasks. For the secondary tasks, subjects were split into two equal groups, with random allocation to each group. The first group was given the mental arithmetic task, while subjects in the second group were given the number matrix task. Each of the tasks was practised individually and control performance measured. Then each of the primary, visually presented tasks was combined with each of the secondary, aurally presented tasks in a counterbalanced order.

In the dual task conditions, the presentation of the letter span or visual span task started first. Presentation of the secondary, auditory material commenced about two seconds later. Performance on the secondary task was required throughout presentation and recall of the visually presented material, with the secondary task stopping immediately after the subject had made their response in the span procedure.

Results

We adopted a span procedure for two of our tasks in order to avoid unnecessary variance in the data due to variations in individual subjects' abilities. On these grounds, it was reasonable to assume that the control span scores were adequate measures of the limits of the appropriate cognitive mechanisms under single task conditions. That is, we assumed that the visual span score represented the best possible performance for a given subject with this material, and similarly for the letter span. Thus any decrements due to the introduction of secondary tasks were measured against what we assumed was optimum performance. Also, this effectively equated visual and letter span performance, avoiding problems of scaling differences.

The dual task span scores were transformed into percentages of control performance, and these data entered into an analysis of variance comprising one between-subject factor (imaginal matrix or arithmetic as the secondary task), and one repeated measure (visual span or letter span performance). Mean percentage data are shown in Figure 2. Table 1 shows the mean raw span scores from which these percentages were derived. The ANOVA revealed that the form of secondary task had no differential effect on overall span scores ($F < 1$). There was a marginal difference between visual span and letter span dual-task performance ($F(1, 30) = 3.09$; $0.05 < p < 0.1$). However, the most striking finding, as shown in Figure 2 was a highly significant interaction between the type of secondary task and whether the primary task involved visual or letter span ($F(1, 30) = 116.7$; $p < 0.001$).

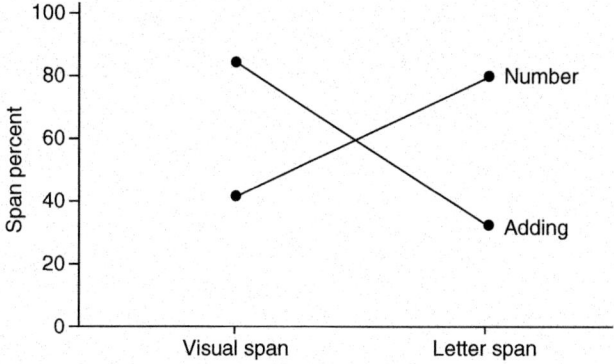

Figure 2 Primary task performance as a percentage of single task performance in experiment 1.

For the secondary tasks, clearly there was a variable number of responses, the number of which depended on the length of time taken to measure span in the dual-task conditions. As such, it was not meaningful to measure secondary task, dual-task performance as a percentage of performance of the secondary task performed alone. Instead we measured number of correct responses as a percentage of the total number of items attempted. This seemed to be the most sensible measure given the variable numbers of data points involved. However, we have reported in Table 1 the mean raw number correct expressed as a fraction of the mean number attempted in that condition. Percentage performance as calculated above, is also shown in Table 1.

Table 1 Primary and secondary dual-task percentage performance in experiment 1. Mean raw scores are given in parentheses

	Primary task performance			
	Control	Arithmetic	Control	Number matrix
Visual	100% (15.5)	85.5% (13.0)	100% (18.2)	42.8% (7.7)
Letter	100% (9.4)	34.0% (3.2)	100% (8.4)	79.9% (6.7)

	Secondary task performance		
	Control	+ Visual	+ Letter
Number	100% (18/18)	69% (11/15)	79% (19/24)
Arithmetic	95% (9.5/10)	83% (22/26)	74% (12/16)

Note
Small discrepancies in percentages are due to rounding errors when averaging.

ANOVA on the percentage data revealed that the span tasks did not differentially affect overall secondary task performance ($F < 1$), nor was there an overall difference in the level of matrix or arithmetic performance ($F < 1$). However, type of span score differentially affected the type of secondary task ($F(1, 30) = 7.32$; $p < 0.05$).

Discussion

These results indicated that there was a small overall impairment in performance associated with a requirement to perform two tasks simultaneously. Specifically, concurrent arithmetic did produce a small impairment (15%) with 12 of the 16 subjects showing a drop from baseline span performance ($p < 0.05$, Sign test) in visual span. However, the most striking feature was the differential nature of the interference with the type of secondary task: Visual span was much more substantially impaired by the secondary number matrix task (57%). In contrast, letter span was somewhat impaired (20%) by the number matrix task with all subjects showing a drop from optimum scores, but letter span was much more substantially impaired (66%) by the mental arithmetic task.

The data are consistent with the Phillips and Christie finding that a secondary arithmetic task produces an impairment of performance in retention of matrix patterns. However, the magnitude of this disruption is relatively small. This supports the Broadbent and Broadbent (1981) suggestion that the Phillips and Christie one-item recency effect is possibly rather more sensitive to general disruption than items earlier in the series. It also appears to be more sensitive to disruption than is performance on the visual span task.

One possible criticism of this study is that the visual span procedure may not use the same cognitive resources as have been traditionally studied in the context of a specialized visuo-spatial memory system. It is a relatively new task and as such we would need to ensure that it has characteristics similar to those tasks used previously. As such, we devised a second experiment designed to investigate the degree of disruption of the visual span and letter span tasks by tasks that are well established as involving primarily visuo-spatial or verbal resources, namely those originally described by Brooks (1967).

A second potential criticism is that in experiment 1, the type of secondary task was a between-groups factor. This should not cause a problem since the subjects were drawn from the same pool, and allocated randomly to the two groups. However, to ensure that the results were not in part due to inadvertent group differences, we used a completely repeated measures design in experiment 2.

Experiment 2

Method

Subjects

Subjects were 5 male and 11 female members of the Applied Psychology Unit subject panel. Their mean age was 45 years.

Materials

The visual span and letter span tasks used in experiment 1 were again used in experiment 2.

The secondary tasks were adapted from procedures developed originally by Brooks (1967). These tasks have been described extensively in the literature on this topic (e.g. Baddeley and Lieberman 1980). The visual task involved asking subjects to visualise a 4×4 matrix of squares. The square on the second row and in the second column was designated as the starting square. The subject's task was to repeat back a sequence of sentences that describe movements (up, down, left or right) around the matrix. The subject was instructed to remember the sequence of movements by visualizing them in the imagined matrix. The contrasting verbal task consists of using the same sentences as for the visual task, but replacing the words 'up', 'down', 'left' and 'right' by the words 'good', 'bad', 'slow' and 'quick'. Again the subject's task was to repeat the sequence of sentences as presented. However, they were instructed to remember the sequence by means of verbal rote rehearsal.

The visuo-spatial task involved a sequence of six sentences. However, performance is generally poorer on the verbal version, and a standard procedure is to present fewer of the nonsense sentences (e.g. Baddeley et al. 1975b). In this case, the verbal version involved four sentences. In both versions of the task, the sentences were read aloud by the experimenter and the subject responded vocally.

Design and procedure

Subjects were tested individually, and all subjects completed all conditions. First, control visual and letter spans were measured. For half of the subjects, visual span was measured first. For the remaining subjects, letter span was measured first. Next, subjects were given two practice trials on one of the Brooks tasks. If a subject was unclear as to the procedure, further practice trials were given prior to four control trials. For half of the subjects, the visuo-spatial material was presented first. In this case, the control trials on the visuo-spatial material were followed by combining the Brooks visuo-spatial task with the visual or letter span tasks. Half of these subjects first combined the Brooks visuo-spatial task with the visual span task. The remaining subjects in this group first combined the Brooks' visuo-spatial task with the verbal span task. The second group of subjects followed a similar procedure except that the Brooks verbal task was given first.

In the dual-task conditions, the experimenter commenced a presentation of the Brooks material as soon as the first item appeared on the screen for the visual or letter span task. The secondary task was performed continuously throughout the visual or letter span procedure. Thus several sets of Brooks' stimuli would be presented during the span procedures. It was never the case that the Brooks' material finished before the span procedure was completed.

The Brooks sentences were presented at a rate of one every three seconds, but this was not synchronized with primary task presentation.

Results

As with experiment 1, the control span scores were taken as representing baseline performance against which to measure the disruptive effects of the secondary tasks. Thus, the visual span and letter span scores for the dual-task conditions, expressed as a percentage of control span scores were entered into an analysis of variance. Mean percentage data along with mean raw span scores are shown in Table 2. There was no main effect of whether the secondary task was the Brooks visuo-spatial or verbal task ($F(1, 15) < 1$). However there was an overall difference between visual span and letter span such that the visual span performance was less impaired by the presence of a secondary task ($F(1, 15) = 29.32$; $p < 0.001$). In addition, as Table 2 clearly shows there was an interaction between visual and letter span, and the type of secondary task ($F(1, 15) = 57.74$; $p < 0.001$). This indicated that the visual span task was affected more severely by the Brooks matrix task than was the letter span task. Conversely, the letter span performance was more impaired by the Brooks verbal task than was visual span performance.

As with the secondary tasks used in experiment 1, the procedure necessarily resulted in different numbers of responses in the different conditions. We adopted the same procedure as before, expressing number of individual sentences correct as a percentage of the number attempted in a given condition. Table 2 shows percentage performance on the secondary tasks as calculated above along with the mean raw data expressed as a fraction.

Analysis of variance on secondary task data, revealed that there was better performance overall with the Brooks' verbal material than with the spatial material ($F(1, 15) = 26.2$; $p < 0.001$), but no overall difference as a function of the type of primary task ($F(1, 15) = 1.26$). There was an interaction between type of primary and secondary task ($F(1, 15) = 6.60$; $p < 0.05$) such that the Brooks spatial task was more disrupted by visual span, while the Brooks verbal task was more disrupted by letter span.

Table 2 Primary and secondary dual task performance in experiment 2 as raw mean scores. Percentage performance is given in parentheses

	Primary task performance		
	Control	Br. spat	Br. verb
Visual span	100% (14.9)	36% (5.3)	63% (9.1)
Letter span	100% (6.9)	47% (3.3)	18% (1.0)
	Secondary task performance		
	Control	V. span	L. span
Br. spat	92% (22/24)	56% (24/43)	61% (26/43)
Br. verb	88% (14/16)	75% (18/25)	63% (45/71)

Note
Small discrepancies in percentages are due to rounding errors when averaging.

Discussion

The differential disruptive effects are clear in this experiment. The visual span task was more severely affected by performing concurrently the Brooks spatial task, while letter span was more severely affected by the secondary verbal task. This is the result that we might expect on the basis of a model which proposed two specialized mechanisms; one for verbal material, the other for visual material. Further, the effect does not appear to be due to a 'trade-off' in resources between the primary and secondary tasks. The differential impairment is mimicked in performance on the Brooks tasks.

This result is unsurprising, but establishes that the visual span procedure is likely to involve the same cognitive resources as have been studied previously, thus reinforcing our conclusions from experiment 1. One observation is that there is a fairly large overall disruption due to the secondary tasks, and this overall effect is much larger than that obtained in experiment 1. This supports other evidence that there is a substantial general processing load involved in performing the Brooks memory tasks, in addition to their specialized processing requirements (e.g. Logie et al. 1989). The issue of task difficulty is tackled in the general discussion, where it is argued that these results are handled by suggesting the involvement of both specialized and general purpose resources in visuo-spatial short-term memory and processing function.

General discussion

What do these results add to our existing knowledge about visual short-term memory? Phillips' (1983) view was that retention of matrix patterns probably relies heavily on general purpose resources, since an ostensibly non-visual task (arithmetic) seemed to interfere with visual retention. However, the results reported here seem to weaken that argument. Arithmetic does indeed appear to disrupt visual retention. However, the degree of the disruption is far outweighed by the differential disruption produced by a secondary task with a strong visual processing component. The differential nature of the disruption rules out an explanation in terms of task difficulty. That is, that a difficult secondary task may produce more interference than an easy secondary task. Such a model would not predict the clear cross-over interactions obtained.

One objection to the results might be that virtually all of the tasks, except visual pattern span, studied in these experiments require some form of semantic processing in addition to any specialized, modality-specific processing. There are two problems with this argument. First, it is highly arguable whether letter span involves semantic processing, since the large literature on serial ordered retention of visually presented letter sequences suggests that the sequences are retained in a phonological form (e.g. Conrad 1964). Second, Avons and Phillips (1987) have argued that short-term memory for matrix patterns may sometimes involve semantic coding. Third, even if it were true that letter span involved semantic coding under some circumstances, we would expect interference between semantic processing tasks but that visual pattern span would be relatively unaffected by such tasks. The cross-over interactions obtained demonstrate that this argument is not applicable to the present set of results.

A second objection might be that the procedure of always testing the single-task performance before the dual-task performance might lead to exaggerated interference. This again is easily dealt with. This fixed order is true for all tasks in the experiments, and any effect of practice or fatigue would be spread equally, affecting overall levels of performance. It could not account for the clear cross-over interactions obtained.

An alternative view might be that the results reflect interference between similar tasks rather than the operation of specialized resources. Thus we might postulate a general resource pool which could be applied to any task, whether it involves verbal or visual processing. Thus a secondary visual task interferes because the general pool already is involved in visual processing. A complimentary argument would hold for verbal processing. However, what of the much smaller disruption which obtains when a visual primary task is combined with a secondary verbal task, and vice versa? Here one must assume that the general pool has some means of dealing with the two tasks more or less independently. It is difficult to distinguish this view logically and functionally from one which postulates separate specialized systems. In addition, the notion of a general resource pool is rapidly becoming acknowledged as being theoretically arid (e.g. Wickens 1984).

A further difficulty with an explanation in terms of a general resource pool is that there is now very strong evidence for a specialized verbal short-term memory system (e.g. Baddeley and Hitch 1974; Shallice and Warrington 1970; Vallar and Baddeley 1984). It seems therefore very compelling to interpret the differential nature of the interference in the verbal domain as involving both a specialized verbal processing system and a general purpose resource, perhaps acting at a supervisory level. Given that the pattern of results in the visual domain is the exact converse of that for the verbal primary task, an analogous argument to that for a specialized verbal resource seems appropriate.

The results are certainly highly consistent with the hypothesis of a specialized short-term visual storage system, that is independent of an analogous system for short-term verbal storage. The differential interference can be thought of as reflecting the operation of two such independent systems. The general disruption observed can be thought of as reflecting the involvement in visual processing of the 'supervisory process' or the central executive of working memory.

What of the obvious alternative, that the results obtain from the operation of just two systems? If we accept the argument for a specialized verbal system, could we not explain the results in terms of the specialized verbal resource and a general purpose resource which was also responsible for visual processing? One difficulty encountered by such a model is the contrast between the general disruption produced by the secondary tasks, and the striking differential disruption observed. Were it the case that visual retention involved only general purpose resources, one might expect that the general disruption by any secondary task would be considerably larger for visual than for verbal processing. This is patently not the case in the present experiments. For example, the Brooks tasks appear to involve a high demand on general purpose resources in addition to their differential processing requirements (e.g. Logie et al. 1989). If we examine Figure 1, Table 1 and Table 2, it is clear that in all cases the reverse is true; the general disruption is greater for the letter span task than it is for the matrix span task. For example, in Figure 1, letter span is disrupted by concurrent number imaging at least as much, if not more than is the visual span disrupted by concurrent arithmetic. In Table 2, the visual span task is disrupted by the Brooks verbal task much more than is letter span affected by the Brooks spatial task. A model based on the use of general purpose resources for visual short-term storage would have considerable difficulty in accounting for these results. A more rational explanation appears to argue for a separate, specialized visual short-term memory system, which also requires a certain amount of monitoring by a general purpose resource.

There is a possible further, logical objection to this interpretation. The strict logic of double dissociation as used here works well if we consider only two systems. However, there are at least three mechanisms involved in the working memory hypothesis; an articulatory loop, a visuo-spatial sketch pad and a central executive. To demonstrate the existence of three mechanisms, the dissociation technique has to be rather more sophisticated. One approach would be to attempt a triple dissociation, in an experiment with nine contrasting conditions. However, the central executive is very different in complexity and in nature to the two 'slave' systems. As such, a triple dissociation would be extremely difficult if not impossible to achieve.

A rather more viable research strategy is to provide evidence from a variety of sources and experimental manipulations that will converge on a coherent account. This approach has been particularly successful with the development of the articulatory loop (see Baddeley 1986), involving effects of phonological similarity (Conrad 1964), word length (Baddeley et al. 1975a), irrelevant speech (Salamé and Baddeley 1982), and patients with very specific deficits of verbal short-term memory (e.g. Vallar and Baddeley 1984; Shallice and Warrington 1970). The articulatory loop also appears to have some involvement in the production of speech (e.g. Ellis 1980; Logie et al. 1989), in some aspects of reading (Baddeley and Lewis 1981; Besner 1987) and in counting (Logie and Baddeley 1987).

We would argue that the results reported here contribute to a body of converging evidence for the concept and nature of a specialized short-term visual memory system. Specifically, the results provide additional support for the view that visual and verbal short-term memory functions are conceptually dissociable, and that both appear to be dissociable from a general purpose cognitive resource. The visual span procedure used in the present study suggests that visual short-term memory is limited by pattern complexity, and the disruption by secondary imagery tasks suggests that it is involved in constructing and retaining a visual image. These results support and extend findings reported previously (Baddeley and Lieberman 1980; Baddeley et al. 1975b; Farmer et al. 1986).

What other evidence is there for the characteristics of such a system? The Phillips and Christie (1977a, 1977b) and Broadbent and Broadbent (1981) studies suggest that visual short-term memory is limited by the number of patterns of a given complexity. In the case of Phillips and Christie, the limit is just one pattern. Broadbent and Broadbent suggest that the limit may be several items and that there may be an effect of visual similarity of the items for retention. Some more recent work by Hue and Erickson (1988) has shown visual similarity effects in immediate recall of unfamiliar Chinese characters. These results suggest that information in these tasks is coded visually, with confusions among visually similar items contained in the store.

Logie (1986) reported that there appears to be an effect of irrelevant visual material, suggesting direct access to visual short-term memory with visual presentation. A variety of studies suggest that there may also be some involvement of a visual short-term store in tracking tasks (e.g. Baddeley and Lieberman 1980, Morris 1986, 1987), and it is possible that the effects of tracking reflect a more general involvement of visual short-term memory in the control of movement (Johnson 1982; Quinn and Ralston 1986).

We have explored the notion of a specialized visuo-spatial scratch pad within the context of a working memory system, and the general notion of a working memory is proving to be increasingly fruitful (Baddeley 1986). To what extent do the ideas developed in this context fit with other influential, contemporary approaches? Two such approaches are immediately relevant; theories of mental imagery (e.g. Kosslyn 1980; Farah 1985) and resource theory (e.g. Navon and Gopher 1979; Wickens 1984).

Visual imagery

Kosslyn's (1980) theory of visual imagery comprises a number of processes that act on a short-term 'visual buffer' and long-term visual memory. The buffer is thought to hold visual images that can be generated from long-term memory or as a result of encoding through visual perception. This overlap between visual perception and visual imagery has received strong support from studies by Finke (1980, 1985) and by Farah (1985, 1988). Some very recent evidence is provided by Farah (1988) who reviews neuropsychological studies demonstrating that visual perception and visual imagery tasks involve the same neuroanatomy. In addition, Farah draws attention to reports of a number of single case studies of neurological patients with cortical blindness who also have difficulty in carrying out tasks that are commonly thought to require visual imagery.

The idea of a visual buffer fits well with the notion of a VSSP. In particular, the studies reviewed by Farah (1988) are consistent with Logie's (1986) report that unattended and irrelevant visually presented material interferes with visual imagery tasks. Kosslyn (1980: 139–141) suggests that the visual buffer has a limited capacity, that contents are subject to decay over time unless they are refreshed, presumably by a process of 'visual rehearsal' (Watkins et al. 1984), and that material is prone to interference from any new material that may enter the buffer.

It is clear from these various sources of evidence that there is considerable support for a visual short-term memory system that is independent of verbal temporary storage, and is involved both in generation of visual images and in visual perception.

Multiple resources

A further alternative approach is within the framework of multiple resources. This refers to a number of specialized resources that may act independently of one another or in concert to meet task demands (see Wickens (1984) for a more detailed discussion of multiple resource theory). Such a view is also entirely consistent with the working memory model, and within the context of multiple resource theory, Wickens and Weingartner (1985) have demonstrated dissociations of verbal and visual processing that are very similar to those reported in the working memory literature.

One common theme in multiple resource theory is that in dual tasks there is commonly a 'cost of concurrence' (Navon and Gopher 1979), or a general processing load resulting from the necessity to coordinate the activities of two or more resources. Such coordinating activity is referred to as an 'executive time-sharer' (Hunt and Lansman 1981; Moray 1967; McLeod 1977; Taylor et al. 1967). This view bears a close similarity to the interpretation of our own results, where the general disruption was attributed to the operation of the central executive, while the differential disruption was attributed to independent, specialized resources. In our case, the specialized resources were referred to as a visuo-spatial scratch-pad and an articulatory loop.

Concluding remarks

A more detailed discussion of the extent to which the ideas from these differing contemporary approaches interrelate is beyond the scope of this paper, and a more detailed treatment is under development (Logie, in preparation). However, it is clear that there is considerable support for the concept of a specialized visual working memory system, to be drawn from these differing literatures, and the attraction of the working memory model lies in its ability to cope with a wide variety of seemingly disparate findings. It appears that the process of converging operations, obtaining evidence from a variety of paradigms rather than relying solely on the results of a single study, offers considerable scope for further enhancing our understanding of the characteristics of visual short-term memory function.

Note

* These experiments were conducted while Robert H. Logie was employed by the Medical Research Council Applied Psychology Unit (MRC APU) Cambridge, UK. Experiment 1 was conducted while G. Zucco was a visiting student at MRC APU, supported by a grant from the Department of General Psychology, University of Padova, Italy. Finally we would like to thank two anonymous referees for their comments on an earlier draft of this article.

Requests for reprints should be sent to R.H. Logie, Dept. of Psychology, University of Aberdeen, Aberdeen AB9 2UB, Scotland.

References

Avons, S. and W.A. Phillips, 1987. Representation of matrix patterns in long- and short-term visual memory. Acta Psychologica 65, 227–246.
Baddeley, A.D., 1986. Working memory. Oxford: Oxford University Press.
Baddeley, A.D. and G.J. Hitch, 1974. 'Working memory'. In: G. Bower (ed.), The psychology of learning and motivation, Vol VIII. New York: Academic Press, pp. 47–90.
Baddeley, A.D. and V.J. Lewis, 1981. 'Inner active processes in reading: The inner voice, the inner ear and the inner eye'. In: A.M. Lesgold and C.A. Perfetti, Interactive processes in reading. Hillsdale, NJ: Erlbaum. pp. 107–129.
Baddeley, A.D. and K. Lieberman, 1980. 'Spatial working memory'. In: R. Nickerson (ed.), Attention and performance VIII. Hillsdale, NJ: Erlbaum. pp. 521–539.
Baddeley, A.D., N. Thomson and M. Buchanan, 1975a. Word length and the structure of short-term memory. Journal of Verbal Learning and Verbal Behavior 14, 575–589.

Baddeley, A.D., W. Grant, E. Wight and N. Thomson, 1975b. 'Imagery and visual working memory'. In: P.M.A. Rabbitt and S. Dornic (eds.), Attention and performance V. London: Academic Press, pp. 205–217.

Besner, D., 1987. Phonology, lexical access in reading, and articulatory suppression: A critical review. The Quarterly Journal of Experimental Psychology 39A, 467–478.

Broadbent, D.E. and M.H.P. Broadbent, 1981. Recency effects in visual memory. Quarterly Journal of Experimental Psychology 33A, 1–15.

Brooks, L.R., 1967. The suppression of visualization by reading, Quarterly Journal of Experimental Psychology 19, 289–299.

Conrad, R., 1964. Acoustic confusions in immediate memory. British Journal of Psychology 55, 75–84.

Ellis, A.W., 1980. Errors in speech and short-term memory: The effect of phonemic similarity and syllable position. Journal of Verbal Learning and Verbal Behavior 19, 624–634.

Farah, M.J., 1985. 'The neurological basis of mental imagery: A componential analysis'. In: S. Pinker (ed.), Visual cognition. Cambridge, MA: MIT Press, pp. 295–271.

Farah, M.J., 1988. Is visual imagery really visual? Overlooked evidence from neuropsychology. Psychological Review 95, 307–317.

Farmer, E.W., J.V.F. Berman and Y.L. Fletcher, 1986. Evidence for a visuo-spatial scratch-pad in working memory. Quarterly Journal of Experimental Psychology 38A, 675–688.

Finke, R.A., 1980. Levels of equivalence in imagery and perception. Psychological Review 87, 113–132.

Finke, R.A., 1985. Theories relating mental imagery to perception. Psychological Bulletin 98, 236–259.

Hue, C. and J.R. Erickson, 1988. Short-term memory for Chinese characters and radicals. Memory and Cognition 16, 196–205.

Hunt, E. and M. Lansman, 1981. 'Individual differences in attention'. In: R. Sternberg (ed.), Advances in the psychology of intelligence, Vol. 1. Hillsdale, NJ: Erlbaum.

Johnson, P., 1982. The functional equivalence of imagery and movement. Quarterly Journal of Experimental Psychology 34A, 349–365.

Kosslyn, S.M., 1980. Image and mind. Harvard, MA: Harvard University Press.

Logie, R.H., 1986. Visuo-spatial processing in working memory. Quarterly Journal of Experimental Psychology 38A, 229–247.

Logie, R.H., in preparation. Visual short-term memory.

Logie, R.H. and A.D. Baddeley, 1987. Cognitive processes in counting. Journal of Experimental Psychology: Learning, Memory and Cognition 13(2), 310–326.

Logie, R.H. and A.D. Baddeley, 1990. 'Imagery and working memory'. In: P.J. Hampson, D.F. Marks and J.T.E. Richardson, Imagery: Current developments. London: Routledge. pp. 103–128.

Logie, R.H., A.D. Baddeley, A. Mané, E. Donchin and R. Sheptak, 1989. Working memory in the acquisition of complex cognitive skills. Acta Psychologica 71, 53–87.

Logie, R., R. Cubelli, S. Della Sala, M. Alberoni and P. Nichelli, 1989. 'Anarthria and verbal short-term memory'. In: J. Crawford and D. Parker (eds.), Developments in clinical and experimental neuropsychology. New York: Plenum Press, pp. 203–211.

McLeod, P., 1977. A dual task response modality effect: Support for multi-processor models of attention. Quarterly Journal of Experimental Psychology 29, 651–667.

Moray, N., 1967. Where is attention limited. A survey and a model. Acta Psychologica 27, 84–92.

Morris, N., 1986. Working memory constellations. Unpublished PhD Thesis, University of Durham.

Morris, N., 1987. Exploring the visuo-spatial scratch pad. Quarterly Journal of Experimental Psychology 39A, 409–430.

Navon, D. and D. Gopher, 1979. On the economy of the human processing system. Psychological Review 86, 214–255.

Phillips, W.A., 1983. Short-term visual memory. Philosophical Transactions of the Royal Society of London B302, 295–309.

Phillips, W.A. and D.F.M. Christie, 1977a. Components of visual memory. Quarterly Journal of Experimental Psychology 29, 117–133.

Phillips, W.A. and D.F.M. Christie, 1977b. Interference with visualization. Quarterly Journal of Experimental Psychology 29, 637–650.

Quinn, J.G. and G.E. Ralston, 1986. Movement and attention in visual working memory. Quarterly Journal of Experimental Psychology 38A, 689–703.

Salamé, P. and A.D. Baddeley, 1982. Disruption of short-term memory by unattended speech: Implications for the structure of working memory. Journal of Verbal Learning and Verbal Behavior 21, 150–164.

Shallice, T. and E.K. Warrington, 1970. Independent functioning of verbal memory stores: A neuropsychological study. Quarterly Journal of Experimental Psychology 22, 261–273.

Taylor, M.N., P.M. Lindsay and S.M. Forbes, 1967. Quantification of shared capacity processing in auditory and visual discrimination. Acta Psychologica, 27, 223–229.

Vallar, G. and A.D. Baddeley, 1984. Fractionation of working memory: Neuropsychological evidence for a phonological short-term store. Journal of Verbal Learning and Verbal Behavior 23, 151–161.

Watkins, M.J., Z.F. Peynircioglu and D.J. Brems, 1984. Pictorial rehearsal. Memory and Cognition 12, 553–557.

Wickens, C.D., 1984. 'Processing resources in attention'. In: R. Parasuraman and D.R. Davies (eds.), Varieties of attention. New York: Academic Press, pp. 63–102.

Wickens, C.D. and A. Weingartner, 1985. 'Process control monitoring: The effects of spatial and verbal ability and concurrent task demand'. In: R.E. Eberts and C.G. Eberts (eds.), Trends in ergonomics/human factors II. Amsterdam: North-Holland, pp. 25–32.

Wilson, J.T.L., J.H. Scott and K.G. Power, 1987. Developmental differences in the span of visual memory for pattern. British Journal of Developmental Psychology 5, 249–255.

Part V

The central executive

Some 10 years after the original paper, the opportunity arose to summarise the current state of the working memory model in a book (Baddeley, 1986) which also allowed me to attempt to link our theoretical framework to its potential applications. All went well until I realised that I had said virtually nothing about the central executive, a clear case of Hamlet without the prince! I did of course frequently refer to the central executive although principally to account for processes that were beyond the remit of the phonological loop and visual spatial sketchpad, effectively using it as a homunculus, a little man who sits above the model and conveniently carries out all the processes that we are unable to explain.

In our defence, this strategy allowed us to focus on those aspects of working memory that we were currently investigating while at the same time accepting that this was not the whole story, a position taken by Attneave (1960) in his paper entitled, In defense of homunculi. The first chapter in this section develops this justification in response to prior neuropsychologically-based criticism of the concept of a central executive. However, if the model was to develop it would clearly need to move beyond this essentially defensive position. In search of inspiration, I began to look for models of attentional control. I found a large and sophisticated literature on the role of attention in *perception*, but little on the control of action, with one exception. There was the model proposed by Norman and Shallice (1986) who suggested that action was controlled in two ways. The first was based on existing habits together with a sophisticated form of conflict resolution when more that one way forward was possible. When a satisfactory resolution could not be reached in this way, the second system took control. This later was termed the Supervisory Attentional System (SAS). Unlike the habit-based system it was capable of creating new plans based on external evidence that would allow it to solve the current problem. A good example might be driving to work on a familiar route. For the most part this could be achieved relatively automatically, often resulting in arriving with no memory of the journey, despite having avoided other cars and stopped appropriately at traffic lights. On the other hand, if a block from road works was encountered, then the SAS would be called upon to work out an alternative route based on additional information from long-term memory, together with evaluation of the potential costs and benefits of possible actions. Norman was particularly interested in the habitual system since he was concerned to understand memory lapses and errors of judgement that might come from the inappropriate reliance on habit, while Shallice was interested in explaining the attentional deficits of patients with bilateral damage to the frontal lobes. The SAS system seemed to provide a plausible candidate for the central executive and has continued to do so.

However, while Shallice has focused principally on a more detailed specification of the underlying system, together with its representation in a computational model, I myself took a rather broader pragmatic approach, asking the question of just what functions would a central executive require in order to operate effectively. The first chapter in this section describes this approach. It was delivered as a Bartlett Lecture to the Experimental Psychology Society, and it will be clear that it left much unfinished business. It still does, although I think it can be argued that our homunculus has in fact proved useful in asking and sometimes answering relevant questions about the executive. While he is still far from retirement, at least some of his capacities are being split off and potentially explained.

The final chapter in this section illustrates the application of the concept of an executive that comprises separable subcomponents to the practical problem of understanding the cognitive deficits resulting from Alzheimer's disease. We discovered that Alzheimer's disease patients have a particular difficulty in performing two tasks at the same time, even when difficulty level is titrated to match control participants, a result that has now been replicated many times. It has both clinical significance as an early marker of Alzheimer's disease (Della Sala, Cocchini, Logie, Allerhand & MacPherson, 2010) and theoretical relevance to the analysis of the central executive (Cocchini, Logie, Della Sala, MacPherson & Baddeley, 2002). My efforts reinforced our earlier conclusion that the central executive raised a range of very tough questions, one of which concerned its link to long-term memory. Our assumption at the time was that it was a purely attentional system and did not possess storage capacity. If so, how was it involved in prose recall?

References

Attneave, F. (1960). In defense of homunculi. In W. Rosenblith (Ed.), *Sensory communication*. (pp. 777–782). Cambridge, MA: Holt, MIT Press.
Baddeley, A. D. (1986). *Working memory*. Oxford: Oxford University Press.
Cocchini, G., Logie, R. H., Della Sala, S., MacPherson, S. E., & Baddeley, A. D. (2002). Concurrent performance of two memory tasks: Evidence for domain-specific working memory systems. *Memory & Cognition, 30*, 1086–1095.
Della Sala, S., Cochini, G., Logie, R. H., Allerhand, M., & MacPherson, S. E. (2010). Dual task during encoding, maintenance and retrieval in Alzheimer's disease. *Journal of Alzheimer's Disease, 19*, 503–515.

Norman, D. A., & Shallice, T. (1986). Attention to action: Willed and automatic control of behaviour. In R. J. Davidson, G. E. Schwarts, & D. Shapiro (Eds), *Consciousness and self-regulation: Advances in research and theory* (Vol. 4, pp. 1–18). New York: Plenum Press.

Papers

1 Baddeley, A. D. (1998). The central executive: A concept and some misconceptions. *Journal of the International Neuropsychological Society, 4*, 523–526.
2 Baddeley, A. D. (1996). Exploring the central executive. *Quarterly Journal of Experimental Psychology, 49A*, 5–28.
3 Baddeley, A. D., Logie, R., Bressi, S., Della Sala, S., & Spinnler, H. (1986). Dementia and working memory. *Quarterly Journal of Experimental Psychology, 38A*, 603–618.

14 The central executive
A concept and some misconceptions

A. D. Baddeley

Parkin's criticisms of the central executive are based on a series of misconceptions. The central executive is not an organ that might or might not exist, but a scientific concept. Part of its function is to separate the analysis of executive processes from the question of their anatomical location. Like other components of working memory, it is fractionable into subsystems. How the subsystems interrelate and how they map onto the anatomical substrate are empirical questions under active current investigation. (*JINS*, 1998, *4*, 523–526.)

Introduction

As Parkin points out, the concept of a central executive was proposed by Baddeley and Hitch (1974) as part of their expansion of the earlier concept of a unitary short-term memory into a multicomponent working memory. It remained an undeveloped component of the model until Baddeley (1986) suggested that it might be useful to conceptualize the executive in terms of the SAS model developed by Norman and Shallice (1986). Despite the complexity of the problem of executive control, the model has proved fruitful and is continuing to show considerable development. It is therefore unfortunate that Parkin has chosen to create and criticize an amalgam of the working memory and SAS models as they were 10 years ago. It is, however, clear from many discussions that such misinterpretations are by no means uncommon, and I therefore welcome the opportunity of attempting to correct them. My comments will be confined to discussion of the concept of working memory, but I suspect that Shallice would share my concern about several of the misconceptions implicit in Parkin's critique and would like to refer the reader to a recent paper (Shallice & Burgess, 1996) in which, far from proposing a unitary module of the type criticized, they differentiate no fewer than eight executive sub-processes.

Parkin's criticism rests on his assumption that the central executive reflects a modular system that is coterminous with the frontal lobes. His criticism focuses on the assumption of a unitary system whose validity depends upon the capacity to map it onto a single anatomical location. I agree that these are implausible assumptions, and have in fact spent a good deal of time over the last decade attacking them (Baddeley, 1996a; Baddeley & Wilson, 1988). I suggest that a number of important misconceptions underlie Parkin's criticisms, misconceptions that are likely to present problems for any approach to executive function. For that reason I believe they are worth spelling out in greater detail. They concern

1. the basic scientific philosophy underlying Parkin's approach;
2. the tendency to assume that a coherent functional concept must have a unitary anatomical location;
3. the assumption that, because a system functions as a whole, it is therefore modular and nonfractionable; and finally,
4. the assumption that a system with homunculus-like properties is likely to be scientifically sterile.

I conclude by briefly outlining some evidence for the value of the working memory approach to executive control in terms of what it has so far achieved.

Some pretheoretical assumptions

As a scientist, I assume that my task is to develop concepts and theories that give an economical account of what we already know, in ways that facilitate the acquisition of further knowledge. Like any other concept, I assume that the central executive should be judged by its usefulness in accounting for what we know and facilitating further research. I do *not* regard it as a hypothesis that requires a yes/no answer if it is to be useful, nor do I regard it as an internal organ whose existence depends upon an exact mapping of function to precise anatomical location. Hence, to deny the existence of the central executive is to commit a category error; the concept of the central executive certainly does exist. The important scientific issue is whether it is a *useful* concept. Note that to be useful, a concept does not have to be in any absolute sense "true" or "correct." The concept in physics of the atom as an irreducible unit of matter has been superseded by fractionation into progressively more detailed components, but that does not invalidate the earlier usefulness of the concept in forwarding the development of physics.

Function and anatomy

The study of executive processes has in the past been bedeviled by the tendency to conflate the study of executive deficits with their hypothetical location within the frontal lobes. Baddeley and Wilson (1988) argued against this approach on the grounds that

A executive processes need not be unitary;

B   the frontal lobes represent a large and multi-faceted area of the brain, which is unlikely to be unitary in function;
C   executive processes are likely to involve links between different parts of the brain and hence are unlikely to be exclusively associated with a frontal location;
D   consequently patients may conceivably have executive deficits without clear evidence of frontal damage; and
E   patients with frontal lesions will not always show executive deficits.

The concept of a "dysexecutive syndrome" was proposed explicitly to allow the discussion of function to be separated from the question of the anatomical location of such functions. It is therefore somewhat ironic to find Parkin using the absence of a simple mapping of executive function onto anatomy as an argument against the concept of a central executive.

The central executive as a unitary system

As Parkin observes, the central executive was represented in the original model as a large unfilled ovoid. The reason that it was unfilled was that we knew virtually nothing about its functions. My own theoretical style is to allow the model to be constrained by the data, not to postulate complex hypothetical mechanisms that in the absence of constraining evidence will almost certainly be wrong. Given the way in which the simpler phonological and visuospatial slave systems have fractionated into subcomponents as we understood them further (Baddeley, 1996b), it seems inconceivable that the central executive will not also fractionate into subsystems. Over the last decade we have therefore been gradually attempting to find ways of breaking the system up into components, beginning with the study of dual task performance (Baddeley et al., 1986) and subsequently postulating other executive processes involving the focusing of attention, attention switching together with a system concerned with the control of long-term memory (Baddeley, 1996a). Such an approach does not of course yield a simple empirically testable yes/no hypothesis, but rather represents a way of progressively investigating an important but extremely complex set of processes. Having established a *prima facie* case for a particular function, it then makes sense to attempt to localize it anatomically, as was done successfully in the case of the capacity for dual task performance (D'Esposito et al., 1995). Such an approach leaves open the question of whether the resulting executive processes will prove to operate in a hierarchical way with one particular process being of overwhelming importance, or whether the central executive system may be better considered as an alliance amongst processes of approximate equal weight.

This might be a good point to correct the impression given by Parkin that the BADS (Wilson et al., 1996) was developed on the assumption that it was measuring a single unitary executive. The test was influenced by the concept of a dysexecutive syndrome, in proposing to separate the executive functions from the question of their anatomical localization, but was not specifically associated with the working memory model; it certainly did not assume a single unitary executive system (B.A. Wilson, personal communication, 1997). The whole purpose of including a range of different tests, originating from different theoretical traditions, was to maximize the likelihood of detecting executive deficits of any kind. In short, the aims of the BADS test were clinical and pragmatic, and its theoretical background eclectic.

When a homunculus is helpful

As Parkin points out, the central executive in the original working memory model is little more than a homunculus, a little man who takes all the decisions that are beyond the capacity of the slave systems. Viewed as a testable hypothesis, a homunculus is clearly a thoroughly bad thing. However, as a way of partitioning a complex area, even a homunculus can be useful. In the first instance, the homunculus allowed us to concentrate on the simpler and more tractable slave systems, while still acknowledging that they are capable of being influenced by more complex strategic factors. By labeling such additional factors as the central executive, we implied a commitment to investigate them, a commitment on which we are now busily engaged (Baddeley, 1996a; Baddeley et al., 1997).

But how should one deal with a homunculus? The first task is to specify the jobs that it performs. In the case of the central executive we have already suggested the need for several executive functions, including dual task performance, attentional focusing, attention switching, and interfacing with LTM. Such executive processes can then be tackled one at a time, with more detailed analysis being followed by a stage in which the separability of the specified executive subprocesses is tested and their interrelationship explored. One hopes eventually to reach a point at which each of the homunculus' tasks has understood, making the homunculus redundant (see Baddeley, 1996a; Baddeley & Della Sala, 1996, for further discussion).

How useful is the concept of a central executive?

The concept of a central executive represents just one of a number of possible approaches to the analysis of executive processes. Such processes are enormously important, but are probably the most complex aspects of human cognition and, as such, are unlikely to be fully understood in the near future. I would however suggest that the central executive concept has already proved useful in the following ways. First, in separating out the complex aspects of attentional control from the slave systems, it has facilitated the understanding of both phonological and visuospatial short-term memory. Secondly, the concept of a dysexecutive syndrome has proved useful in disentangling the functional analysis of the executive processes from the important but separate question of their anatomical location.

A good example of this is the application of the working memory model to Alzheimer's disease (AD). Analysis of the cognitive deficit in AD patients suggested a central executive impairment (Morris & Baddeley, 1988; Spinnler et al., 1988). Tasks were developed to test this hypothesis, leading to evidence for a differential disruption (Baddeley et al., 1991), while subsequent application of the tasks to patients with frontal lobe damage produced evidence both for a fractionation of executive processes and for the dependence of dual-task control on the frontal lobes (Baddeley & Della Sala, 1996; Baddeley et al., 1997). Further support for the control of dual task performance by the frontal lobes was independently obtained by D'Esposito et al. (1995) in a PET-based functional imagery study.

Finally, at a conceptual level, the increasing use of the term "dysexecutive syndrome" by investigators who are not specifically concerned with the working memory model suggests that the concept is serving a useful function. The use of a functional term rather than the anatomically based term "frontal syndrome" does not of course mean that data concerning the anatomical localization of executive processes are unimportant. But it does imply that it is pragmatically useful to be able to separate the analysis of function from the question of anatomical localization.

Having identified potential executive processes and having tested them across a range of different tasks and material, anatomical localization can then be used as a possible source of further evidence for a distinct and separable process. It is important to bear in mind, however, that specifying a function and investigating its characteristics is not dependent on that function having a unitary anatomical localization.

Conclusion

The central executive certainly exists, but as a concept—not as a modular organ coterminous with the frontal lobes. As in the case of other components of working memory, it is proving to fractionate into subcomponents and, in doing so, provides a useful basis for studying the complexities of executive control and identifying subprocesses, which may then be mapped on to their anatomical substrate.

Acknowledgments

This work was funded by MRC Grant G9423916. I am grateful to Chris Jarrold for comments and discussion.

References

Baddeley, A.D. (1986). *Working memory*. Oxford: Oxford University Press.
Baddeley, A.D. (1996a). Exploring the central executive. *Quarterly Journal of Experimental Psychology, 49A,* 5–28.

Baddeley, A.D. (1996b). The fractionation of working memory. *Proceedings of the National Academy of Sciences USA, 93*, 13468–13472.

Baddeley, A.D., Bressi, S., Della Sala, S., Logie, R., & Spinnler, H. (1991). The decline of working memory in Alzheimer's Disease: A longitudinal study. *Brain, 114*, 2521–2542.

Baddeley, A.D. & Della Sala, S. (1996). Working memory and executive control. *Proceedings of the Royal Society London, B, 351*, 1397–1404.

Baddeley, A.D., Della Sala, S., Papagno, C., & Spinnler, H. (1997). Dual task performance in dysexecutive and non-dysexecutive patients with a frontal lesion. *Neuropsychology, 11*, 187–194.

Baddeley, A.D. & Hitch, G. (1974). Working memory. In G.A. Bower (Ed.), *The psychology of learning and motivation* (Vol. 8, pp. 47–89). New York: Academic Press.

Baddeley, A.D., Logie, R., Bressi, S., Della Sala, S., & Spinnler, H. (1986). Dementia and working memory. *Quarterly Journal of Experimental Psychology, 38A*, 603–618.

Baddeley, A.D. & Wilson, B. (1988). Frontal amnesia and the dysexecutive syndrome. *Brain and Cognition, 7*, 212–230.

D'Esposito, M., Detre, J.A., Alsop, D.C., Shin, R.K., Atlas, S., & Grossman, M. (1995). The neural basis of the central executive system of working memory. *Nature, 378*, 279–281.

Morris, R.G. & Baddeley, A.D. (1988). Primary and working memory functioning in Alzheimer-type dementia. *Journal of Clinical and Experimental Neuropsychology, 10*, 279–296.

Norman, D.A. & Shallice, T. (1986). Attention to action: Willed and automatic control of behavior. In R.J. Davidson, G.E. Schwartz, & D. Shapiro (Eds.), *Consciousness and self-regulation: Advances in research and theory* (Vol. 4, pp. 1–18). New York: Plenum Press.

Shallice, T. & Burgess, P. (1996). The domain of supervisory processes and temporal organization of behaviour. *Philosophical Transactions of the Royal Society of London B, 351*, 1405–1411.

Spinnler, H., Della Sala, S., Bandera, R., & Baddeley, A.D. (1988). Dementia, ageing and the structure of human memory. *Cognitive Neuropsychology, 5*, 193–211.

Wilson, B.A., Alderman, N., Burgess, P., Emslie, H., & Evans, J. (1996). The behavioural assessment of the dysexecutive syndrome. Bury St Edmunds: Thames Valley Test Company.

15 Exploring the central executive

A. D. Baddeley

The central executive component of working memory is a poorly specified and very powerful system that could be criticized as little more than a homunculus. A research strategy is outlined that attempts to specify and analyse its component functions and is illustrated with four lines of research. The first concerns the study of the capacity to coordinate performance on two separate tasks. A second involves the capacity to switch retrieval strategies as reflected in random generation. The capacity to attend selectively to one stimulus and inhibit the disrupting effect of others comprises the third line of research, and the fourth involves the capacity to hold and manipulate information in long-term memory, as reflected in measures of working memory span. It is suggested that this multifaceted approach is a fruitful one that leaves open the question of whether it will ultimately prove more appropriate to regard the executive as a unified system with multiple functions, or simply as an agglomeration of independent though interacting control processes. In the meantime, it seems useful to continue to use the concept of a central executive as a reminder of the crucially important control functions of working memory.

In the 20 years since its publication, the model of working memory proposed by Baddeley and Hitch (1974) has continued to be useful in stimulating further research, although the three subcomponents have differed markedly in both the amount of further research evoked and also in their apparent success in accounting for available results. The most extensively explored and probably the simplest is the phonological loop (see Gathercole & Baddeley, 1994 for a review). The visuo–spatial sketchpad has proved less tractable, but it has continued to see steady progress through work that is often linked into the related question of factors underpinning visual imagery (for a recent review see Logie, 1995). Meanwhile, the central executive component of working memory remains the least studied, despite the fact that it is almost certainly the most important component in terms of its general impact on cognition.

In the early years of the model, the neglect of the central executive was intentional, as it seemed better to concentrate efforts on the more tractable problems of the two slave systems. By the mid 1980s, however, the degree of neglect had become an embarrassment to the model, and the attempt began to redress the balance (Baddeley, 1986). The intriguing problem of working out the functional and possible evolutionary significance of the phonological loop provided a potent distractor, but in recent years, I believe, we have begun to make progress. The evidence for such a claim is described below.

Strategies for analysing the central executive

The executive as ragbag

It is probably true to say that our initial specification of the central executive was so vague as to serve as little more than a ragbag into which could be stuffed all the complex strategy selection, planning, and retrieval checking that clearly goes on when subjects perform even the apparently simple digit span task. This still seems a sensible way of starting to explore working memory, as it accepts the complexities and the ultimate need to explain them, while concentrating on analysing the simpler and presumably more tractable slave systems. The Baddeley (1986) version of the model finally began the attempt to specify the central executive in greater detail, relying heavily on the Supervisory Activating System (SAS) component of Norman and Shallice's (1980) model of attentional control. This model has the advantage that it relates working memory to Shallice's study of frontal lobe patients (Shallice, 1982) and to Norman's concern with slips of action. However, the working memory model was still open to the objection that the central executive was just a convenient homunculus—a little man who sits in the head and in some mysterious way makes the important decisions.

The central executive and the frontal lobes

In recent years there have been at least two dominant approaches to attempting to understand the processes underlying executive control—one principally stemming from neuropsychology, the other being rooted in the psychometric tradition—and a number of attempts have been made to combine the two. In the area of neuropsychology, there is abundant evidence that disorders of executive control are associated with damage to the frontal lobes (Shallice, 1982, 1988). One approach to understanding executive processes, therefore, is to attempt to study the functions of the frontal lobes (e.g. Duncan, 1986; Duncan, Johnson, Swales, & Freer, in preparation; Shallice & Burgess, 1991, 1993). We could therefore define the central executive anatomically, as that system that resides in the frontal lobes.

However, although I regard the neuropsychological evidence as highly relevant, this does not seem to me to be the most fruitful line for the working memory model to take. The model is principally a functional model that would exist and be useful even if there proved to be no simple mapping on to underlying neuroanatomy. To add a component that was defined neuroanatomically rather than functionally would therefore be anomalous. It would also be likely to be unhelpful, bearing in mind the fact that the frontal lobes constitute an extremely large area that almost certainly has multiple functions that are as yet poorly understood. It is entirely possible that, although the frontal lobes are often involved in many executive processes, other parts of the brain may also be involved in executive control. If we identify the central executive exclusively with frontal function, then we might well find ourselves excluding from the central executive processes that are clearly executive in nature, simply because they prove not to be frontally located. Equally, we would be in danger of describing functions as executive simply because they were based on the frontal lobes. Note that this is not an argument against exploring the role of the frontal lobes, merely against defining the functional concept of a central executive in terms of a specific anatomical location.

Extending this argument, Baddeley and Wilson (1988) suggested that it was desirable to extend this proposed separation of functional and anatomical concepts into neuropsychology. In the case of most neuropsychological functions, this separation already occurs: a patient is referred to as suffering from an *amnesic syndrome* rather than a *temporal lobe* or *hippocampal syndrome*, and it would be generally agreed that it is more profitable to talk about a patient's *dyslexia*, or *acalculia* rather than attempt to specify the dysfunction purely in terms of hypothetical anatomical locus of damage. For that reason, we suggested that the commonly used concept of a *frontal syndrome* should be replaced by the functional concept of a *dysexecutive syndrome*. The anatomical substrate of the dysexecutive syndrome represents an important question, but it does not form a component of its definition. In what follows, I shall make extensive use of neuropsychological evidence, much of it from patients who have damage to the frontal lobes, but I do not propose to use anatomical localization as a *defining* criterion for the central executive.

Psychometric approaches to the central executive

A second and related approach to the analysis of executive processes has been through the study of individual differences. This approach is reflected in two separate but related research streams, one based on the traditional concept of intelligence, and the other more directly influenced by the assumption made by Baddeley and Hitch (1974) that working memory involves the simultaneous storage and manipulation of material. This latter approach, which has been particularly influential in North America, is considered later. The more traditional psychometric approach has been based upon the assumption that intelligence measures reflect the operation of a central cognitive processor, which could potentially be identified with the central executive of working memory. This, in turn, leads to the question of whether intelligence is better considered as reflecting a single general factor or capacity, for example Spearman's *g*, or whether, as Thurstone

(1938) proposed, g can usefully be broken down into a number of subprocesses. This general issue continues to be pursued actively using populations of normal subjects (Kyllonen & Christal, 1990), of neuropsychological patients (Burgess & Shallice, in press; Della Sala, Gray, Spinnler, & Trivelli, in preparation; Duncan, 1986), and of normal elderly subjects (Rabbitt, 1983; Salthouse, 1991). The results of such studies are clearly relevant to a concept such as working memory, but they depend crucially upon exactly which tests are included in the study, and what subject groups are tested. Although there is some encouragement for the clustering of tests, this is by no means always so. Furthermore, the clusters obtained do not at present show any consistent pattern. Some neuropsychological studies suggest a clustering of classic "frontal" tasks (Della Sala et al., in preparation), but others do not (e.g. Burgess & Shallice, in press; Duncan et al., in preparation). In the case of studies using normal subjects, the pattern is again unclear, as reflected in the present volume: Lehto finds correlations between some aspects of "frontal" tasks and some working memory tasks, and Waters and Caplan obtain patterns of correlations that appear to be largely specific to type of material and method of processing.

The homunculus: friend or foe?

Given that we adopt neither the anatomical nor the classic psychometric approach, what other methods are open? One possibility is to accept the homunculus with all its limitations, but to argue that such a concept is not only useful in defining the scope of our attempts to understand the subsidiary slave systems, but may also be productive, provided we systematically attempt to analyse the functions performed by the homunculus. If we can separate and explain just one part of the role currently attributed to the homunculus, then we shall have made progress. If we can define and analyse a range of several executive processes, then we would be in a position to begin to ask whether they are better regarded as individual and separable functions, or whether a unitary account could be given.

Using this gradualist approach, we can hope eventually to account for many—and indeed one hopes all—of the executive functions that at present are performed by our rather unsatisfactory homunculus. We would then be in a position to use psychometric methods to ask whether these are better regarded as reflecting the various operations of a single unitary controller, or whether they might be better regarded as an executive committee of interacting but independent administrators. This is clearly an ambitious and lengthy enterprise, which may prove ultimately unsuccessful. Even if this proves to be the case, however, the attempt to fractionate and understand a range of executive processes is likely to prove fruitful in raising new questions and providing new findings that will need to be accommodated by any adequate model of executive control. The sections that follow give an account of the progress we have so far made along this particular path.

Although presented as a formal strategy, in actual practice it has evolved gradually and pragmatically, starting with an attempt to capture just one necessary function of a central executive—namely, that of co-ordinating the two proposed WM slave systems. This component of the programme was initially stimulated by the attempt to test the hypothesis that patients suffering from Alzheimer's Disease showed particular impairment in the operation of the central executive, but it has more recently been extended to a broader range of patients, notably including those with frontal lobe lesions.

A second strand of research stemmed from the attempt to link the central executive to the SAS system postulated by Norman and Shallice, by means of the task of random generation (Baddeley, 1986). A third component of our approach is much more recent, being based on the assumption that one important role of the central executive should be to act as an attentional controller, selecting certain streams of incoming information and rejecting others. We have just begun to investigate this aspect of the executive's function, concentrating initially on the effects of ageing.

The final executive capacity to be considered is the ability to select and manipulate information in long-term memory. Although this is the least developed aspect of our own work, it has formed an important component of the work on individual differences in working memory that have been studied intensively elsewhere, notably by Carpenter, Just, and their colleagues and by Engle and his associates. Having reviewed progress on these four approaches, I will return to the question of whether this strategy seems sufficiently fruitful to merit continuation. Should we sack the homunculus, or continue in our long-term aim of whittling away his functions until he eventually becomes unnecessary, and can gracefully retire?

Approaches to fractionating the executive

Dual-task performance

Our first attempt to devise a measure of executive function stemmed from research on the memory deficit accompanying Alzheimer's Disease. A preliminary study indicated the expected substantial deficit in episodic long-term memory, but also suggested a degree of impairment extending across aspects of working memory, both verbal and visuo-spatial (Spinnler, Della Sala, Bandera, & Baddeley, 1988). We hypothesized that this might reflect the common central executive contribution to the two slave systems of working memory and attempted to devise a series of tasks that would allow us to test this more directly. One necessary feature of the model is its capacity to co-ordinate information from the two slave systems. We therefore set up a number of tests in which Alzheimer's Disease (AD) patients, normal elderly control and young control subjects were tested on individual tasks reflecting the operation of the relevant slave systems and were then required to combine performance on two such tasks (Baddeley, Logie, Bressi, Della Sala, & Spinnler, 1986).

For this first series of studies, we used pursuit tracking as the visuospatial task. The subjects were required to keep a light pen in contact with a moving spot of light on a VDU. We varied rate of movement of the spot so as to ensure that all subjects were performing at an equivalent level, 70% time on target. This was combined with each of three other tasks. The first involved articulatory suppression, the second reaction time to a tone, and the third digit span. Length of digit sequence was dependent on each subject's digit span and hence varied between groups, ensuring that error level was equivalent across subjects when the memory task was performed alone.

The results of this study indicated that articulatory suppression did not significantly impair performance in any of the three groups, although an earlier study by Morris (1986) and a later follow-up study observed that patients were significantly impaired, whereas the elderly and young controls were unaffected by suppression. When combined with the concurrent RT task, tracking in the AD patients showed a significantly greater decrement than was found in either of the two control groups. This was not simply a trade-off effect, as performance on both speed and accuracy of the reaction time task showed a similar pattern of significantly greater impairment in the AD patients. Similar results were also shown for the concurrent digit span task, with both tracking and memory performance showing a significantly greater decrement in the AD patients than in either of the control groups, who did not differ. Our results were therefore consistent with the hypothesis that the capacity to combine performance on two tasks—a capacity that, we have argued, is a necessary function of the central executive—is particularly impaired in AD patients. However, although our results are certainly consistent with this view, other interpretations need to be considered.

One possibility that has been suggested is that our results might simply reflect impaired performance on the constituent peripheral tasks rather than the cost of their co-ordination. We think this is unlikely, for two reasons: (1) We carefully titrated the level of difficulty in the case of our digit span and tracking tasks to a point at which all groups were performing at the same level. It is of course still possible to argue that achieving this level demanded more attention from the AD patients, but there is no reason to assume that any of the groups were doing anything other than devote their full available resources to the task. A second reason for rejecting the specific peripheral deficit interpretation is that the dual task deficit in AD patients found in this and other studies using other combined tasks (Greene, Baddeley, & Hodges, in press), is not found in the case of normal ageing. Despite the fact that age tends to impair performance on the constituent tasks, provided level of performance on the tasks is age-adjusted, combining them does not lead to an enhanced effect of age in this or other studies (Salthouse, 1991). As some aspects of WM do appear to decline with age (Salthouse, 1992; Welford, 1958), this pattern of results supports the view that the executive can be fractionated (see Gick, Craik, & Morris, 1988; Morris, Gick, & Craik, 1988).

A second potential explanation of our results is to suggest that the decrement observed in AD patients reflects some overall deficit, such as one in general intelligence or g. Rabbitt (1983) has suggested, for example, that much of the cognitive deficit observed in ageing is attributable to reduced general intelligence. I am unhappy with this interpretation, for two reasons. First of all, if both AD and ageing represent a simple reduction in g, which is reflected in the capacity to combine tasks, then one would expect dual task performance to be sensitive to age. As suggested above, this is not the case.

The second problem with a concept such as intelligence is that it replaces one problem, the nature of the executive, with another, the nature of intelligence. Most established tests of intelligence, such as the WAIS, Heim's AH4 or Cattell's Culture Fair Test, are based on the performance of a range of subtasks, each of which probably reflects a number of different processes. Whereas it might well be informative to study the contribution of simpler and potentially purer measures to performance on intelligence tests, it is far from clear that a general concept such as intelligence can throw useful light on either the functioning of the central executive or, indeed, on the neuropsychology of Alzheimer's Disease.

A similar problem of lack of specificity occurs in the case of an account of our results in terms of task difficulty. Dual-task performance clearly *is* difficult for AD patients; the role of the central executive hypothesis is to predict *what* will be difficult, rather than simply using the general concept of difficulty to label the tasks on which AD patients fail. Such an argument gains strength from a second experiment, wherein we varied difficulty within a task without increasing the demand for dual-task coordination (Baddeley, Bressi, Della Sala, Logie, & Spinnler, 1991).

The task in question was that of classifying words as belonging to one or more semantic categories. Earlier work by Yntema and Mueser (1960) had shown that the time to categorize a word increases as a function of the number of potential categories from which it is selected. We therefore presented the category judgement task to our patients and controls, studying the influence of number of simultaneously presented categories on time to decide whether a presented word was or was not a category member. We tested performance both immediately and after a six-month interval during which the disease had progressed.

As expected, both patient and control groups took longer with larger numbers of categories, and control subjects were consistently faster and more accurate than were AD patients, whose performance declined as the disease progressed. There was, however, no interaction between difficulty and disease stage, hence arguing against the hypothesis that simply increasing difficulty will make a task more sensitive to the progress of the disease. This contrasted with dual-task performance, where combined performance deteriorated substantially more rapidly than did performance on the constituent tasks when performed alone. It could, of course, be argued that our categorization task had varied the wrong kind of difficulty. However, such an objection simply highlights the inadequacy of a concept as poorly specified as "difficulty". We have proposed a rather specific executive process, which we have shown to be more disrupted in AD patients than either performance on its constituent parts or an alternative task that varied in difficulty. This is clearly not sufficient to settle the issue, but was enough to encourage us to investigate dual-task performance more extensively in both AD patients and in other groups of patients.

Our task seemed to be a useful and interesting one, but it was logistically far from convenient. As a standard piece of laboratory equipment, the pursuit rotor is almost extinct, and although it is possible to develop computerized tracking tasks, we found that lack of standardization was a major problem in purchasing equipment in different countries, even when purchasing machines of apparently identical specification. We therefore decided to develop a paper-and-pencil version of the test, and we began a series of pilot studies in search of a simple paper-and-pencil tracking task that could be scored easily and explained readily to AD patients. After a range of unsuccessful attempts, we ended up with a task in which a chain of 0.5-cm-square boxes is laid out to form an irregular path on a sheet of paper. The subject's task is to start at one end of the chain of boxes, placing a cross in each box in turn and working as rapidly as possible until the chain is complete. A second sheet is then presented, containing rather more boxes, and the subject is asked to fill in the chain of boxes as rapidly as possible for a period of 2 min. After establishing the subject's digit span using standard procedures, sequences of digits at that length are presented and tested continuously for a 2-min period, noting how many are completed and how many correct. The final stage involves combining box-crossing and digit span, again for a 2-min period (Della Sala, Baddeley, Papagno, & Spinnler, in press).

We validated the new method using a sample of 13 AD patients and 12 controls. The results were clear, with all but one AD patient showing clear impairment under dual-task conditions, whereas none of the controls showed a marked decrement. The one exception subsequently proved to have been initially misdiagnosed, proving not to have a progressive dementia, and was hence excluded from the data, which then showed a clear separation, with no overlap between the two groups in susceptibility to dual-task interference.

Although we were pleased that our modification was successful, it still left us with many theoretical questions, one being that of the neuroanatomical basis of performance on this executive task, and in particular its possible link with frontal-lobe function. We therefore gave the task to a sample of 24 neurological patients who were free of AD, but who all showed clear neuroradiological evidence of lesions somewhere in the frontal lobes. All 24 patients were assessed in three ways. Two of these involved standard cognitive measures that are generally assumed to be disrupted following frontal-lobe damage. The first of these is verbal fluency in which subjects attempt to generate as many words as possible in 90 sec, from a specified category such as animals. Poor performance on this task is typically associated with left frontal damage

(Perret, 1974), and the task itself is typically assumed to be associated with the executive process of search and retrieval from long-term memory (Baddeley, Lewis, Eldridge, & Thomson, 1984; Engle, in press). The second classic frontal measure was the Wisconsin Card Sorting Test (Nelson, 1976), in which subjects are required to form concepts as to ways in which cards containing a range of features may be sorted. Once the initial concept (e.g. colour) has been achieved, the experimenter switches to another concept (e.g. number), until eventually the six possible conceptualizations have all been tested. Performance is measured in terms of number of concepts achieved and number of errors, with perseverative errors being particularly characteristic of frontal-lobe damage. Although no task is uniformly and unambiguously associated with frontal lobe damage, these are probably the most frequent and generally accepted measures in the field (Lezak, 1983).

Our third measure attempted to capture the characteristically disinhibited behaviour that often accompanies frontal-lobe function and is typically the aspect of behaviour that makes some frontal-lobe patients fail to cope independently, even when many aspects of cognition that are frequently disturbed following frontal-lobe damage are preserved (see Shallice & Burgess, 1993, for a discussion). The behavioural syndrome is readily recognizable, although difficult to characterize in detail. Rylander (1939, page 20) suggested that it involves "disturbed attention, increased distractibility, a difficulty in grasping the whole of the complicated state of affairs ... while able to work along old routine lines ... but cannot learn to master new types of task". Such patients are often inclined to facetiousness, tend to perseverate in conversation, and may show signs of confabulation. In order to identify this aspect of cognitive dysfunction, we used clinical judgement of two independent assessors, both experienced neurologists and neuropsychologists. One judge categorized the patients as showing dysexecutive behaviour or not on the basis of their performance during the test session; the second independently based his judgement on the existing patient notes. The two assessments agreed for all except two of the 24 patients, and these two were dropped from further analysis.

Normative data for the Wisconsin Card Sorting Task and verbal fluency indicate that our group were significantly impaired, though to varying degree. The assessment of behavioural dysfunction was based on a simple dichotomy, which resulted in approximately equal numbers of patients in each group. There was a clear tendency for patients showing behavioural disorder to demonstrate a clear decrement in performance when the box-crossing and digit span tasks were combined, an effect that showed up particularly strongly on the memory component of the test, with behaviourally disturbed subjects showing a drop from 88% to 65% correct sequences on adding tracking to span, whereas the behaviourally undisturbed group showed no significant decrement (88% to 84%). The behaviourally normal and behaviourally disordered groups did not, however, differ significantly either in verbal fluency or Wisconsin Card Sorting performance.

This set of findings clearly requires replication and extension, but taken at face value it appears to be most consistent with a view of the central executive that involves a number of subcomponents, possibly associated with the functioning of different aspects of the frontal lobes. More specifically, it implies that the disordered and disinhibited behaviour that is sometimes found in frontal lobe patients is associated with difficulty in distributing attention.

How should this pattern of results be interpreted? One option might be to conclude that, because of its association with behavioural manifestations of the dysexecutive syndrome, our dual task is a better measure of executive function than the more established tests such as verbal fluency. We do however, have other reasons for regarding verbal fluency, for example, as a good measure of at least some aspects of the central executive. We found that it was particularly sensitive to the effects of a concurrent heavy digit load (Baddeley et al., 1984), and recent work by Engle (in press) finds that verbal fluency performance relates closely to working memory span measures of the type devised by Daneman and Carpenter (1980). A much more plausible interpretation would seem to be that the two tests simply measure different executive processes, with the two processes being differentially associated with the behavioural problems commonly observed in dysexecutive patients. For example, impaired verbal fluency may be associated with retrieval problems, resulting in disruption in autobiographical memory, either associated with extreme poverty of recollection, as in so-called "dynamic aphasic" patients who have great difficulty in initiating retrieval, or in the apparently opposite pattern also found in dysexecutive cases, where recollection is fluent but inaccurate, resulting in confabulation (Baddeley & Wilson, 1986).

The association between impaired dual-task performance and behavioural disturbance was not predicted in advance of our study. It is, however, reminiscent of results reported by Alderman (in preparation), who studied a group of severely brain-damaged patients who were subjected to a rehabilitation regime based on a token economy system in which socially acceptable behaviour earned points that could subsequently be cashed to purchase small luxury items. The system is usually highly successful in helping seriously behaviourally disturbed patients adapt to living with others. A small subgroup of patients, however, fail to respond to the treatment, and Alderman was concerned to understand what characterized these atypical patients. He tested all patients on a wide range of neuropsychological tests, including most of the standard tests of "frontal" function, together with a number of variants on our dual-task performance test. The standard "frontal" tests showed a weak association with behavioural disturbance, and performance on each of the range of dual-task tests was strongly associated with failure to benefit from the token economy.

The evidence from two sources of a strong association between poor dual-task performance and behavioural problems, therefore, seems to merit further investigation. It is unclear what might be the mediating mechanism, but one possibility is that effective social behaviour requires simultaneous monitoring of one's own interests and desires while attending to the potentially conflicting concerns of those around, a form of dual-task or indeed multi-task activity. However, before proceeding further along such speculative lines, it is necessary to replicate these results, preferably in conjunction with more careful and quantitative monitoring of the behaviour of the dysexecutive and control patients.

Meanwhile dual-task performance is beginning to be used somewhat more widely. Dalrymple-Alford and his associates have found a mild degree of impairment on dual-task performance in patients suffering from Parkinson's disease (Dalrymple-Alford, Kalders, Jones, & Watson, 1994). A study by Greene, Baddeley, and Hodges (in press) also observed impairment in mild AD patients, although the pattern of deficit is somewhat different from that observed in the previously described AD study, in that Greene et al.'s patients decreased their performance on box-crossing, rather than showing a decrement based on increased error rate on the memory test. This raises an important methodological problem that concerns any dual-task performance measure—namely, that of how to combine the two components so as to give a single score. Such a score is necessary if one is to compare the performance of subjects who may have a somewhat different trade-off between performance on the two components of the test. This problem, which is discussed in greater detail by Baddeley et al. (in press), will clearly have to be tackled if the test is to become suitable for general clinical use.

One final point to be noted from these more recent studies is the evidence for dual-task performance as a capacity that extends across a range of different tasks. In the original AD studies, broadly similar results were obtained whether performance on a pursuit rotor was combined with articulatory suppression, reaction time to a tone, or digit span, and our later studies indicated that changing the tracking task to one involving writing crosses in boxes did not diminish the sensitivity of dual-task performance to AD. Alderman also obtained similar results from a range of combinations of tasks, and the study by Greene et al. (in press) observed a correlation of 0.49 between performance on the combined box-crossing and digit span task and the dual-task component of the Test of Everyday Attention (Robertson, Ward, & Ridgeway, 1994). In this subtest, subjects search a list resembling a page of the telephone directory, at the same time as listening for a specified message. Again characteristically, correlation of these two tests with other tests of executive function such as verbal fluency was significant but low ($r=0.33$ and 0.31 for the box-crossing and search tasks, respectively).

In conclusion, the capacity to carry out two tasks simultaneously appears to be a candidate for one separable feature of executive function. However, it would be premature to regard our results as conclusive, bearing in mind problems of allowing for possible differences in speed–error trade-off and of the statistical scaling problems that are inevitably found when comparing impaired and non-impaired populations. We do, however, regard progress as sufficiently encouraging to justify further attempts to tackle these thorny methodological problems.

Random generation

The second area of investigation of executive function was prompted by the problem of explaining a set of results published many years earlier, in which subjects had been encouraged to generate sequences of letters, making the order as random as possible (Baddeley, 1966). The results were highly consistent and broadly fitted the conceptualization that the process of random generation depends on a system of limited informational capacity—hence the more rapid the rate, the less random the output, and the larger the set of selection alternatives, the slower the maximum generation rate. The proposed limited-capacity system appeared to have features in common with the system that limited performance in reaction-time studies, as when random generation was combined with a card-sorting task paced at a rate of 2 sec per response, the greater the number of response alternatives, the less random and more stereotyped was the concurrent sequence of letters generated.

Despite its lawfulness, this pattern of results remained difficult to explain until the arrival of the Norman and Shallice (1980) model. It will be recalled that this involves two sources of control of action, schemata that channel behaviour into well-learnt habitual patterns, together with the Supervisory Attentional System, an attentional controller that is capable of overriding habitual response patterns when it is necessary to initiate new behaviour. Looked at from this viewpoint, random generation could be seen as reflecting a series of habitual letter-retrieval schemata that were based on processes such as alphabetic recitation or the production of common acronyms. The requirement to make the sequence random, however, demanded the constant intervention of the SAS in order to break up these stereotyped sequences. As the SAS was presumably also required for the decision process involved in sorting cards into different categories, card sorting interfered with the randomness of the letter sequence generated (Baddeley, 1986). This initial suggestion has led to the adoption of random generation as a secondary task that might be assumed to disrupt the operation of the central executive. Disruption of a range of tasks, from the learning of simple contingencies (Dienes, Broadbent, & Berry, 1991) through the disruption of strategic thought in chess (Robbins et al., in press), suggests that this assumption is a reasonable one. However, although there have over the years been many speculations as to what underpins the task of random generation, there was until recently little attempt to relate the task more clearly to the functioning of a hypothetical executive component of working memory. An opportunity to remedy this came with a joint grant with Duncan to investigate executive processes.

An important concept in the study of attention over the last 20 years has been that of automaticity. Shiffrin and Schneider (1977) argued that the repeated pairing of a specific stimulus with the same response will gradually reduce the attentional demand of responding to the stimulus, up to a point at which the stimulus will evoke the response virtually automatically. Random generation could be regarded as the opposite end of this continuum from automaticity, the aim being to generate a response that is minimally associated with what went before, hence producing a task that continues to demand attention even after much practice. The pattern of results produced by Baddeley (1966) is consistent with the idea that a common mechanism is involved in studies of reaction time and of random generation but leaves the underlying mechanism unexplained. We decided to attempt a more detailed analysis, which, we hoped, would lead to an explanation.

The vast majority of studies of random generation have used verbal output, typically involving letters or numbers. This is logistically somewhat cumbersome, because it requires copying down the subject's output and then keying these into a computer for analysis of randomness. Number and letter generation also lend themselves to a range of simplifying strategies, reciting telephone numbers or dates, spelling words, or providing the initial letter of names of objects around the room. Furthermore, having only a verbal output mode limits the use of random generation as a secondary task. We therefore decided to explore the possibility of using random keypressing as an alternative generation procedure (Baddeley, Emslie, Kolodny & Duncan, in preparation).

Our first study required subjects to generate random sequences of a hundred numbers at each of three rates—0.5, 1, or 2 sec per response. One condition required subjects to generate random sequences of spoken digits. In a second condition, subjects were seated at a keyboard containing 10 keys, one located under each finger or thumb. Subjects were required to generate a random sequence of presses, again at each of the three rates. Results indicated that the degree of randomness was somewhat less for keypressing than for digits but showed an equivalent decline as generation speed increased. We also observed characteristic stereotyped response sequences, which, as in the case of alphabetic stereotypes, became more frequent as rate of generation increased. In short, key-pressing appeared to be broadly equivalent to digit generation. From this point on we used keyboard generation almost exclusively, typically selecting a rate of one response per second.

Our next study explored further the assumption that random generation reflects the limited capacity of a general purpose executive system. We combined keypressing with performance on a memory span task in which subjects recalled sequences ranging in length from one item to eight. If performance depends upon a general purpose system, then there should be interference between the verbal memory task and the visuo-spatial generation task. Furthermore, if the system reflects a limited-capacity working memory, then the degree of disruption of random generation should increase with concurrent memory load. As Figure 1 shows, this is what we observed.

We went on to investigate the influence on keyboard generation of a range of further tasks, selected on the basis of their expected loading on the central executive component of working memory. We found that articulatory suppression—for example, counting repeatedly from 1 to 6—had no significant effect on random generation, which was, however, substantially

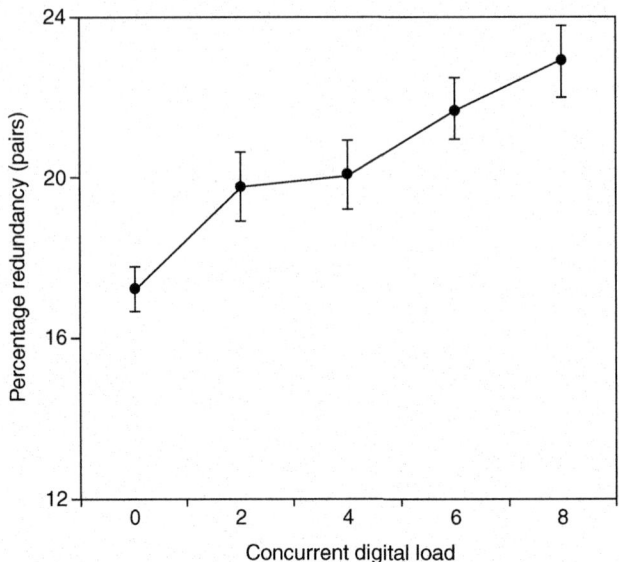

Figure 1 The effect of concurrent digit-span load on the randomness of key pressing.

disrupted by a category generation task in which subjects attempted to produce as many items as possible from a specified semantic category such as *animals* or *fruit*. Such verbal fluency tasks do seem to depend relatively heavily on executive resources, as evidenced both by their susceptibility to concurrent digit span (Baddeley et al., 1984), and to impairment in patients suffering from the dysexecutive syndrome (Baddeley & Wilson, 1988). An even greater degree of impairment was produced by a concurrent requirement to perform the AH3, a demanding measure of fluid intelligence (Heim, 1975). Duncan (1993) has argued that performance on such intelligence tests is an index of executive function and depends upon the operation of the frontal lobes. The overall pattern of results, therefore, was broadly consistent with the assumption that random generation competes for the same limited capacity as is necessary for performance of a range of tasks that depend to a greater or lesser extent on central executive functioning.

However, although this pattern of results was broadly supportive of the initial hypothesis, none of the results so far described places any major constraints upon the possible models of the underlying process. An exception to this was provided by one condition in which we asked our subjects to combine random number generation with random key-pressing, both being required at a 1/sec rate. What might one expect? In comparison to performing either of the generation tasks alone, by combining them we were asking for double the amount of information to be generated. Given that we are moving from a single-task to a dual-task mode, this seemed likely to create even more problems, and we suspected that subjects would simply be unable to perform the task. In fact they coped very well; concurrent digit generation reduced the randomness of keypressing by about the same amount as concurrent category generation or holding a sequence of digits of span length. The reciprocal effect of concurrent keypressing on digit generation was rather less.

In order to explain these results, we had to think in greater detail about the possible processes underlying random generation, and we came up with a model that is somewhat similar to a simplified version of retrieval in Raaijmakers and Shiffrin's (1981) SAM model, as represented, not entirely sympathetically, by Roediger (1993). It involves setting up a retrieval plan, running it, and checking the output, which, if judged to be suitably random, is then emitted at the appropriate time. We assumed that the decrease in randomness at higher speeds occurs principally because of the time taken to shift from one retrieval plan to another. If this were not time-limited, then the subject could presumably switch every time and would not need to bother checking the randomness of the output. On the other hand, if the same retrieval plan is used repeatedly, then the stream of responses is likely to be stereotyped and non-random. Anything that interferes with the capacity to switch retrieval plans will tend to increase the degree of redundancy. Note that although such a system may be quite sensitive to concurrent activity, degradation is gradual rather than catastrophic; the subject simply makes fewer switches of retrieval plan, resulting in more stereotyped responses. In the case of combined keyboard and digit generation, if the performance limitation reflects the switching process, then the result will simply be that generation will continue, but with fewer switches of retrieval plan. Provided the retrieval plan for keypressing is separate from that for digit generation and can operate simultaneously, the outcome will be a general reduction in randomness rather than catastrophic breakdown.

Our final two experiments attempted to test the switching hypothesis directly. If the need to switch retrieval plans is the source of the disruption in performance, then it should be possible to devise a task that places minimal load on memory and other executive processes but has heavy switching demands. We were encouraged in our search by evidence from neuropsychology, where the Trails Test from the Halstead–Reitan battery is regarded as sensitive to frontal-lobe damage. The test measures speed of performance on a task in which subjects use a pencil to join together a series of numbered squares (Trails A). A second subtest (Trails B) requires subjects to alternate between numbered and lettered squares, to produce a sequence that connects squares A–1–B–2–C–3, etc. Patients with frontal-lobe damage are particularly disrupted on this alternating test (Lezak, 1983).

As our subjects were pressing keys, it would obviously not be appropriate to give them the Trails Test, so we developed our own verbal equivalent. In the first study, subjects pressed keys at a rate of 1/sec, either alone or in combination with one of three concurrent tasks, involving either reciting the alphabet at a 1/sec rate, counting at the same rate, or alternating letters and numbers (A–1–B–2–C–3, etc.). Whereas neither counting nor reciting the alphabet had a detectable effect on the randomness of keypressing, the concurrent alternation task markedly reduced randomness. We were therefore encouraged to carry out a replication, this time starting the subjects on each trial with a different initial number–letter pair, for example F–9, and requiring them to repeat and then continue (F–9–G–10–H–11–I–12, etc.). Despite the minimal memory load imposed by this task and its entirely predictable nature, it led to a substantial reduction in redundancy of keypressing and was itself disrupted, as indicated by the number of occasions on which generation ceased and had to be prompted by the experimenter.

We have provisionally concluded therefore that random generation disrupts the operation of the central executive by its demand for the constant switching of retrieval plans. We believe that our simple alternation task performs a similar function and propose to explore it further in studies using both normal subjects and neuropsychological patients.

Although our results are encouraging, we are still some way from an adequate model of random generation. It is, for example, still unclear to what extent the load imposed by generation stems from (1) the need to switch strategies, (2) the problem of accessing new strategies, or (3) the monitoring of the response output. The initial concept of a limited-capacity general processor responsible for both input and retrieval is also clearly over-simplified. Baddeley et al. (1984) observed that a concurrent sorting task had a marked effect on the learning of a list of words but little or no effect on retrieval, suggesting a degree of automaticity in the retrieval process. A subsequent series of experiments by Craik (1995) replicated the learning effect but did in addition find a consistent but smaller effect of a secondary RT task on number of words retrieved. However, when the RT task itself was studied, the effect of concurrent word retrieval on mean RT was substantial. RT did not, however, vary as a function of number of items retrieved per unit time, suggesting a general effect rather than a direct effect of item retrieval on checking or production. Furthermore, RT was not affected by whether subjects were instructed to focus principally on retrieval or on RT. Craik suggests that whereas learning depends on amount of attention available, retrieval behaves differently, depending on the cognitive demand of operating in a general retrieval mode. Craik suggested that this difference is linked to data from recent PET studies that suggest differential cortical localization of encoding and retrieval processes in the left and right frontal regions, respectively.

In conclusion, although our understanding of the role of WM in random generation is far from complete, it seems likely that further analysis may have important implications for the more general question of the role of attentional processes in retrieval.

Selective attention

The third proposed component of executive processing resulted from speculation as to what further capacities might be likely to be required by a general executive processing system, in addition to the capacity to timeshare and to switch retrieval plans. One obvious candidate is the capacity to attend selectively to one stream of information while discarding others—the classic phenomenon of selective attention. This was investigated in a series of experiments carried out in collaboration with Duncan Godden, Elizabeth Maylor, Ian Robertson, and Tony Ward. Middle-aged and elderly subjects were studied, on the assumption that age is a variable that influences executive processes in ways that are important but far from straightforward.

The suggestion that the elderly may have an impairment in working memory extends back at least as far as Welford (1958), and data from studies of ageing were incorporated in initial discussion of the concept of a central executive in working memory (Baddeley, 1986). Available evidence, however, is somewhat equivocal and appears to depend, in ways that are not fully understood, on the exact nature of the task. For example, studies by Morris et al. (1988) and Gick et al. (1988) used the technique developed by Baddeley and Hitch in which a concurrent digit load accompanies a secondary task—in this, case sentence verification. Working memory load was manipulated in two ways, either by increasing the concurrent load or by increasing the syntactic complexity of the sentence being processed. Both of these impaired performance, but whereas syntactic complexity had a greater effect in the elderly than in the young, there was no such interaction between age and concurrent digit load. This inconsistency characterizes much of the literature (see Craik, Anderson, Kerr, & Li, 1995, for a review), suggesting that ageing may be an interesting and productive variable to study within the context of working memory.

One problem in attempting to carry out theoretically driven studies on ageing stems from the fact that almost every physical and cognitive function shows some decline. Consequently, showing that the elderly perform poorly on any given task cannot be regarded as evidence for the task's peculiar vulnerability to ageing, unless other factors are ruled out. This has led to a series of attempts to capture the essence of the ageing deficit in terms of a single measure such as general intelligence (Rabbitt, 1983), speed of processing (Salthouse, 1991), or reduced capacity for inhibition (Hasher & Zacks, 1988). We therefore adopted a strategy of measuring the first two of these and using them as part of a multi-variate analysis that allowed us to ask whether any of our working memory conditions led to age-related deficits that were not attributable to general intelligence or simple speed of processing. If we were to find such additional effects, they would argue in favour of a multi-component executive rather than a monolithic system based on a single underlying capacity.

Our basic paradigm was one in which subjects were instructed to press a key as rapidly as possible whenever a specified stimulus occurred. In the first two of our experiments, subjects were required to count the stimuli and report the total when cued, thereby, we felt, providing a more demanding task than simple reaction time (Wilkins, Shallice & McCarthy, 1987). However, exclusion of counting from the third experiment did not change the pattern of results, so the counting requirement will not be discussed further. We used two manipulations to increase the attentional demand of the task. The first of these was to present irrelevant stimuli, which the subject was instructed to ignore; in Experiment 1 these were always in a different sensory modality, while in Experiments 2 and 3 stimuli within the same modality were also included. Our second attentional manipulation was produced by an occasional instruction, which required the subject to switch from responding to signals in one modality to those in the other. We measured both the overall effect of this and its short-term effect, as Allport, Styles, and Hsieh (1994) have presented evidence suggesting that this initial response is particularly markedly slowed.

The first experiment, therefore, was one in which subjects watched a VDU and responded by pressing a key when a circle appeared. All subjects were tested on four conditions in which visual detection was performed (1) alone, (2) with irrelevant tones, (3) with instruction to respond to *both* circles and tones, and finally (4) with a requirement to switch between circles and tones on a given cue. We tested 24 middle-aged subjects (mean age = 42, range 35 to 50) and 24 elderly subjects (mean age = 72, range 66 to 83), all of whom were members of the APU Subject Panel and all of whom were required to perform a paper-and-pencil test of fluid intelligence, the Cattell Culture Fair Test (Cattell & Cattell, 1960).

The results were disappointingly straightforward. Reaction time was slowed by the presence of an irrelevant signal on the other dimension and by the instruction to switch channels—an effect that, as Allport et al. reported, is particularly marked in the case of the first response after switching. The elderly subjects were consistently slower than the young and, as expected, were lower in performance on the fluid IQ test. When IQ was taken out as a covariate, the age differences disappeared—a result that mirrors closely many similar findings reported by Rabbitt (1983).

Our second experiment utilized the same group of subjects and a design that was similar, apart from the inclusion of conditions in which irrelevant stimuli occurred within the *same* dimension as the target stimuli. Hence, subjects responded to circles but ignored triangles, or responded to low tones but ignored high.

Comparing Experiments 1 and 2 indicated that subjects were slower in responding when they had to ignore irrelevant stimuli, particularly when these occurred within the same sensory dimension. Switching from one modality to another slowed responding, particularly on the initial stimulus on the new modality, and again the elderly responded consistently more slowly than did their middle-aged counterparts. When the analysis was performed using IQ as a covariate in Experiment 2, the age difference disappeared when subjects were ignoring a stimulus on the other dimension or were switching from one dimension to the next. There did, however, remain a significant age effect for the condition in which subjects were required to ignore irrelevant stimuli within the same dimension as the targets.

Bearing in mind the potential theoretical interest of the last finding, we decided to replicate, using auditory stimuli that varied on dimensions other than simple pitch, adding differences in timbre to produce two stimuli that could best be described as a squeak and a grunt. For visual stimuli we used a square and a diamond. In order to simplify the design, we omitted the condition in which subjects switch modalities; this had consistently failed to show a differential sensitivity to age. This is broadly consistent with the findings of Allport et al. (1994), which seem to point to a phenomenon based on negative priming rather than a reflection of some active limited-capacity attentional control mechanism.

As expected, there were overall main effects of age and of whether the irrelevant stimulus was on the same or a different sensory dimension. A multivariate analysis was then performed in which the relevant factors were age, intelligence, and speed of processing, measured by using the mean reaction time from the condition in which the irrelevant stimulus was presented on the unattended dimension. Age continued to have a significant impact on time to respond while ignoring a stimulus in the same dimension, even when the influence of IQ differences and speed of processing were partialled out.

It is clearly important to extend and replicate these findings before drawing firm conclusions about the cognitive effects of ageing. However, if they do prove replicable, this pattern of results appears to suggest that the effects of ageing go beyond the hypothesized simple slowing in general speed of information processing or, indeed, decline in fluid intelligence, because neither of these can account for the remaining age difference in the capacity to ignore irrelevant stimuli in the target dimension. On the other hand, this result would seem to be consistent with the proposal by Hasher and Zacks (1988) that age limits the capacity for utilizing inhibition to sharpen attentional focus and limit distraction. Note, however, that our results do not support the idea of a general reduction in inhibition, since the differential effect does not occur when the irrelevant stimuli are presented on a different sensory dimension.

The nature of the inhibition process is at this point purely speculative, but if one uses its physiological counterpart as a model, then one might imagine something like an attentional focus on the features specifying the target, represented by a bell-shaped distribution, with the target at its centre. The nearer a stimulus is in characteristics to the focal-point, the more attention it will receive, and the more rapidly it will be processed. Stimuli falling outside this distribution will be ignored, and those falling on the border will require processing before rejection.

A simple assumption that age leads to a less highly peaked attentional focus might offer a simple account of our results. For older subjects, with a broader distribution of attention, the irrelevant stimulus will demand and receive more processing than for the middle-aged group. In the case of stimuli on a completely different dimension, then, such stimuli will fall outside the focus of attention for both groups.

It is, however, important not to place too much reliance on these preliminary results. At this stage, their principal implication is in the support they lend to the view that the capacity for focused selective attention provides a promising further component of any complete specification of the central executive.

Activation of long-term memory

One feature that must surely be possessed by the central executive but which has been totally ignored by my own work is the capacity for the temporary activation of long-term memory. Consider the case of KJ, a highly intelligent patient with a dense but pure amnesic syndrome (Wilson & Baddeley, 1988). When presented with a paragraph relating a brief story such as that involved in the logical memory subtest of the Wechsler Memory Scale, his immediate recall was above average (12 idea units), although half an hour later, his recall score was zero. He had not only forgotten the story but even forgotten that he had been told a story. KJ's good initial recall is, of course, by no means typical, because many amnesic patients perform extremely badly even on immediate test. The question remains, however, as to how KJ achieves his excellent initial performance.

My own assumption is that, as Johnson-Laird (1983) suggests, the process of comprehension involves setting up some form of mental model, and that this, in turn, demands working memory capacity. It is conceivable that this process operates entirely within the two slave systems, but it seems unlikely that they would be capable of reflecting the semantic complexity, and capacity to utilize earlier learning that seems to lie at the root of comprehension. A more plausible assumption might be to assume that such models represent the temporary activation of components of long-term memory. Such a view has recently been developed by Ericsson and Kintsch (1995) who illustrate this aspect of WM extensively using examples from prose comprehension and from the performance of mnemonic experts.

The idea that working memory might represent the selective activation of representations in long-term memory is not, of course, a novel one, and in North America at least, could probably be regarded as the modal view (e.g. Crowder, 1993; Roediger, 1993). At one level the view is comparatively uncontroversial. Given that memory span for non-words that resemble English is higher than that for those differing from English in their phonotactic structure (Adams & Gathercole, 1995), it is clear that even the phonological loop is not a *tabula rasa*, but, rather, a system that has developed on the basis of the phonological experience of the rememberer. The phonological store, therefore, depends on activation of a system that itself reflects long-term memory. The problem with the view that STM simply represents currently activated LTM, however, is that such a view is so general as to be theoretically sterile, unless an attempt is made to specify in detail the processes involved. At this point, it is likely to become a matter of theoretical taste whether one chooses to emphasize this single very general feature of dependence on some aspect of LTM or to concentrate on those characteristics that differentiate different memory subsystems. In general, I myself have been more interested in the differences, for example, between the characteristics of the phonological loop and the visuo-spatial sketchpad, or in contrasting both of these with the rich multimodality that characterizes coding in semantic or episodic memory.

Returning to the role of LTM in WM, one way of conceptualizing a hypothetical general retrieval system might be through the concept of a central executive. Such a system should be able to encode and retrieve information both from the slave systems and from temporarily activated components of long-term memory. The Baddeley and Hitch working memory model has already considered one rather specialist form of retrieval, that underlying the recency effect, where an implicit priming mechanism is assumed to be usable across both slave systems and long-term memory (Baddeley & Hitch, 1993). However, we have almost completely neglected the possible role of the central executive in setting up, maintaining, and retrieving temporary representations in long-term memory.

Fortunately, others have been less negligent. In recent years the dominant theme in North American research on working memory has been that stemming from Daneman and Carpenter's (1980) development of the measure they term *working memory span*. This combines the simultaneous requirement to process and store information, and, in its most typical form, involves presenting the subject with a series of sentences, each of which must be processed, and the last word stored. At the end of the sequence, the subject is required to recall the terminal words of each sentence. The maximum number of sentences that can be processed while retaining the final word is the subject's working memory span, and it typically ranges between two and five. The measure has proved very useful in studying the hypothetical role of working memory in a range of complex cognitive tasks, extending from measures of reading comprehension to predicting performance on computer programming courses (Just & Carpenter, 1994; Shute, 1991). When scores on a number of such tasks are combined, the resultant measure correlates very highly with performance on standard intelligence tests (Kyllonen & Christal, 1990), suggesting that it is indeed measuring some capacity of general importance to cognitive functioning. Exactly what this capacity reflects is, however, considerably more controversial, as reflected for example in the contributions to the present volume of Lehto and of Waters and Caplan.

Some of the most careful analytic work in this area has been carried out by Engle, whose recent chapter (in press) gives an excellent overview of his extended research programme. Early work was concerned with the issue of whether the working memory system assumed to underlie the Daneman–Carpenter task was limited to language processing or reflected a more broadly based limited-capacity system. In demonstrating equivalent working memory span effects when arithmetic operations were used instead of sentence verification, Turner and Engle (1989) provided evidence for a more general limited-capacity system—evidence that is, of course, also consistent with our own random generation studies and with Shute's (1991) work on individual differences in working memory span. Subsequent work has been concerned to specify in greater detail the nature of the limitation in capacity that gives rise to individual differences in working memory span.

Cantor and Engle (1993) suggested that working memory might reflect the temporary activation of areas of long-term memory, with high-span subjects being able to activate more extensive regions of long-term memory. Following the assumptions and techniques developed by Anderson (1974), they investigated this hypothesis by using the *fan effect*. This term refers to the fact that time to verify a statement that has previously been presented will be greater if the subject or object of that statement has also been linked to a range of other statements. Hence, a subject who has mastered the set of statements,

> *The vicar is in the canoe*
> *The vicar has red hair*
> *The sailor is in the supermarket*
> *The vicar is in Scotland*

will take less time to verify that the sailor (one proposition) is in the supermarket than that the vicar (four propositions) is in the canoe. Anderson explains this pattern of results by assuming that a limited amount of activation automatically spreads from one unit of the sentence to its associated features. As the amount of activation is limited and the vicar is associated with four different features, each link will be weaker than the equivalent link between sailor and supermarket. Cantor and Engle propose that high-working-memory-span subjects simply have more activation available, and, in accordance with this view, they demonstrate that the slope relating set size to verification time is steeper for subjects with a low working memory span, as measured by a variant of the Daneman and Carpenter task.

A later study by Rosen and Engle (cited by Engle, in press) studies the capacity to generate items from a semantic category such as animal names, demonstrating that performance is substantially higher in high-working-memory-span subjects. A third series of experiments explores the relation between working memory span, Anderson's fan effect, and the closely related demonstration by Sternberg (1966) of a linear relationship between the time it takes to decide whether a probe item comes from a set that has just been presented, and set size. Sternberg himself attributed this effect to an internal memory scanning mechanism, but the phenomenon is also open to a wide range of alternative interpretations, of which Anderson's model provides one.

An ingenious series of experiments by Conway and Engle (1994) is based on the link between the Sternberg and fan effects. Subjects are first of all taught groups of two, four, six, or eight letters, to a point at which, when asked for that group, they can provide the constituent letters perfectly. When given the group and a particular letter, subjects are thus able to indicate whether or not that letter falls within that group. As predicted, reaction time increases linearly with number of items within the probed group, with the slope being steeper for subjects with low working memory span. By specifying the set first (e.g. the four-letter set) but delaying the presentation of the probe letter, Conway and Engle are able to separate out the time it takes to access a given set, from the time to check it for the presence of the probe letter. They find that working memory span does not influence the time to access the set, only the time to verify the presence of the probe. They conclude from this that the former retrieval process is relatively automatic and does not depend on limited-capacity working memory, whereas the latter involves an active search process that depends upon the limited-capacity system available.

The data described so far are captured well by a model that assumes that individual differences in working memory reflect differences between subjects in the amount of activation available. However, despite this supportive evidence, Engle reports two further observations that cause him to abandon this hypothesis. The first of these concerns the detailed analysis of the performance of high- and low-span subjects on the various tasks. Although the effects I have described all hold for high-span subjects, many of them do not do so for subjects with low spans (Engle, in press). For example, combining the category-generation task with an attention-demanding concurrent activity reduces the performance of high-span subjects but has no effect on those with low spans. Having subjects learn a subset of categorized items, which they are instructed to ensure are *not* included in their category generation, has no effect on low-span subjects but severely inhibits the performance of those with high spans. Somewhat surprisingly, a marked effect occurs even when the items to be excluded come from a category (e.g. *furniture*) that is totally unrelated to the generation category (e.g. *animals*). Such findings have important implications for the general utilization of the working memory span measure, because they suggest that rather than providing a measure of some continuously varying capacity across subjects, the measures differentiate between subjects who are using different strategies. Although the differences in strategies may well result from underlying differences in some form of processing capacity, this needs to be tapped more directly if the measure is to be used as anything more than a convenient way of dichotomizing groups into good and poor performers.

The second observation is even more problematic for the simple excitation hypothesis. It resulted from Engle's concern that his experiments based on the Sternberg paradigm had required subjects to learn sets of items that were not mutually exclusive—hence the letter *K* could appear in both Set Size 2 and Set Size 6. In two final experiments, this confounding was avoided, with no item occurring in more than one set. Under these circumstances, the difference in slope between high- and low-working-memory-span subjects disappeared, and the linear relationship between set size broke down around length 8. It therefore appears to be the case that a crucial feature of the previous results was the overlap of items across categories. Engle abandons his earlier model in favour of something that more closely resembles the Baddeley and Hitch concept of a central executive, but an executive that is limited principally in its capacity to inhibit irrelevant information. As Engle points out, such a view is very close to that of Hasher and Zacks; it is, of course, consistent with our own much less extensive findings from the previously described age and selective attention studies.

The presence of individual differences in inhibitory capacity does not, of course, rule out the possibility that excitatory processes also differ across individuals. However, Engle's results suggest at the very least that we need to look very carefully at claims for such differences. Engle's results ask some searching questions about three of the major phenomena of cognitive psychology, namely the Sternberg effect, the fan effect, and the Daneman–Carpenter measure of working memory span. As such, they have substantial implications for understanding the role of working memory in retrieval from long-term memory.

Conclusion: should we sack the homunculus?

To what extent is it still useful to talk about the central executive as if it were a unitary system? My own view is that it continues to be a useful concept that is able to focus on both the attentional characteristics of working memory and its more traditional links with short-term visual and verbal memory. In short, I think the homunculus can be useful, given two provisos: (1) the continued recognition that it constitutes a way of labelling the problem, not an adequate explanation; (2) a continued attempt to understand the component processes that are necessary for executive control, gradually stripping away the various functions we previously attributed to our homunculus, until eventually it can be declared redundant. Whether we will then be left with a single coordinated system that serves multiple functions, a true executive, or a cluster of largely autonomous control processes—an executive committee—remains to be seen.

References

Adams, A.M., & Gathercole, S.E. (1995). Phonological working memory and speech production in preschool children. *Journal of Speech and Hearing Research, 38*, 403–414.

Alderman, N. (in preparation). *Maximising the learning potential of brain injured patients.*

Allport, A., Styles, E.A., & Hsieh, S. (1994). Shifting international set: Exploring the dynamic control of tasks. In C. Umilta & M. Moscovitch (Eds.), *Attention and performance XV* (pp. 421–452). Cambridge, MA: MIT Press.

Anderson, J.R. (1974). Retrieval of propositional information from long-term memory. *Cognitive Psychology, 6*, 451–474.

Baddeley, A.D. (1966). The capacity for generating information by randomization. *Quarterly Journal of Experimental Psychology, 18*, 119–129.
Baddeley, A.D. (1986). *Working memory.* Oxford: Oxford University Press.
Baddeley, A.D., Bressi, S., Della Sala, S., Logie, R., & Spinnler, H. (1991). The decline of working memory in Alzheimer's Disease: A longitudinal study. *Brain, 114,* 2521–2542.
Baddeley, A.D., Emslie, H., Kolodny, J., & Duncan, J. (in preparation). Random generation and the executive control of working memory.
Baddeley, A.D., & Hitch, G.J. (1974). Working memory. In G.H. Bower (Ed.), *The psychology of learning and motivation, Vol. 8* (pp. 47–89). New York: Academic Press.
Baddeley, A.D., & Hitch, G.J. (1993). The recency effect: Implicit learning with explicit retrieval? *Memory and Cognition, 21,* 146–155.
Baddeley, A.D., Lewis, V., Eldridge, M., & Thomson, N. (1984). Attention and retrieval from long-term memory. *Journal of Experimental Psychology: General, 113,* 518–540.
Baddeley, A.D., Logie, R., Bressi, S., Della Sala, S., & Spinnler, H. (1986). Dementia and working memory. *Quarterly Journal of Experimental Psychology, 38A,* 603–618.
Baddeley, A.D., & Wilson, B. (1986). Amnesia, autobiographical memory and confabulation. In D.C. Rubin (Ed.), *Autobiographical memory* (pp. 225–252). New York: Cambridge University Press.
Baddeley, A.D., & Wilson, B. (1988). Frontal amnesia and the dysexecutive syndrome. *Brain and Cognition, 7,* 212–230.
Burgess, P.W., & Shallice, T. (in press). Fractionnement du syndrome frontale. *Revue de Neuropsychologie.*
Cantor, J., & Engle, R.W. (1993). Working memory capacity as long-term memory activation: An individual differences approach. *Journal of Experimental Psychology: Learning, Memory, and Cognition, 5,* 1101–1114.
Cattell, R.B., & Cattell, A.K.S. (1960). *Handbook for the individual or group Culture Fair Intelligence Test.* Champaign, IL: Testing Inc.
Conway, A.R.A., & Engle, R.W. (1994). Working memory and retrieval: A resource-dependent inhibition model. *Journal of Experimental Psychology: General, 123,* 354–373.
Craik, F.I.M. (1995). *Should PET change our views on the frontal lobes and memory?* Presented at Memory Disorders Research Society Meeting, King's College, August.
Craik, F.I.M., Anderson, N., Kerr, S.A., & Li, K.H. (1995). Memory changes in normal ageing. In A.D. Baddeley, B.A. Wilson, & F.N. Watts (Eds.), *Handbook of memory disorders* Chichester: John Wiley.
Crowder, R.G. (1993). Systems and principles in memory theory: Another critique of pure memory. In A.F. Collins, S.E. Gathercole, M.A. Conway, & P.E. Morris (Eds.), *Theories of memory* (pp. 139–161). Hove: Lawrence Erlbaum Associates, Ltd.
Dalrymple-Alford, J.C., Kalders, A.S., Jones, R.D., & Watson, R.W. (1994). A central executive deficit in patients with Parkinson's disease. *Journal of Neurology, Neurosurgery and Psychiatry, 57,* 360–367.
Daneman, M., & Carpenter, P.A. (1980). Individual differences in working memory and reading. *Journal of Verbal Learning and Verbal Behavior, 19,* 450–466.
Della Sala, S., Baddeley, A., Papagno, C., & Spinnler, H. (in press). Dual-task paradigm: A means to examine central executive. *Annals of the New York Academy.*
Della Sala, S., Baddeley, A., Papagno, C., & Spinnler, H. (in preparation). *Dual task performance in dysexecutive and non-dysexecutive patients with a focal frontal lesion.*

Della Sala, S., Gray, C., Spinnler, H., & Trivelli, C. (in preparation). *The riddle of executive testing*.

Dienes, Z., Broadbent, D.E., & Berry, D.C. (1991). Implicit and explicit knowlege bases in artificial grammar learning. *Journal of Experimental Psychology: Learning, Memory, and Cognition, 17*, 875–887.

Duncan, J. (1986). Disorganisation of behaviour after frontal lobe damage. *Cognitive Neuropsychology, 3*, 271–290.

Duncan, J. (1993). Selection of input and goal in the control of behaviour. In A. Baddeley & L. Weiskrantz (Eds.), *Attention: Selection, awareness and control* (pp. 171–187). Oxford: Clarendon Press.

Duncan, J., Johnson, R., Swales, M., & Freer, C. (in preparation). *Frontal lobe deficits after head injury: Unity and diversity of function*.

Engle, R.W. (in press). Working memory and retrieval: An inhibition-resource approach. In J.T.E. Richardson (Ed.), *Working memory in human cognition*. New York: Oxford University Press.

Ericsson, K.A., & Kintsch, W. (1995). Long-term working memory. *Psychological Review, 102*, 211–245.

Gathercole, S.E., & Baddeley, A.D. (1994). *Working Memory and Language*. Hove: Lawrence Erlbaum Associates, Ltd.

Gick, M.L., Craik, F.I.M., & Morris, R.G. (1988). Task complexity and age differences in working memory. *Memory and Cognition, 16*, 353–361.

Greene, J., Baddeley, A.D., & Hodges, J. (in press). Autobiographical memory and executive function in early dementia of Alzheimer type. *Neuropsychologia*.

Hasher, L., & Zacks, R.T. (1988). Working memory, comprehension, and aging: A review and a new view. In G.H. Bower (Ed.), *The psychology of learning and motivation, Vol. 22* (pp. 193–225). New York: Academic Press.

Heim, A. (1975). *AH2 and AH3*. Windsor, Berks: NFER Publishing.

Johnson-Laird, P.N. (1983). *Mental models*. Cambridge: Cambridge University Press.

Just, M.A., & Carpenter, P.A. (1994). Unpublished paper presented at the Working Memory Conference, Cambridge.

Kyllonen, P.C., & Christal, R.E. (1990). Reasoning ability is (little more than) working-memory capacity. *Intelligence, 14*, 389–433.

Lezak, M.D. (1983). *Neuropsychological assessment*. New York: Oxford University Press.

Logie, R.H. (1995). *Visuo-spatial working memory*. Hove: Lawrence Erlbaum Associates, Ltd.

Morris, R.G. (1986). Short-term forgetting in senile dementia of the Alzheimer's type. *Cognitive Neuropsychology, 3*, 77–97.

Morris, R.G., Gick, M.L., & Craik, F.I.M. (1988). Processing resources and age differences in working memory. *Memory and Cognition, 16*, 362–366.

Nelson, H.E. (1976). A modified card-sorting task sensitive to frontal lobe defects. *Cortex, 12*, 313–324.

Norman, D.A., & Shallice, T. (1980). *Attention to action: Willed and automatic control of behavior*. University of California at San Diego, CHIP Report 99.

Perret, E. (1974). The left frontal lobe of man and the suppression of habitual responses in verbal categorical behaviour. *Neuropsychologia, 12*, 323–330.

Raaijmakers, J.G.W., & Shiffrin, R.M. (1981). Search of associative memory. *Psychological Review, 88*, 93–134.

Rabbitt, P.M.A. (1983). How can we tell whether human performance is related to chronological age? In D. Samuel, S. Alegeri, S. Gershon, V.E. Grimm, & G. Teffano (Eds.), *Aging of the Brain* (pp. 9–18). New York: Raven Press.

Robbins, T.W., Anderson, E.J., Barker, D.R., Bradley, A.C., Fearneyhough, C., Gillespie, P.H., Henson, R., Hudson, S.R., & Baddeley, A.D. (in press). Working memory in chess. *Memory and Cognition*.

Robertson, I.H., Ward, T., & Ridgeway, V. (1994). *The Test of Everyday Attention*. Flempton: Thames Valley Test Company.

Roediger, H.L. (1993). Learning and memory: Progress and challenge. In D.E. Meyer & S. Kornblum (Eds.), *Attention and performance XIV: Synergies in experimental psychology, artificial intelligence, and cognitive neuroscience* (pp. 509–528). Cambridge, MA: MIT Press.

Rylander, G. (1939). Personality changes after operation on the frontal lobes. *Acta Psychiatrica Neurologica* (Supplement No. 30).

Salthouse, T.A. (1991). Mediation of adult age differences in cognition by reductions in working memory and speed of processing. *Psychological Science, 2*, 179–183.

Salthouse, T.A. (1992). *Mechanisms of age-cognition relations in adulthood*. Hillsdale, NJ: Lawrence Erlbaum Associates, Inc.

Shallice, T. (1982). Specific impairments of planning. *Philosophical Transaction of the Royal Society London, B, 298*, 199–209.

Shallice, T. (1988). *From neuropsychology to mental structure*. Cambridge: Cambridge University Press.

Shallice, T., & Burgess, P.W. (1991). Deficits in strategy application following frontal lobe damage in man. *Brain, 114*, 727–741.

Shallice, T., & Burgess, P. (1993). Supervisory control of action and thought selection. In A. Baddeley & L. Weiskrantz (Eds.) *Attention: Selection, awareness and control* (pp. 171–187). Oxford: Clarendon Press.

Shiffrin, R.M., & Schneider, W. (1977). Controlled and automatic human information processing: II. Perceptual learning, automatic attending and a general theory. *Psychological Review, 84*, 127–190.

Shute, V.J. (1991). Who is likely to acquire programming skills? *Journal of Educational Computing Research, 7*, 1–24.

Spinnler, H., Della Sala, S., Bandera, R., & Baddeley, A.D. (1988). Dementia, ageing and the structure of human memory. *Cognitive Neuropsychology, 5*, 193–211.

Sternberg, S. (1966). High speed scanning in human memory. *Science, 153*, 652–654.

Thurstone, L. (1938). Primary mental abilities. *Psychometric Monograph, 1*.

Turner, M.L., & Engle, R.W. (1989). Is working-memory capacity task-dependent? *Journal of Memory and Language, 28*, 127–154.

Welford, A.T. (1958). *Ageing and human skill*. London: Oxford University Press.

Wilkins, A.J., Shallice, T., & McCarthy, R. (1987). Frontal lesions and sustained attention. *Neuropsychologia, 25*, 359–365.

Wilson, B.A., & Baddeley, A.D. (1988). Semantic, episodic and autobiographical memory in a post-meningitic amnesic patient. *Brain and Cognition, 8*, 31–46.

Yntema, D.B., & Mueser, G.E. (1960). Remembering the present states of a number of variables. *Journal of Experimental Psychology, 60*, 18–22.

16 Dementia and working memory

*A. D. Baddeley, R. Logie, S. Bressi,
S. Della Sala and H. Spinnler*

This study explored the hypothesis that patients suffering from dementia of the Alzheimer type (DAT) are particularly impaired in the functioning of the Central Executive component of working memory, and that this will be reflected in the capacity of patients to perform simultaneously two concurrent tasks. DAT patients, age-matched controls and young controls were required to combine performance on a tracking task with each of three concurrent tasks, articulatory suppression, simple reaction time to a tone and auditory digit span. The difficulty of the tracking task and length of digit sequence were both adjusted so as to equate performance across the three groups when the tasks were performed alone. When digit span or concurrent RT were combined with tracking, the deterioration in performance shown by the DAT patients was particularly marked.

Introduction

Dementia was defined by the Committee of Geriatrics of the Royal Society of Physicians as follows:

> Dementia is a global impairment of higher cortical functions, including memory, the capacity to solve the problems of everyday living, the performance of learned perceptual motor skills and the correct use of social skills and control of emotional reactions, in the absence of gross clouding of consciousness.
>
> (Anon., 1981)

The condition is characteristically irreversible and progressive. Although dementia may result from a range of causes, the most common is Alzheimer's disease. One of the clearest and most sensitive indicators of Alzheimer's disease is provided by memory performance, where patients show both lapses in everyday memory and grossly impaired performance on a range of laboratory tasks (Miller, 1977). The present study attempts to provide a more detailed analysis of the breakdown in memory. It is hoped that such an analysis may be of interest at three distinct levels: (1) it may provide a better understanding of the disease, (2) it may provide new and more sensitive tests for the early detection of dementia, and (3) it may cast light on the functioning of normal memory, as has proved the case in other studies of memory deficits.

The memory deficit shown by patients suffering from Alzheimer's disease resembles that of both the classic amnesic and the head-injury patient in showing poor long-term learning but differs from these groups in the extent of the working memory deficit. Performance of demented patients is impaired on digit span (Miller, 1977), suggesting a superficial resemblance to STM deficit patients. However, demented patients differ from STM patients in showing a decrement in performance on the Corsi non-verbal memory span task (Spinnler, Della Sala, Bandera and Baddeley, submitted). This latter study also indicated that DAT patients show comparatively unimpaired performance on the recency component of free recall, an observation also made by Wilson, Baker, Fox and Kaszniak (1983), in contrast to STM patients who typically show impaired recency (Shallice and Warrington, 1970).

One framework for attempting to explain this pattern of results is offered by the working memory model of Baddeley and Hitch (1974). This suggests that working memory comprises a controlling Central Executive system aided by a number of slave systems, with access to passive storage as reflected in the recency effect. It has been suggested (Baddeley, 1986) that such demented patients may be particularly impaired in the functioning of the Central Executive. Such a deficit would be likely to impair learning ability and also lead to poorer performance on span tasks that are assumed within the model to depend at least in part upon the adequate functioning of the Central Executive. It would also lead to an impairment in the capacity for scheduling two or more concurrent tasks. On the other hand, the recency effect would be much less likely to be impaired by a Central Executive deficit as it is based on passive storage and is comparatively uninfluenced by concurrent load (Baddeley and Hitch, 1977).

A recent study by Morris (1984) suggests that DAT patients show every sign of using the Articulatory Loop component of working memory, as indicated by the presence of clear effects of phonological similarity, word-length and articulatory suppression on span performance, coupled with an overall impairment in immediate memory span. Subsequent experiments by Morris (1986) indicate impaired performance on the Peterson and Peterson (1959) task in which subjects are required to maintain verbal material over a brief delay filled by a secondary distracting task. The capacity to store information while simultaneously processing a heavy cognitive load is assumed to be one of the functions of working memory, a function that would be expected to depend on the Central Executive component.

While a case can certainly be made for the view that patients suffering from Alzheimer's disease show a particular decrement in Central Executive function, testing this hypothesis is far from easy, largely because the detailed functioning of the Central Executive has not been adequately explored in normal subjects. Baddeley (1986) has suggested that the attentional control model posed by Norman and Shallice (1980) and applied to the study of frontal lobe deficits by Shallice (1982) might prove a suitable candidate for a model of the Central

Executive in working memory. The Norman and Shallice model assumes that behaviour is controlled at two levels, the first of these involving a series of ongoing programmes or schemata that typically run in parallel, with contention scheduling procedures available to resolve conflicts. Such programmes can, however, be initiated, terminated or modified by a higher-level Supervisory Activating System, which is necessary for initiating new behaviour, for making changes in ongoing activity and for resolving major conflicts that may occur in the concurrent performance of two or more activities. Such a system can, of course, be defective at a number of different points, and it is suggested that normal ageing, damage to the frontal lobes and dementia all affect the Central Executive in somewhat different ways (Baddeley, 1986).

It is, however, clearly the case that the sort of detailed analysis that is possible within the Articulatory Loop component of working memory is simply not yet available for investigating hypothetical deficits of the Central Executive. One response to this state of affairs is to delay the application of the executive component of the working memory model to the analysis of neuropsychological problems until it is more adequately established using normal subjects. This approach has the drawback that it would lead to the postponement of neuropsychological investigation of any but the simplest cognitive functions until sometime in the distant future when available models are regarded as sufficiently well developed. An alternative is to use the existing working memory model, despite its limitations, as a framework to explore the nature of the neuropsychological deficit. Provided that tests of the model are not too narrowly limited, they are likely to provide data that will shape the current model and at the same time yield information that is likely to be useful regardless of the ultimate fate of the working memory model.

The study of dementia encounters two further problems, that of specificity of diagnosis of dementia, and that of its degree of severity. Any sample of patients referred for diagnosis is likely to contain patients with a range of different aetiologies in addition to those suffering from Alzheimer's disease. These include rarer dementias such as Huntingdon's disease and normotensive hydrocephalus, as well as patients suffering from multi-infarct dementia, which results from multiple minimal brain lesions caused by a series of minor strokes. In addition to having different causes, such dementias may well produce different dysfunction, with the result that any clear analysis is likely to depend crucially on adequate diagnosis. This is far from easy, demanding information from a range of sources together with the skills of an experienced clinical neurologist. Even under such circumstances, a diagnosis is typically only finally confirmed after the patient has died, and his brain has been found to yield the characteristic pattern of neurofibrilary tangles and plaques; hence patients are typically diagnosed in terms of the somewhat tentative classification of "dementia of the Alzheimer's type" (DAT).

A second important variable in studying dementia is that of severity. The disease is typically progressive, developing from a pattern of initial deficits that may possibly be quite specific to one of massive and general deterioration.

Severe cases are likely to show decrement on virtually any test used and, indeed, are likely to have great difficulty in even understanding instructions. Such cases are unlikely to provide insight into the detailed nature of the underlying psychological deficit, and for that reason studies are likely to be more informative if they focus on mild to moderate cases. The present study therefore compares the performance of mild to moderate dements with a sample of non-demented patients of a similar age and background. A third young control group was also tested so as to allow a comparison between the effects of dementia and those of normal ageing.

The focus of the study was on the capacity for successfully coordinating performance on two parallel tasks. We chose pursuit tracking as our primary task. This allowed us to adjust the difficulty of the task so as to match performance across our groups and also allowed a continuous monitoring of performance. We combined tracking with each of three secondary tasks. The first of these was articulatory suppression, selected as being a task that would make minimal demands on the Central Executive. As suppression relies primarily on the Articulatory Loop, it would be unlikely to produce substantial direct interference with tracking, a task that is assumed to depend primarily on the Visuo-spatial Sketchpad component of working memory (Baddeley, 1986). However, although the information load of suppression is minimal, combining it with tracking does require the concurrent performance of two activities, and hence it is conceivable that performance would be impaired in DAT patients.

We chose as a more demanding secondary task simple reaction time to a tone. While simple reaction time is a relatively undemanding task, there is evidence that it will reflect the attentional capacity available to a subject (Posner, 1978), hence allowing the concurrent measurement of performance on both tasks, something that is hard to achieve in the case of articulatory suppression. While simple reaction time allows performance on the secondary task to be measured, it does not allow us to control task difficulty across groups, leaving open the possible objection that the reaction time is simply more difficult for the DAT patients than for the controls. We therefore opted for a third condition, in which task difficulty could be matched across groups. The task we used was concurrent digit span, a task that earlier work suggests does make significant demands on both the Central Executive of working memory and the Articulatory Loop system (Baddeley and Hitch, 1974).

Method

Experimental subjects

DAT patients

A total of 28 DAT patients were tested. They form part of the larger sample described by Della Sala, Nichelli and Spinnler (1986), referred to the Neurological Service in Milan over a three-year period. Of the 224 people referred, 129 were provisionally diagnosed as DAT patients on the basis of clinical history and neurological examination, combined with CT scan and laboratory data, which were used to exclude other possible dementing illnesses. Of the 129, 63 were excluded as being too severe. Severity was assessed on the basis of conventional cut-off points on the following pre-tests: Temporal orientation score on the Benton, Van Allen and Fogel (1964) test should be at least 50%. Orientation in terms of personal identity using the test devised by Della Sala and Spinnler (in press) should be at least 50%. Ability to provide appropriate information on family members should be at least 70% on the test devised by Della Sala, Nespoli, Ronchetti and Spinnler (1984). Finally, score on a scale of everyday coping ability should be at least 70%.

A second feature of dementia is that it should be progressive; hence patients were included only if they showed unequivocal evidence of deterioration as measured by neurological and psychological assessment over a period of at least six months. Other criteria for inclusion included availability and willingness to be tested and the capacity to read and write as measured informally. Patients with a history of other neurological or psychiatric diseases were excluded, as were patients with evidence of chronic progressive liver, kidney, lung or heart failure, history of alcohol abuse or evidence of having within the last 48 hours taken drugs affecting CNS functions. Finally, patients were excluded if they were not currently living in a family setting without need of special care, or if they did not live within Milan or its hinterland. When these criteria were applied, the initial sample of 224 referred for diagnosis reduced to a total of 28 DAT patients comprising 12 men and 16 women with a mean age of 64.9 years (SD = 7.0, range 51–80) and a mean of 9.5 years of education (SD = 4.3, range 5–17). The mean length of illness was 2.0 years (SD = 1.7, range 1–4).

The overall level of intellectual performance of the patients was further assessed using Raven's Matrices (Raven, 1954) to measure overall intelligence, the Token Test (De Renzi and Faglioni, 1977) as a measure of language performance, and the Street Gestalt Completion Test as a measure of perceptual performance (Street, 1931). Normative data are available for all of these, based on an ongoing study involving a sample of 321 normal subjects. Performance of the DAT patients on Raven's Matrices gave a mean of 18.0 (SD = 11.1), whereas the median score for normals was 28.5, and the fifth percentile cut-off point is 15.0. In the case of the Token Test, the patients scored a mean of 24.8 (SD = 7.8) as compared to a normal median of 33.0 and a fifth percentile cut-off of 26.5. Mean performance on the Street Test was 4.8 (SD = 3.2) for the patients, as against a median for normal subjects of 7.0 and a fifth percentile cut-off of 2.25. The profile of impairment across intelligence language and perception was relatively homogeneous in 15 cases, while 13 showed a more selective pattern of dysfunction. While all patients showed evidence of memory impairment, 12 showed additional evidence of a language deficit and 1 a particular impairment of spatial cognition.

Of the 28 DAT patients, 5 had a CT scan that was within the expected range for their age, 9 showed gross evidence of atrophy, and 14 showed minimal-to-moderate atrophy. Two patients showed atrophy that was more marked in the left hemisphere. On the standard neurological examination, 15 patients presented so-called release signs (chiefly tonic mouth reflex) or paratonia. Some degree of motor impersistence (Joynt, Benton and Fogel, 1962) occurred in 17 of the patients, while 2 patients presented with some degree of extrapyramidal rigidity.

Control groups

Two control groups were used, one matched on age and educational background with the DAT patients, the other comprising young subjects. All controls were free of evidence of present or past nervous, organic or physical disease that might be expected to impair cognitive performance and were free of alcohol or any other substance that might influence CNS performance. The 28 age-matched controls comprised 13 men and 15 women with mean age 64 years (SD = 4.5, range 57–72) and a mean of 9.2 years of education (SD = 3.7, range 5–17). The young controls comprised 9 men and 11 women with a mean age of 24.3 years (SD = 2.8, range 20–31). Their mean educational level was 15.8 years (SD = 1.9, range 12–17). This is, of course, substantially in excess of the two elderly groups, a difference that is due at least in part to the increase in number of statutory years of education required since the early years of this century. The absence of matching on this variable should be borne in mind in considering the performance of the young control group.

Procedure

Adaptive tracking

The basic tracking task involved presenting subjects with a 2 × 2 cm white square on a colour monitor. The square moved in random directions around the screen, and the subject's task was to follow the movement of the square by means of a light-sensitive pen. Whenever the pen moved off the square, the colour of the square changed to orange. When the pen was replaced on the square, the colour returned to white. In the adaptive version of the task, the square initially moved relatively slowly, and the speed gradually increased until the subject was unable to maintain the pen on target for more than 60% of the time. The speed of movement was always constant for a period of 20 sec before it was increased. The total time required on the task for this pretesting phase was variable as it depended upon the performance of each individual subject, but averaged about 4 min. The movement of the square and the monitoring of the light pen were controlled by an Apple II computer.

Primary tracking task

The basic tracking task was as for adaptive tracking, except that the speed at which the target moved around the screen was stable. Before continuing with the experiment, subjects were given three 20-sec periods of tracking with the level of difficulty set as described above. If subjects improved their performance above 60% time on target, the difficulty level was increased (by increasing the speed of movement of the target), and three further 20-sec trials were given at this new level. This procedure continued until the subject's performance appeared to stabilize between 40% and 60% time on target. This level of difficulty was then used for the primary tracking task throughout the remaining conditions and acted as a baseline level of performance on the tracking task performed alone. The tracking task was performed for 2 min on each trial, and the subject's performance was measured in terms of percentage time on target. In pilot tests it was found that tracking on a vertical screen for periods of more than a few seconds was physically tiring, particularly with the elderly. Thus the monitor was set into a table at an angle of 30° from the horizontal.

Adaptive tracking and tracking alone at a stable rate were completed prior to all other conditions. The subject was then introduced to each of the secondary tasks. However, the order of the tasks was counterbalanced within each subject group.

Secondary tasks

Articulatory Suppression. In this task, the subject was required to count from one to five repeatedly at a regular rate. Subjects were encouraged to maintain a rate of two per second throughout the testing period. The experimenter demonstrated this rate, and then subjects were given practice. This task was then combined with the primary tracking task for a 2-min session. The rate of suppression was not formally recorded, but where the suppression rate slowed noticeably, the experimenter encouraged the subject to speed up.

Reaction Time to Tones. In this task, the subject was presented with a series of tones from a loudspeaker. The task involved pressing a footswitch as rapidly as possible after commencement of the tone. The inter-tone interval was randomly varied between a maximum of 4 and a minimum of 6 sec, in order to ensure that the subject was unable to use the rhythmic cues associated with a regular rate. The number of missed tones and the correct response times were recorded by an Apple II computer. Subjects were given a few tones for practice, and then the task was performed for two 2-min trials, one alone and the other combined with the tracking task. Because of the random inter-tone interval, the number of tones in each 2-min period ranged between 23 and 25 tones, with a mean over all subjects of 24 tones.

Memory Span. Here subjects were first tested using a standard digit span procedure. The subject was presented with a list of digits at one digit per second and was asked for immediate ordered spoken recall. Initially only one digit was presented, and the number in each list was gradually increased by one item, with three lists at each length. Presentation ceased when the subject was unable to recall two of the three sequences at a given length, and span was taken to be the previous list length. Subjects were then presented with digit lists at their own span for a period of 2 min alone and for 2 min combined with the primary tracking task. Performance was measured by the percentage of sequences that were recalled completely correctly. The number of lists presented in 2 min was dependent on the individual span of each subject and ranged from 11 to 15 sequences.

Results

The primary interest of the study lies in the comparison between DAT patients and age-matched controls, hence this will be considered first. The comparison of the performance of the normal elderly and young groups provides useful background information but is of secondary importance and will be considered later. In comparing the various groups, we shall begin by considering overall performance on the primary tracking task, asking two preliminary questions: (1) does tracking performance vary as a function of secondary task, and (2) if so, do such differences interact with subject group? As explained in the introduction, a working-memory hypothesis would predict both of these, namely a disruption of tracking by secondary task and an enhancement of this disruption in the case of the DAT patients. Should these predictions by upheld in the overall analysis, then a more detailed examination of performance on both primary and secondary tasks will be made in turn for the effects of suppression, concurrent reaction time and concurrent digit span.

One further point should be considered before discussing the analysis of results. Performance of our three groups of subjects on the tracking task was equated by varying the speed at which the target moved. This avoids the problem of comparing performance at markedly different time-on-target scores but introduces the further problem as to whether a 5% decrement at the fast rate is equivalent to a 5% drop in time-on-target at the slow rate achieved by the DAT patients. We make the assumption that it is, since we see no way of comparing our groups meaningfully without such an assumption. In theory, our assumption could be tested by plotting performance operating characteristic curves for our three groups. In practice this is not feasible, for two reasons. First, the amount of data necessary to established reliable functions would place unrealistic demands on the attentional capacity of our DAT patients. Secondly, the plotting of such functions assumes that the same strategy will be used across the function, an assumption that we feel is highly questionable in the case of our demented patients (for a further discussion of this point see Baddeley, Eldridge, Lewis and Thomson, 1984).

We shall therefore argue as follows: If our DAT patients under control conditions show a time-on-target performance equivalent to, or better than, the control groups and, when a secondary task is added, decline to a level below that of the controls, then a prima facie case has been established for their greater vulnerability to secondary task interference. While the task concerned is different, it involves a more slowly moving target, which, other things being equal, ought to be less susceptible to interference that the rapidly moving target experienced by the control groups.

Figure 1 shows the mean overall percentage of time on target for the three subject groups as a function of condition. An analysis of variance was carried out involving the elderly and DAT groups and the four conditions. There proved to be no overall effect of subject group ($F<1$) but significant effects of conditions [$F(3, 162) = 22.46$, $p<0.001$] and a significant interaction between subject group and conditions [$F(3, 162) = 4.36$, $p<0.01$]. When the two groups were examined individually using one-way analysis of variance, the normal elderly subjects showed an overall effect of conditions [$F(3, 81) = 4.55$, $p<0.01$], and a further analysis using the Newman–Keuls test indicated that the only significant difference was between performance on

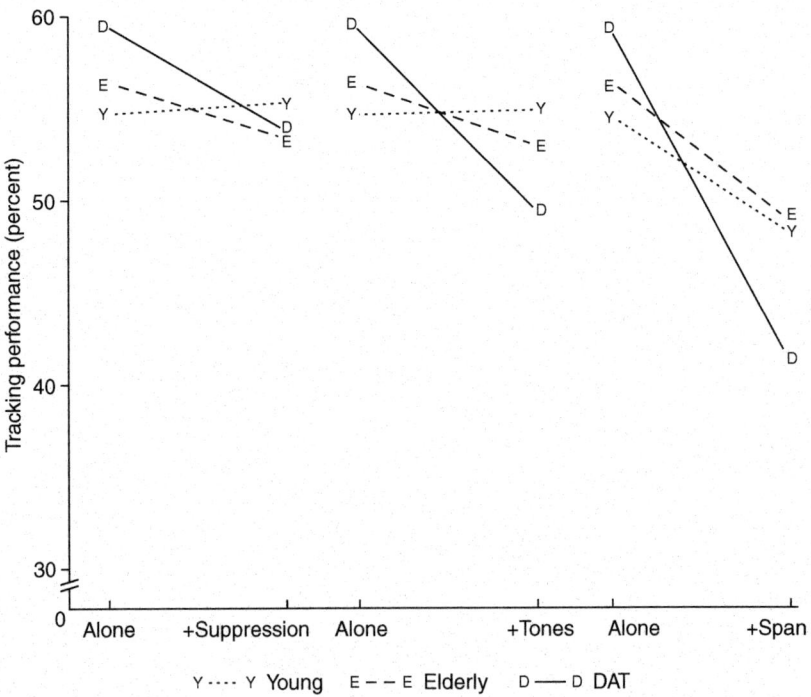

Figure 1 Adaptive tracking performance of senile dementia patients (DAT) and elderly and young controls when tracking is performed alone and when it is combined with concurrent articulatory suppression, RT to a tone or auditory digit span.

the tracking task alone and performance when tracking was combined with span ($p<0.01$). When an equivalent analysis was performed on the DAT patients, there again proved to be a significant effect of conditions [$F(3, 81)=19.28$, $p<0.001$] but in this case tracking alone led to significantly higher performance than tracking with suppression ($p<0.05$) or than tracking with concurrent RT ($p<0.01$) or tracking with concurrent span ($p<0.01$). Having obtained the predicted interaction between dementia and secondary task effects, we felt justified in performing a more detailed analysis of the individual tasks.

Tracking and suppression

Mean tracking performance with and without suppression is shown in Figure 1. An analysis of variance of percentage of time on target for the DAT and age-control patients indicated no overall difference between the groups ($F<1$), but a significant effect of suppression [$F(1, 54)=4.71$, $p<0.05$] and no interaction ($F<1$).

Tones and tracking

Figure 1 shows overall performance on tracking and the effect of adding the RT task. Analysis of variance of time on target on the primary task again indicated no difference between the groups ($F<1$), but a significant effect of the RT task on tracking [$F(1, 54)=23.53$, $p<0.001$] and a significant interaction [$F(1, 54)=6.13$, $p<0.05$]. Reaction time and error data are shown in Table 1. Three of the DAT group missed all of the tones while tracking, and their data were excluded from the reaction time analysis. Analysis of variance of the reaction time data suggested an overall difference in reaction time between the two groups [$F(1, 51)=55.13$, $p<0.001$], together with a significant effect of combining tracking and RT [$F(1, 51)=44.56$, $p<0.001$] and a significant interaction between these two [$F(1, 51)=20.54$, $p<0.001$]. The elderly control subjects made virtually no errors on this task (mean = 0.18) in contrast to the DAT group, who omitted a mean of 2.79 responses under control conditions, and 9.82 when RT was combined with tracking.

Table 1 Effects of concurrent tracking on mean RT to tones[a]

	Young	Elderly controls	DAT patients
RT alone	381.4 (0)	418.0 (0.11)	721.3 (2.79)
RT while tracking	434.6 (0)	479.4 (0.25)	1041.9 (9.82)

Notes
Mean number of omission errors shown in brackets.
[a] In msec.

Tracking and digit span

Figure 1 shows performance on combined tracking and digit span. Analysis of variance on tracking scores indicated no significant overall effect of subject group ($F<1$) but a highly significant effect of condition [$F(1, 54) = 61.47$, $p<0.001$] and a significant interaction between subject group and condition [$F(1, 54) = 11.09$, $p<0.01$]. Performance on the memory task is shown in Table 2. Analysis of variance suggests that the DAT group made significantly more errors in the span task than did the elderly group [$F(1, 54) = 18.71$, $p<0.001$], and both groups made more errors when span was combined with tracking [$F(1, 54) = 219.74$, $p<0.001$]. There was also an interaction between groups and the dual task effect [$F(1, 54) = 34.11$, $p<0.001$].

Effects of normal ageing on performance

Overall analysis of tracking performance for the young and the elderly indicated a significant effect of condition [$F(3, 138) = 6.36$, $p<0.01$], but no overall difference between young and old ($F<1$) and no interaction between age and conditions ($F<1$). This general lack of an age effect on performance was also reflected in the more detailed analysis. There was no significant effect of articulatory suppression on performance ($F<1$), no effect of age ($F<1$) and no interaction ($F<1$). Analysis of the effects of the concurrent RT task indicated no overall effect on tracking of age ($F<1$), no significant overall effect of tones [$F(1, 46) = 1.42$, $p>0.05$] and no interaction between age and concurrent task effects [$F(1, 46) = 2.18$, $p>0.05$]. As Table 1 suggests, mean RT did not differ significantly between the young and the old [$F(1, 46) = 2.6$, $p>0.1$]. RT was, however, increased by concurrent tracking [$F(1, 46) = 33.81$, $p<0.001$], but this did not interact with age [$F(1, 46) < 1$]. Number of tonal signals missed did, however, show a difference between the young and the elderly, with the young never missing a signal, while the elderly had a mean miss rate that was low (0.18 out of 24), but nevertheless significantly greater than the young [$F(1, 46) = 4.74$, $p<0.05$]. Bearing in mind that the incidence is extremely low and that the distribution of responses clearly does not conform to the assumptions appropriate to analysis of variance, this difference should be treated with considerable caution. There was no significant interaction between age and condition [$F(1, 46) = 1.16$, $p<0.1$].

Table 2 Memory span and tracking

	Young	Elderly controls	DAT patients
Span task alone	0	2.42	1.79
Span task with tracking	13.76	20.39	43.11

Note
Mean percentage of erroneous sequences as a function of subject group.

Analysis of the tracking and concurrent span data indicated a significant overall effect of span on tracking [$F(1, 46) = 20.07$, $p < 0.001$] but no effect of age and no interaction between age and concurrent task ($F < 1$ in each case). Span performance is shown in Table 2. Analysis of variance indicated a highly significant effect of concurrent tracking on span [$F(1, 46) = 67.85$, $p < 0.001$], together with a small effect of age [$F(1, 46) = 4.28$, $p < 0.05$], but no interaction between age and concurrent tracking [$F(1, 46) = 1.19$, $p > 0.1$]. In general, then, there is clear evidence that combining tasks impairs performance for both groups, together with some minimal evidence for general age effects on performance, but no suggestion of an interaction between age and secondary task.

Discussion

We shall begin by discussing the principal comparison between DAT patients and age-matched controls. It is clear from the initial analysis that dements are more disrupted than controls by the addition of a secondary task, even when pains are taken to ensure that the primary tracking task is set at a comparable level of difficulty for DAT and control patients. However, interpretation of this predicted interaction is by no means unequivocal in some of the cases. In the case of articulatory suppression, the dual task effect is very small, even in the case of the DAT patients. Furthermore, since monitoring performance on the suppression task was not undertaken, we cannot rule out some form of differential trade-off between tasks, though this does not seem a likely explanation of the difference between DAT patients and controls.

The data from combining tracking and RT are somewhat more compelling. Performance on tracking shows a crossover interaction indicating that DAT patients are more dramatically impaired by concurrent RT than are controls. It is clear from performance on the RT task that this interaction does not reflect a differential trade-off between the two groups, since RT and errors both show an interaction between dementia and conditions, indicating that the DAT patients are more vulnerable to the demands of combining two tasks. Interpretation of this result, however, is constrained by the fact that DAT patients perform more poorly than the controls when RT is studied alone. This raises the possible objection that the secondary task was simply more difficult for the dements than the controls. If this is so, then our results may simply indicate that adding a difficult task to tracking impairs performance more than adding an easier task.

Fortunately such an objection cannot be applied to the third condition, since here the digit span task was adjusted so as to ensure that performance on the secondary task was equivalent for the dements and controls. Under these conditions we again observe a crossover interaction between performance on concurrent tracking and subject group, coupled with an equivalent interaction between condition and subject group for the secondary span task. In short, adding two tasks that are equated between the groups for difficulty produces a disproportionately large decrement in performance for the DAT patients. This is, of course, exactly what was predicted by the Central Executive interpretation, which suggests that DAT patients will have particular difficulty in integrating and coordinating two concurrent tasks.

In contrast to the very marked interactions between dementia and secondary task effects, we observed little or no differential disruption of performance in the elderly when compared to the young controls. This is particularly striking since the young controls are of higher educational level and hence might be expected to have some advantage over the elderly, beyond that afforded by differences in age. Two interpretations of this are possible. The first is that our study was, for one reason or another, relatively insensitive. While this is possible, our results certainly indicate that the study was sensitive enough to pick up differences among conditions, and of course clear effects of dementia.

A second interpretation is to conclude that the lack of difference between the young and old control subjects simply indicates that we have been successful in equating the difficulty of the various tasks for our subjects. If so, one might expect to observe differences between the elderly and the young in the level of difficulty on the tracking task that yielded the mean pretest matching score of 40–60% time on target. Examination of the data indicate that the elderly achieved the criterion tracking performance at a reliable lower difficulty level than did the young [$F(1, 46) = 4.54$, $p < 0.05$]. The DAT group, in turn, achieved 40–60% tracking performance at an even lower difficulty level than the elderly [$F(1, 54) = 31.08$, $p < 0.001$]. As Table 1 shows, the elderly also showed longer mean RTs than the young. It appears to be the case, then, that the elderly and the young do differ in cognitive capacity, but that when performance on the two concurrent tasks is equated between young and old, then combining them does not lead to any greater decrement in the elderly than in the young.

Our results are therefore consistent with the initial hypothesis that DAT patients are particularly impaired in the operation of the Central Executive, and that this system is important for integrating the performance of two or more concurrent tasks. As mentioned earlier, such a prediction is not, of course, peculiar to a working memory interpretation of dementia. A sceptic might, for example, suggest that anything that makes a task more difficult will differentially penalize DAT patients. Such an interpretation does, of course, leave open the question of how difficulty should be specified, and as such runs the risk of circularity.

It seems unlikely that such a "difficulty" hypothesis can confidently be rejected on the basis of any single experiment. Consider, for example, the following hypothetical experiment in which a primary task such as tracking is combined with two other tasks, A and B. It might well be the case that when performed alone, A led to shorter RTs than B, but when combined with tracking, it produced more impairment than B. An advocate of the difficulty hypothesis could then argue that the "difficulty" came from the problem of combining tasks rather than from the difficulty of the tasks per se. Such a result would not therefore refute the "difficulty" interpretation but would force it in the direction of beginning to specify the nature of the underlying processes.

Having an existing model such as that of working memory will, we believe, help us to look for patterns of results that will increasingly constrain the interpretation of the effects of dementia on performance, just as using the complex web of results on the role of phonological coding in immediate memory has allowed us to develop and refine the articulatory loop hypothesis. It is, however, only fair to point out that the same argument could be used to defend, for example, Craik's (1984) interpretation of the decline of memory with age in terms of reduced "processing resources". We find a working memory framework useful for conceptualizing our results but cannot yet claim that it gives a substantially better account of dementia than other available models. Few, if any, current models are sufficiently precise to be unequivocally supported, or refuted, by our data.

What, finally, are the possible practical implications of our results? One of the problems facing any attempt to treat dementia is the need to detect it an an early stage. Whatever the underlying cause of dementia, it seems likely that any treatment that is devised will be most effective if provided during the early stages, before the occurrence of profound neural and intellectual deterioration. This calls for the development of tests that are sensitive to dementia and capable of screening out the patients suffering from dementia from those exhibiting the normal signs of ageing. Our secondary task approach seems promising in this respect, since it combines considerable sensitivity to the effects of DAT while being relatively impervious to the effects of normal ageing. However, dual task techniques tend to be inherently more complex logistically than single tasks. It remains to be seen whether sufficient increase in sensitivity occurs to offset these potential logistic costs.

References

Anon. (1981). Organic mental impairment of the elderly. *Journal of the Royal College of Physicians of London*, **15**, 142–167.

Baddeley, A. D. (1986). *Working memory*. London: Oxford University Press.

Baddeley, A. D., Eldridge, M., Lewis, V. and Thomson, N. (1984). Attention and retrieval from long-term memory. *Journal of Experimental Psychology: General*, **113**, 518–540.

Baddeley, A. D. and Hitch, G. J. (1974). Working memory. In G. Bower (Ed.), *Recent advances in learning and motivation, Vol. VIII*. New York: Academic Press, pp. 47–90.

Baddeley, A. D. and Hitch, G. J. (1977). Recency re-examined. In S. Dornic (Ed.), *Attention and performance VI*. Hillsdale, N.J.: Lawrence Erlbaum Associates, pp. 647–667.

Benton, A. L., Van Allen, W. M. and Fogel, M. L. (1964). Temporal orientations in cerebral disease. *Journal of Nervous and Mental Diseases*, **139**, 110–119.

Craik, F. I. M. (1984). Age differences in remembering. In L. R. Squire and N. Butters (Eds.), *Neuropsychology of memory*. New York: Guildford Press, pp. 3–12.

Della Sala, S., Nespoli, A., Ronchetti, E. and Spinnler, H. (1984). Does chronic liver failure lead to chronic mental impairment? In *Advances in hepatic encephalopathy and urea cycle diseases*. Basle: Targer, pp. 448–456.

Della Sala, S., Nichelli, P. and Spinnler, H. (1986). An Italian series of patients with organic dementia. *Italian Journal of Neurological Science*, **7**, 27–41.

Della Sala, S. and Spinnler, H. (1986). Indifference Amnesique in a case of global amnesia following acute brain hypoxia. *European Neurology*, **25**, 98–109.

De Renzi, E. and Faglioni, P. (1977). Normative data and screening power of a shortened version of the Token Test. *Cortex*, **13**, 424–433.

Joynt, R. J., Benton, A. L. and Fogel, M. L. (1962). Behavioral and pathological correlates of motor impersistence. *Neurology*, **12**, 876–881.

Miller, E. (1977). *Abnormal ageing*. New York: Wiley.

Morris, R. (1984). Dementia and the functioning of the articulatory loop system. *Cognitive Neuropsychology*, **1**, 143–157.

Morris, R. (1986). Short-term forgetting in senile dementia of the Alzheimer's type. *Cognitive Neuropsychology*, **3**, 77–97.

Norman, D. A. and Shallice, T. (1980). Attention to action: Willed and automatic control of behavior. *University of California San Diego CHIP Report 99*.

Peterson, L. R. and Peterson, M. J. (1959). Short-term retention of individual verbal items. *Journal of Experimental Psychology*, **58**, 193–198.

Posner, M. I. (1978). *Chronometric explorations of mind*. Hillsdale, N.J.: Lawrence Erlbaum Associates.

Raven, J. C. (1954). *Progressive matrices*. Firenze: Edizione Italiana Organizzazione Speciale.

Shallice, T. (1982). Specific impairments of planning. *Philosophical Transactions of the Royal Society London B*, **298**, 199–209.

Shallice, T. and Warrington, E. K. (1970). Independent functioning of verbal memory stores: A neuropsychological study. *Quarterly Journal of Experimental Psychology*, **22**, 261–273.

Spinnler, H., Della Sala, S., Bandera, R. and Baddeley, A. (submitted). Dementia, ageing and the structure of human memory.

Street, R. F. (1931). *A gestalt completion test*. New York: Teachers' College Press.

Wilson, R. S., Baker, L. D., Fox, J. H. and Kaszniak, A. (1983). Primary memory and secondary memory in dementia of the Alzheimer type. *Journal of Clinical Neuropsychology*, **5**, 337–344.

Part VI

The episodic buffer

By the late 1990s it was becoming increasingly clear that the three component model was encountering problems in specifying the link with long-term memory. This was shown most clearly in accounting for the huge increase in verbal memory span between unrelated words (5) and connected prose (up to 15). This limitation was highlighted by the success in predicting performance on a wide range of cognitive tasks by a measure based on sentence memory, working memory span. This stemmed from a seminal experiment by Daneman and Carpenter (1980) who were interested in the role of working memory in text comprehension. They devised a task that combined storage and processing by requiring participants to read out a series of sentences and then recall the last word of each. This and related measures proved to correlate with a wide range of cognitive capacities including reasoning and intelligence test performance.

Extensive efforts have subsequently been made to understand the nature of this correlation, notably by groups associated with Engle (Engle, Cantor & Carullo, 1992; Engle, Tuholski, Laughlin & Conway, 1999) and with Miyake (Miyake et al., 2000), typically using multivariate correlational techniques. While some progress has been made, it is probably true to say that the issue of how best to characterise the executive system remains unresolved. Of central importance to the three component model, however, is the question of exactly how data from working memory span studies can be incorporated, given that none of the three components is assumed to have sufficient storage capacity to hold even a single sentence unaided. The first chapter in this section describes the problems confronting the original working memory model and proposes a fourth component, the *episodic buffer*. Like our original 1974 paper it was invited in this case following a talk I gave at an international meeting. It is very speculative and would, I suspect have been rejected had I submitted it to a psychology journal. Somewhat to my surprise, it seemed to resonate with concerns of others at the time and received more that 5,000 citations.

However, we were concerned that the concept of an episodic buffer might simply provide a convenient but unproductive way of explaining awkward results. Graham Hitch, Richard Allen and I set out to demonstrate that the concept could be used constructively. We tested the hypothesis that it played a central role in binding information from different sources into episodes, initially proposing that this process would depend heavily on the central executive. We were successful in developing methods to investigate binding, an increasingly popular research area, but found no evidence for its dependence on the central executive, concluding instead that the buffer was an important but passive storage system that allowed already bound information from different sources to be combined into episodes which were then accessible to conscious awareness. Evidence for this view was first presented by Allen, Baddeley and Hitch (2006), and has been followed by a series of later studies, several of which are described in the second chapter in this section.

The final chapter was invited by the Annual Review of Psychology and gives a broad overview of my current views of working memory and its relation to other theories in the field.

References

Allen, R., Baddeley, A. D., & Hitch, G. J. (2006). Is the binding of visual features in working memory resource-demanding? *Journal of Experimental Psychology: General, 135*, 298–313.

Daneman, M., & Carpenter, P. A. (1980). Individual differences in working memory and reading. *Journal of Verbal Learning and Verbal Behaviour, 19*, 450–466.

Engle, R. W., Cantor, J., & Carullo, J. J. (1992). Individual differences in working memory and comprehension: A test of four hypotheses. *Journal of Experimental Psychology: Learning, Memory, and Cognition, 18*, 972–992.

Engle, R. W., Tuholski, S. W., Laughlin, J. E., & Conway, A. R. A. (1999). Working memory, short-term memory, and general fluid intelligence: A latent-variable approach. *Journal of Experimental Psychology: General, 128*, 309–331.

Miyake, A., Friedman, N. P., Emerson, M. J., Witzki, A. H., Howerter, A., & Wager, T. D. (2000). The unity and diversity of executive functions and their contributions to complex 'frontal lobe' tasks: A latent variable analysis. *Cognitive Psychology, 41*, 49–100.

Papers

1 Baddeley, A. D. (2000). The episodic buffer: A new component of working memory? *Trends in Cognitive Sciences, 4*, 417–423.
2 Baddeley, A. D., Allen, R. J., & Hitch, G. J. (2011). Binding in visual working memory: The role of the episodic buffer. *Neuropsychologia, 49*, 1393–1400.
3 Baddeley, A. (2012). Working memory, theories models and controversies. *The Annual Review of Psychology, 63*, 12.1–12.29.

17 The episodic buffer

A new component of working memory?

A. D. Baddeley

In 1974, Baddeley and Hitch proposed a three-component model of working memory. Over the years, this has been successful in giving an integrated account not only of data from normal adults, but also neuropsychological, developmental and neuroimaging data. There are, however, a number of phenomena that are not readily captured by the original model. These are outlined here and a fourth component to the model, the episodic buffer, is proposed. It comprises a limited capacity system that provides temporary storage of information held in a multimodal code, which is capable of binding information from the subsidiary systems, and from long-term memory, into a unitary episodic representation. Conscious awareness is assumed to be the principal mode of retrieval from the buffer. The revised model differs from the old principally in focussing attention on the processes of integrating information, rather than on the isolation of the subsystems. In doing so, it provides a better basis for tackling the more complex aspects of executive control in working memory.

Theoretical structures within cognitive science come in different forms, ranging from detailed mathematical or computational models of narrow and precisely defined phenomena, to broad theoretical frameworks that attempt to make sense of a wide range of phenomena and that leave open much of the more detailed specification. The purpose of such a framework is to represent what is currently known while at the same time prompting further questions that are tractable. This is likely either to extend the range of applicability of the model, or to increase its theoretical depth, subsequently leading to more precisely specified sub-models. The concept of working memory proposed by Baddeley and Hitch[1] provided such a framework for conceptualizing the role of temporary information storage in the performance of a wide range of complex cognitive tasks (see Box 1). It represented a development of earlier models of short-term memory, such as those of Broadbent[2], and Atkinson and Shiffrin[3], but differed in two ways. First it abandoned the concept of a unitary store in favour of a multi-component system, and second it emphasized the function of such a system in complex cognition, rather than memory *per se*.

Over the 25 years since the publication of our initial paper, the concept of working memory (WM) has proved to be surprisingly durable. In one form or another, it continues to be actively used within many areas of cognitive science, including mainstream cognitive psychology[4], neuropsychology[5], neuroimaging[6], developmental psychology[7] and computational modelling[8,9]. However, there have always been phenomena that did not fit comfortably within the Baddeley and Hitch model, particularly in its more recent form. An attempt to come to terms with these has led to a reformulation of the theoretical framework, which will be described below. The reformulation leads to the proposal of a new component of working memory, the 'episodic buffer'.

Problems for the current model

The phonological loop: limits and limitations

The phonological loop gives a reasonably good account of a wide range of data (see Box 2). There are, however, phenomena that do not seem to fit neatly into the picture without serious further modification. Consider, first, the effect of articulatory suppression, whereby the subject continues to utter an irrelevant word such as 'the', while attempting to remember and repeat back a visually presented sequence of numbers. According to the model, suppression should prevent the registration of visual material in the phonological loop, producing a devastating impact on subsequent recall. Suppression does have a significant effect, but by no means devastating; in a typical study, auditory memory span might drop from 7 to 5 digits[10]. Furthermore, patients with grossly impaired short-term phonological memory, resulting in an auditory memory span of only one digit, can typically recall about four digits with visual presentation[11]. How are such digits stored?

An obvious possibility is in terms of the visuospatial sketch-pad. However, the evidence indicates that this system is good at storing a single complex pattern, but not suited to serial recall[12]. Furthermore, if visual coding were involved then one might expect suppression to make recall performance very sensitive to effects of visual similarity. A recent study by Logie et al. does indeed show visual similarity effects[13]. They are, however, small and not limited to conditions of articulatory suppression.

Box 1 The concept of working memory

The term working memory is used in at least three different ways in different areas of cognitive science. It is used here, and in cognitive psychology generally to refer to a limited capacity system allowing the temporary storage and manipulation of information necessary for such complex tasks as comprehension, learning and reasoning (Refs a,b). In the animal learning laboratory the term refers to the storage of information across several trials performed within the same day, as demanded by tasks such as the radial arm maze (Ref. c). In artificial intelligence, production system architectures apply the term to the component, often unlimited in capacity, that is assumed to be responsible for holding the productions (Ref. d).

These three meanings are thus not interchangeable. Performance of rats on a the radial arm maze, for example, probably relies upon long-term memory (LTM), while the unlimited capacity of the working memory component typically assumed by production system architectures differs markedly from the capacity limitation assumed by most of the models proposed within cognitive psychology.

The multi-component model of working memory (WM) that forms the basis of this review developed from an earlier concept of short-term memory (STM), that was assumed to comprise a unitary temporary storage system. This approach was typified by the model of Atkinson and Shiffrin (Ref. e). However, their model encountered problems (1) in accounting for the relationship between type of encoding and LTM (Ref. f), (2) in explaining why patients with grossly defective STM had apparently normal LTM, and (3) in accounting for the effects of a range of concurrent tasks on learning, comprehending and reasoning (Ref. b).

Baddeley and Hitch proposed the three-component WM model (shown in Figure 1a) to account for this pattern of data. The model comprised an attentional control system, the 'central executive', aided by two subsidiary slave systems, the 'phonological loop' and the 'visuospatial sketchpad' (Ref. b). The loop is assumed to hold verbal and acoustic information using a temporary store and an articulatory rehearsal system, which clinical lesion studies, and subsequently neuroradiological studies, suggested are principally associated with Brodmann areas, 40 and 44 respectively. The sketchpad is assumed to hold visuospatial information, to be fractionable into separate visual, spatial and possibly kinaesthetic components, and to be principally represented within the right hemisphere (areas 6, 19, 40 and 47). The central executive is also assumed to be fractionable. Although it is less well understood, frontal lobe areas appear to be strongly implicated. An excellent recent overview of short-term and working memory is given by Gathercole (Ref. g).

Working memory and long-term memory were initially treated as quite separate because patients with clear short-term phonological deficits appear to have intact LTM (Ref. b). Subsequent research has shown that such patients do have specific deficits in long-term phonological learning, for example, learning the vocabulary of a new language (Ref. h). Further evidence based on the link between phonological loop performance and vocabulary level in children, suggests that the loop might have evolved to enhance language acquisition (Ref. h). As predicted by this supposition, patients with phonological loop deficits have great difficulty in acquiring novel vocabulary. It seems likely that a similar function is served by the visuospatial sketchpad, although there is as yet little investigation of this topic. If one accepts the hypothesis of an equivalent function for the sketchpad, possibly in acquiring visuospatial semantics, then the framework is modified to that shown in Figure 1b.

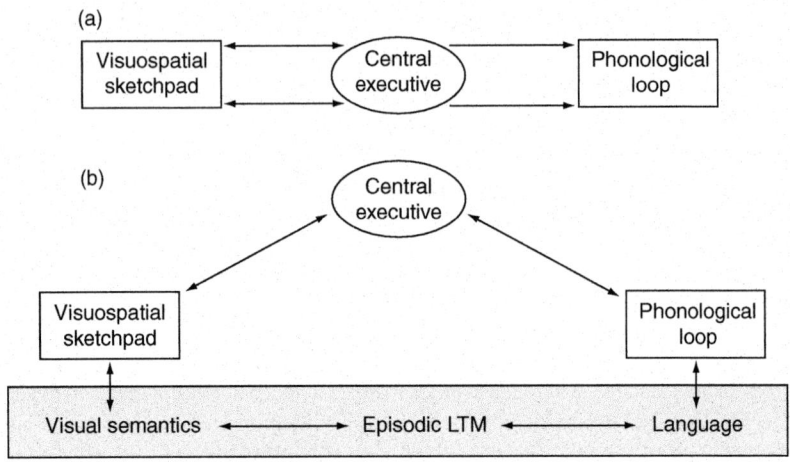

Figure 1 **(a) The initial three-component model of working memory proposed by Baddeley and Hitch (Ref. b)**. The three-component model assumes an attentional controller, the central executive, aided by two subsidiary systems, the phonological loop, capable of holding speech-based information, and the visuospatial sketchpad, which performs a similar function for visual information. The two subsidiary systems themselves form active stores that are capable of combining information from sensory input, and from the central executive. Hence a memory trace in the phonological store might stem either from a direct auditory input, or from the subvocal articulation of a visually presented item such as a letter. **(b) A further development of the WM model**. It became clear that the phonological loop plays an important role in long-term phonological learning, in addition to short-term storage. As such it is associated with the development of vocabulary in children, and with the speed of acquisition of foreign language vocabulary in adults. The shaded areas represent 'crystallized' cognitive systems capable of accumulating long-term knowledge (e.g. language and semantic knowledge). Unshaded systems are assumed to be 'fluid' capacities, such as attention and temporary storage, and are themselves unchanged by learning, other than indirectly via the crystallized systems (Ref. i).

References

a Miller, G.A. *et al.* (1960) *Plans and the Structure of Behavior*, Holt
b Baddeley, A.D. and Hitch, G. (1974) Working memory. In *The Psychology of Learning and Motivation* (Bower, G.A., ed.), pp. 48–79, Academic Press
c Olton, D.S. *et al.* (1980) Hippocampal function: working memory or cognitive mapping? *Physiol. Psychol.* 8, 239–246
d Newell, A. and Simon, H.A. (1972) *Human Problem Solving*, Prentice-Hall
e Atkinson, R.C. and Shiffrin, R.M. (1968) Human memory: a proposed system and its control processes. In *The Psychology of Learning and Motivation: Advances in Research and Theory* (Spence, K.W., ed.), pp. 89–195, Academic Press

f Craik, F.I.M. and Lockhart, R.S. (1972) Levels of processing: a framework for memory research. *J. Verbal Learn. Verbal Behav.* 11, 671–684
g Gathercole, S.E. (1999) Cognitive approaches to the development of short-term memory. *Trends Cognit. Sci.* 3, 410–419
h Baddeley, A.D. *et al.* (1998) When long-term learning depends on short-term storage. *J. Mem. Lang.* 27, 586–595
i Cattell, R.B. (1963) Theory of fluid and crystallized intelligence: a critical experiment. *J. Educ. Psychol.* 54, 1–22

This effect of visual similarity on span for verbal materials presents something of a problem; given that it occurs under standard non-suppressed conditions, it indicates that visual and phonological information are combined in some way. The current WM model has no mechanism that allows this, given that the central executive lacks storage capacity. The data suggest the need for some form of 'back-up store'[14] that is capable of supporting serial recall, and presumably of integrating phonological, visual and possibly other types of information.

Box 2 The phonological loop

The phonological loop is probably the best developed component of the working memory model. It is assumed to comprise a temporary phonological store in which auditory memory traces decay over a period of a few seconds, unless revived by articulatory rehearsal. The loop is assumed to have developed on the basis of processes initially evolved for speech perception (the phonological store) and production (the articulatory rehearsal component). It is particularly suited to the retention of sequential information, and its function is reflected most clearly in the memory span task, whereby a sequence of items such as digits must be repeated back immediately in the order of presentation. Digit span, the maximum number of digits that can be retained perfectly on 50% of occasions (typically about seven), is assumed to be determined jointly by the durability of the memory trace, and the time required to refresh the trace by subvocal rehearsal.

This model gives a simple account of the following phenomena:

(1) The phonological similarity effect
Items such as letters or words that are similar in sound are harder to remember accurately (e.g. the sequence *g, c, b, t, v, p* is harder than *f, w, k, s, y, q*), whereas visual or semantic similarity has little effect (Refs a,b). This implies an acoustic or phonological code.

(2) The word-length effect
Subjects find it easier to recall a sequence of short words (e.g. *wit, sum, harm, bag, top*) than long words (*university, aluminium, opportunity, constitutional, auditorium*). It takes longer to rehearse the polysyllables, and to produce them during recall. This allows more time for the memory trace to deteriorate (Ref. c).

(3) The effect of articulatory suppression

When subjects are prevented from rehearsing the items to be remembered, by being required to recite continuously an irrelevant sound such as the word 'the', performance declines markedly. Suppression also removes the effect of word length; if the words are not verbally rehearsed, it does not matter how long they take to articulate (Ref. c).

(4) Transfer of information between codes

Because of the efficiency of the phonological store in serial recall, adult subjects typically opt to name and subvocally rehearse visually presented items, thereby transferring the information from a visual to an auditory code. Articulatory suppression prevents this: it removes the effect of phonological similarity for visually presented items, but not for auditory, as these are automatically registered in the phonological store (Ref. d).

(5) Neuropsychological evidence

Patients with a specific deficit in phonological STM behave as if their phonological store is defective. The articulatory rehearsal process is defective in aphasic patients with dyspraxia, because they are unable to set up the speech motor codes necessary for articulation (Ref. e). Dysarthric patients whose speech problems are peripheral, however, show a normal capacity for rehearsal, suggesting that it is the central rehearsal code, rather than its overt operation, that is crucial for rehearsal (Ref. f).

References

a Conrad, R. and Hull, A.J. (1964) Information, acoustic confusion and memory span. *Br. J. Psychol.* 55, 429–432

b Baddeley, A.D. (1966) Short-term memory for word sequences as a function of acoustic, semantic and formal similarity. *Q. J. Exp. Psychol.* 18, 362–365

c Baddeley, A.D. *et al.* (1975) Word length and the structure of short-term memory. *J. Verb. Learn. Verb. Behav.* 14, 575–589

d Murray, D.J. (1968) Articulation and acoustic confusability in short-term memory. *J. Exp. Psychol.* 78, 679–684

e Waters, G.F. *et al.* (1992) The role of high-level speech planning in rehearsal: evidence from patients with apraxia of speech. *J. Mem. Lang.* 31, 54–73

f Baddeley, A.D. and Wilson, B.A. (1985) Phonological coding and short-term memory in patients without speech. *J. Mem. Lang.* 24, 490–502

Prose recall

Further problems for the simple phonological loop hypothesis are presented by data on the recall of prose. If asked to recall a sequence of unrelated words, subjects typically begin to make errors once the number of words exceeds five or six. However, if the words comprise a meaningful sentence, then a span of 16 or more is possible[11]. This has of course been known for many years; it presents a good example of what Miller referred to as 'chunking'[15].

Additional information, typically from long-term memory (LTM), is used to integrate the constituent words into a smaller number of chunks with capacity being set by the number of chunks rather than the number of words. This emphasizes once again the question of how information from different sources is integrated, raising the additional question of where the chunks are stored: are they held in the phonological loop, in LTM, or in some third back-up store?

If they reside within the phonological loop, then a patient whose loop capacity is limited to a single word should show no advantage for sentential sequences. In fact, patient 'PV', who has a word span of one word, has a sentence span of five[16]. Note that this is far smaller than the fifteen or so that one would normally expect. Given that her LTM appears to be quite normal, this argues against the idea that chunking is a purely LTM phenomenon. This leaves a back-up store interpretation as the simplest interpretation, although some form of complex interaction between the loop and LTM cannot be ruled out. Such an interpretation would also, however, need to account for data on the recall of longer sequences of prose, where performance is measured in terms of gist rather than by verbatim recall.

The immediate recall of a prose paragraph, typically comprising some 15–20 'idea units', is an important component of many clinical measures of memory. Patients are asked to recall the passage immediately after hearing it, and again after a filled delay of some 20 minutes. Severely amnesic patients are uniformly bad at delayed prose recall. In some cases, however, immediate recall can be virtually normal[17]. Preserved immediate recall appears to demand preserved intelligence and/or the absence of impairment in the functioning of the central executive system[18].

The amount of recalled material far exceeds the capacity of the phonological loop, particularly in view of the fact that the process of recall would be likely to overwrite material already in the phonological store. An account in terms of the visuospatial sketchpad encounters problems, in terms of the limited capacity of the sketchpad, its unsuitability for serial recall, and the lack of any evidence that suggests that only imageable sequences can be recalled. A third possibility might appear to be storage within the central executive. However, this is assumed to be an attentional control system with no intrinsic storage capacity.

Our provisional interpretation of preserved immediate prose recall in densely amnesic patients was in terms of some form of temporary activation of LTM, along lines similar to those suggested by Ericsson and Kintsch[19] who proposed a form of long-term working memory. We agreed with their view that the comprehension of prose passage involves the activation of existing structures within long-term memory, ranging from those at the level of the word or phrase up to conceptual schemas, such as those proposed by Bartlett[20] and subsequently by Schank[21]. It is not clear, however, how simple reactivation of old knowledge is capable of creating new structures, which can themselves be manipulated and reflected upon. I could, for instance, ask you to consider the idea of an ice-hockey-playing elephant, something which you have presumably not encountered too frequently in the past. This would raise the question of how he would hold the stick, and what would be his best position; he could no doubt deliver a formidable body check, but might be even better in goal. In order to solve this team-selection problem, it is necessary to maintain and manipulate the relevant knowledge of elephants and ice hockey. This process could in principle occur in LTM. However, there is no evidence to indicate that patients with normal intelligence combined with grossly impaired LTM have any difficulty in creating the long-term representations necessary for either problem solving or immediate recall, an activity that appears to be limited more by executive processes[18]. Might this process of creating and manipulating mental models also be a useful role for some kind of back-up store?

Although the examples so far have all depended on the recall of prose, similar phenomena have been reported in other domains. For example, Tulving (pers. commun.) reports the claim of a densely amnesic patient to continue to be able to play a good game of bridge. When Tulving tested this, he observed that the patient was not only able to keep track of the contract, but also of which cards had been played; indeed, he and his partner won the rubber. Once again, we appear to have evidence for a temporary store that is capable of holding complex information, manipulating it and utilizing it over a time scale far beyond the assumed capacity of the slave systems of WM.

The problem of rehearsal

One feature of the initial phonological-loop model is its assumption of separable processes of storage and rehearsal, with the latter being broadly equivalent to uttering the material to be recalled subvocally (see Box 2). Strong support for this view came from the effect on serial recall of articulatory suppression and its interactions with the word-length effect (the fact that it is easier to recall a sequence of short words than long words). Alternative accounts of the word-length effect have been presented recently[22], but in my view have difficulty accounting for the interaction of length with articulatory suppression. More problematic for our interpretation of rehearsal in terms of subvocalization are the data suggesting that some kind of rehearsal occurs in children before they have acquired the adult subvocal rehearsal strategy[23]. Even more problematic is the difficulty of giving a good account of rehearsal in the visuospatial sketchpad, and of course, in the suggested back-up store. Hence, while not wishing to abandon the assumption of subvocal rehearsal, the evidence suggests that it may not be typical of other aspects of WM.

Subvocal rehearsal of materials such as a digit sequence offers two probable advantages over rehearsal in other modalities. The first of these concerns the fact that subjects can literally regenerate digits by speaking them, a process that also appears to be possible in a covert form, equivalent to running the speech output programme[24]. Secondly, digits and words involve existing lexical representations, allowing any deterioration in the memory trace to be repaired during the process of rehearsal. Thus, if I am recalling a digit sequence and remember one of the items as *-ive*, then I know the correct item has to be the digit *five*, and not *ive, mive* or *thrive*.

In the case of the visuospatial sketchpad, although there has been speculation as to the possible role of eye movements, there is no firm evidence for a specific output process equivalent to vocalization. In contrast to verbal memory studies, the material in most visual memory experiments does not comprise familiar shapes or objects, as these tend to be nameable, which would allow subjects to recode the material verbally so as to take advantage of the capacity of the phonological loop for storing serial order.

It therefore seems more plausible to assume some form of general rehearsal, which perhaps involves the sequential attention to the component of the material to be recalled. In the case of an abstract pattern, this might involve chunking into a number of subcomponents. In the case of a prose passage, it presumably involves attending to the structure that has been built in order to represent the passage as part of the process of comprehension[19]. Hence, although additional assumptions will need to be made about the process of rehearsal operating within the proposed back-up store, similar assumptions are already necessitated by the question of rehearsal in the sketchpad.

Consciousness and the binding problem

Although the question of conscious awareness was not tackled directly, the WM model was implicitly assumed to play a role in consciousness. For example, the visuospatial sketch-pad was assumed to be involved in the storage and manipulation of visual images, and the phonological loop to play an equivalent role in auditory–verbal imagery[25]. Baddeley and Andrade[26] recently attempted to explore this assumption in a series of experiments in which subjects were required to maintain auditory or visual images and rate their vividness, while at the same time performing tasks selected to disrupt selectively either the visuospatial or phonological slave systems. Our results indicated that the relevant systems were indeed involved in conscious awareness, but suggested in addition a substantial role for LTM and for the central executive. These results reinforced the view that aspects of conscious awareness can be studied empirically, with highly coherent results. In doing so, they supported the WM framework as a useful empirical tool, whilst at the same time exposing its limitations. The model had no means of storing the complex images other than the two slave systems, which clearly play a role in the maintenance of verbally cued images, but one that is not nearly as important as the contribution of LTM. Once again, this suggests the existence of a store that is capable of drawing information both from the slave systems and from LTM, and holding it in some integrated form.

Such a process was indeed implicit in an earlier attempt to give an account of conscious awareness in terms of the WM model. Baddeley[27] suggested that

WM might play an important role in solving the binding problem – the question of how information from a range of separate independent sensory channels is bound together to allow the world to be perceived as comprising a coherent array of objects. Such a process requires the integration of information regarding the perceived location, colour, movement, smell and tactile features of objects. It was suggested that the central executive played a crucial role in such integration, although neglecting the fact that the executive contained no short-term multi-modal store capable of holding such complex representations.

To summarize, although the visual and verbal slave systems of the WM model do offer a plausible account of a wide range of data, evidence from patients with short-term memory deficits, from the resistance in serial recall to articulatory suppression, and from the recall of prose, all suggest the need to assume a further back-up store. Evidence for the integrated storage of information from different modalities and systems comes from the small but significant impact of visual similarity on verbal recall, and from the very substantial impact of meaning on the immediate recall of sentences and prose passages. There is a clear need, therefore, to assume a process or mechanism for synergistically combining information from various subsystems into a form of temporary representation. Such a representation also offers a possible solution to the binding problem and the role of consciousness. The term 'episodic buffer' is proposed for this suggested fourth component of the working memory model.

The episodic buffer

The episodic buffer is assumed to be a limited-capacity temporary storage system that is capable of integrating information from a variety of sources. It is assumed to be controlled by the central executive, which is capable of retrieving information from the store in the form of conscious awareness, of reflecting on that information and, where necessary, manipulating and modifying it. The buffer is episodic in the sense that it holds episodes whereby information is integrated across space and potentially extended across time. In this respect, it resembles Tulving's concept of episodic memory[28]. It differs, however, in that it is assumed to be a temporary store that can be preserved in densely amnesic patients with grossly impaired episodic LTM. It is, though, assumed to play an important role in feeding information into and retrieving information from episodic LTM. The model of WM incorporating the episodic buffer is illustrated in Figure 2.

The component proposed is a buffer in that it serves as an interface between a range of systems, each involving a different set of codes. It is assumed to achieve this by using a common multi-dimensional code. The buffer is assumed to be limited in capacity because of the computational demand of providing simultaneous access to the necessarily wide range of different codes[29].

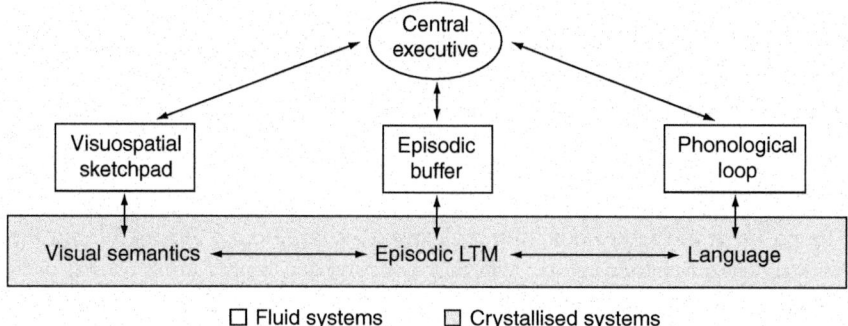

Figure 2 **The current version of the multi-component working memory model.** The episodic buffer is assumed to be capable of storing information in a multi-dimensional code. It thus provides a temporary interface between the slave systems (the phonological loop and the visuospatial sketchpad) and LTM. It is assumed to be controlled by the central executive, which is responsible for binding information from a number of sources into coherent episodes. Such episodes are assumed to be retrievable consciously. The buffer serves as a modelling space that is separate from LTM, but which forms an important stage in long-term episodic learning. Shaded areas represent 'crystallized' cognitive systems capable of accumulating long-term knowledge, and unshaded areas represent 'fluid' capacities (such as attention and temporary storage), themselves unchanged by learning.

The episodic buffer can be accessed by the central executive through the medium of conscious awareness. The executive can, furthermore, influence the content of the store by attending to a given source of information, whether perceptual, from other components of working memory, or from LTM. As such, the buffer provides not only a mechanism for modelling the environment, but also for creating new cognitive representations, which in turn might facilitate problem solving.

How is the buffer implemented biologically?

Of the various speculations as to the biological mechanism of binding, I would regard the process of synchronous firing as providing one promising hypothesis[30]. I would not expect the buffer to have a single unitary anatomical location; given its putative importance, some redundancy would be biologically useful in making the system more robust. However, frontal areas are very likely to be important for both the central executive and the episodic buffer, as evidenced by behavioural data on the operation of executive processes in general[31], together with fMRI studies implicating areas within the right frontal lobes in the capacity to combine two separate tasks. A recent fMRI study by Prabhakaran *et al.*[32] compared the retention of verbal and spatial information held in an integrated or unintegrated form. The results showed greater right frontal activation for integrated information, with unintegrated retention showing more posterior activation of areas previously implicated in verbal and spatial WM. Prabhakaran et al. conclude that 'The present fMRI results provide evidence for another buffer, namely, one that allows for temporary retention of integrated information' (Ref. 32, p. 89).

So what's new?

The proposal of the episodic buffer clearly does represent a change within the working memory framework, whether conceived as a new component, or as a fractionation of the older version of the central executive. By emphasizing the importance of coordination, and confronting the need to relate WM and LTM, it suggests a closer link between our earlier multi-component approach and other models that have emphasized the more complex executive aspects of WM. The revised framework differs from many current models of WM in its continued emphasis on a multi-component nature, and in its rejection of the suggestion that working memory simply represents the activated portions of LTM (Ref. 33). It also rejects the related view that the slave systems merely represent activations within the processes of visual and verbal perception and production[34,35]. Although WM is intimately linked both to LTM and to perceptual and motor function, it is regarded as a separable system involving its own dedicated storage processes.

Some outstanding issues

What problems are raised by the proposed new component of WM? I would suggest, first of all, that there is an immediate need to investigate its boundaries. Why not assign all memory to the episodic buffer, for example? I suggest that the existing evidence for the fractionation of WM, including that from neuropsychology and neuroradiology, indicates that this would be a thoroughly retrograde step[6]. It is less clear, however, whether one can draw a clear line between the two slave systems of WM and the episodic buffer. Fortunately, tools already exist for answering this kind of question, making use of neuropsychological cases with specific phonological or visuospatial memory deficits, and by using dual-task interference procedures. Such procedures are, of course, rarely process-pure, typically having at least some central executive component. However, by using the same secondary tasks to study both visuospatial and phonological memory, it is possible to tease apart the specific contribution from each slave system from the contribution of the executive component[26].

Separation of the episodic buffer from episodic LTM presents another problem. It seems likely that the study of selected neuropsychological patients will prove most productive, contrasting cases with executive deficit but preserved LTM with pure amnesic patients. Here and elsewhere, it seems likely that neuroradiological imaging could provide an additional valuable tool in association with the study of single case studies, and experimental conditions in which the LTM and WM component can be differentially controlled.

The episodic buffer emphasizes the integration of information, in contrast to earlier approaches to WM, which have focused on separating the various components. We need now to know much more about the role of executive processes in chunking. In connection with this, we need to be able to separate the relatively automatic binding of properties that occur in the processes of normal perception from the more active and attentionally demanding integrative processes that are assumed to play such an important role in the episodic buffer[29]. As Prabhakaran et al. have shown[32], neuroradiological measures have considerable potential for investigating the processes underlying this capacity, and the study of patients with executive deficits following frontal lobe damage also offers a promising line of investigation[31].

The suggestion that the episodic buffer forms the crucial interface between memory and conscious awareness places it at the centre of the highly active line of research into the role of phenomenological factors in memory and cognition. Tulving[28], for example, defines his concept of episodic memory explicitly in terms of its associated phenomenological experience of remembering. Although not all theorists would wish to place phenomenological experience so centrally, there is increasing evidence to suggest that conscious monitoring of the evidence supporting an apparent memory plays a crucial role in separating accurate recall from false memory, confabulation and delusion[36]. If the episodic buffer does indeed provide the storage, and the central execu-tive the underlying processing for episodic memory, then unravelling their complexities is likely to provide a fruitful and potentially tractable activity for many years to come.

Acknowledgements

This paper was based on a presentation at the Psychonomics Society meeting, Los Angeles, November 1999. I wish to thank Fergus Craik, Sergio Della Sala, Graham Hitch, Susan Gathercole, Christopher Jarrold, Endel Tulving and the anonymous referees for comments on an earlier draft. The support of MRC Grant G9423916 is gratefully acknowledged.

References

1 Baddeley, A.D. and Hitch, G.J. (1974) Working memory. In *The Psychology of Learning and Motivation* (Bower, G.A., ed.), pp. 47–89, Academic Press
2 Broadbent, D.E. (1958) *Perception and Communication*, Pergamon Press
3 Atkinson, R.C. and Shiffrin, R.M. (1968) Human memory: a proposed system and its control processes. In *The Psychology of Learning and Motivation: Advances in Research and Theory* (Spence, K.W., ed.), pp. 89–195, Academic Press
4 Hitch, G.J. and Logie, R.H. (1996) Working memory (Special Issue). *Q. J. Exp. Psychol.* 49A, 1–266

5 Becker, J.T. (1994) Working memory (Special Issue). *Neuropsychology* 8, Issue 4
6 Smith, E.E. and Jonides, J. (1996) Working memory in humans: neuropsychological evidence. In *The Cognitive Neurosciences* (Gazzaniga, M.S., ed.), pp. 1009–1020, MIT Press
7 deRibaupierre, A. and Hitch, G.J. (1994) The Development of Working Memory. *Int. J. Behav. Devel.* (Special Issue) 17, 1–200
8 Burgess, N. and Hitch, G.J. (1992) Towards a network model of the articulatory loop. *J. Mem. Lang.* 31, 429–460
9 Brown, L.D.A. and Hulme, C. (1996) Non-word repetition, STM and age-of-acquisition: a computational model. In *Models of Short-term Memory* (Gathercole, S.E., ed.), pp. 129–148, Psychology Press
10 Baddeley, A.D. *et al.* (1984) Exploring the articulatory loop. *Q. J. Exp. Psychol.* 36, 233–252
11 Baddeley, A.D. *et al.* (1987) Sentence comprehension and phonological memory: some neuropsychological evidence. In *Attention and Performance Vol. XII: The Psychology of Reading* (Coltheart, M., ed.), pp. 509–529, Lawrence Erlbaum
12 Phillips, W.A. and Christie, D.F.M. (1977) Components of visual memory. *Q. J. Exp. Psychol.* 29, 117–133
13 Logie, R.H. *et al.* (2000) Visual similarity effects in immediate serial recall. *Q. J. Exp. Psychol.* 53A, 3, 626–646
14 Page, M.P.A. and Norris, D. (1998) The primacy model: a new model of immediate serial recall. *Psychol. Rev.* 105, 761–781
15 Miller, G.A. (1956) The magical number seven, plus or minus two: some limits on our capacity for processing information. *Psychol. Rev.* 63, 81–97
16 Vallar, G. and Baddeley, A.D. (1984) Fractionation of working memory: neuropsychological evidence for a phonological short–term store. *J. Verbal Learn. Verbal Behav.* 23, 151–161
17 Wilson, B.A. and Baddeley, A.D. (1988) Semantic, episodic and autobiographical memory in a post–meningitic amnesic patient. *Brain Cognit.* 8, 31–46
18 Baddeley, A.D. Levels of working memory. In *Perspectives on Human Memory and Cognitive Aging: Essays in Honor of Fergus Craik* (Naveh Benjamin, M. *et al.*, eds), Psychology Press (in press)
19 Ericsson, K.A. and Kintsch, W. (1995) Long-term working memory. *Psychol. Rev.* 102, 211–245
20 Bartlett, F.C. (1932) *Remembering*, Cambridge University Press
21 Schank, R.C. (1982) *Dynamic memory*, Cambridge University Press
22 Neath, I. and Nairne, J.S. (1995) Word-length effects in immediate memory: overwriting trace-decay theory. *Psychonomic Bull. Rev.* 2, 429–441
23 Gathercole, S.E. and Hitch, G.J. (1993) Developmental changes in short-term memory: a revised working memory perspective. In *Theories of Memory* (Collins, A. *et al.*, eds), pp. 189–209, Lawrence Erlbaum
24 Hickok, G. and Poeppel, D. (2000) Towards a functional neuroanatomy of speech perception. *Trends Cognit. Sci.* 4, 131–138
25 Baddeley, A.D. and Logie, R.H. (1992) Auditory imagery and working memory. In *Auditory Imagery* (Reisberg, D., ed.), pp. 179–197, Lawrence Erlbaum
26 Baddeley, A.D. and Andrade, J. (2000) Working memory and the vividness of imagery. *J. Exp. Psychol. Gen.* 129, 126–145
27 Baddeley, A.D. (1993) Working memory and conscious awareness. In *Theories of Memory* (Collins, A.F. *et al.*, eds), pp. 11–28, Lawrence Erlbaum

28 Tulving, E. (1989) Memory: performance, knowledge and experience. *European J. of Cog. Psychol.* 1, 3–26
29 Hummel, J. (1999) The binding problem. In *The MIT Encyclopedia of the Cognitive Sciences* (Wilson, R.A. and Keil, F.C., eds), pp. 85–86, MIT Press
30 Singer, W. (1999) Binding by neural synchrony. In *The MIT Encyclopedia of the Cognitive Sciences* (Wilson, R.A. and Keil, F.C., eds), pp. 81–84, MIT Press
31 Roberts, A.C. et al. (1996) Executive and cognitive functions of the prefrontal cortex. *Philos. Trans. R. Soc. London Ser.* B 351, 1387–1527
32 Prabhakaran, V. et al. (2000) Integration of diverse information in working memory within the frontal lobe. *Nat. Neurosci.* 3, 85–90
33 Cowan, N. (1999) An embedded-processes model of working memory. In *Models of Working Memory* (Miyake, A. and Shah, P., eds), pp. 62–101, Cambridge University Press
34 Jones, D.M. (1993) Objects, streams and threads of auditory attention. In *Attention: Selection, Awareness and Control* (Baddeley, A.D. and Weiskrantz, L., eds), pp. 87–104, Clarendon Press
35 Allport, D.A. (1984) Auditory–verbal short-term memory and conduction aphasia. In *Attention and Performance Vol. X: Control of Language Processes* (Bouma, H. and Bouwhuis, D.G., eds), pp. 313–326, Lawrence Erlbaum
36 Conway, M.A. et al. (1996) Recollections of true and false autobiographical memories. *J. Exp. Psychol. Gen.* 125, 69–95

18 Binding in visual working memory
The role of the episodic buffer
A. D. Baddeley, R. J. Allen and G. J. Hitch

The episodic buffer component of working memory is assumed to play a central role in the binding of features into objects, a process that was initially assumed to depend upon executive resources. Here, we review a program of work in which we specifically tested this assumption by studying the effects of a range of attentionally demanding concurrent tasks on the capacity to encode and retain both individual features and bound objects. We found no differential effect of concurrent load, even when the process of binding was made more demanding by separating the shape and color features spatially, temporally or across visual and auditory modalities. Bound features were however more readily disrupted by subsequent stimuli, a process we studied using a suffix paradigm. This suggested a need to assume a feature-based attentional filter followed by an object based storage process. Our results are interpreted within a modified version of the multicomponent working memory model. We also discuss work examining the role of the hippocampus in visual feature binding.

The programme of research that we describe was motivated by an attempt to further develop our earlier multicomponent approach to working memory and in particular to explore the concept of an episodic buffer. The experiments that follow do not depend on such a framework for their relevance and validity, but a brief account of the broad theoretical framework will help to explain the particular questions we asked and our subsequent interpretations.

The term working memory is used very widely. In our own case we use it principally to refer to a broad framework of interacting processes that involve the temporary storage and manipulation of information in the service of performing complex cognitive activities. We take as our starting point the three component model initially proposed by Baddeley and Hitch (1974), together with the later proposal of a fourth component, the episodic buffer (Baddeley, 2000). The original model assumed a limited capacity controller, the central executive, supported by two temporary storage systems, the phonological loop and the visuo-spatial sketchpad. This three-component system has provided a useful framework for detailed investigations for many years, although the nature of the central executive has continued to demand further exploration. Our initial version assumed a system that could both store and manipulate information, effectively a homunculus, potentially so powerful that it was difficult to investigate.

We subsequently attempted to tackle this problem by abandoning the assumption that the central executive had the capacity to store information (Baddeley & Logie, 1999). However, it soon became clear that the temporary storage provided by the slave systems would not be sufficient to account for a range of functions that were assumed to be central to the operation of working memory. These include memory for prose, and the capacity for individual differences in working memory span to account for differences in performance levels across a wide range of capabilities, ranging from prose comprehension through the acquisition of programming skills to performance on standard intelligence test questions (Daneman & Carpenter, 1980; Engle & Kane, 2004). This led to the postulation of a fourth component, the episodic buffer (Baddeley, 2000, see Figure 1).

The episodic buffer was assumed to be a temporary multidimensional store that forms an interface between the subsystems of working memory, long-term memory and the central executive. As a multidimensional buffer, it allows a range of different sub-systems to interact, despite their being based on different codes, with a major function of the buffer being to bind together different sources of information to form integrated chunks. Its capacity is assumed to be limited by the number of episodes or chunks that it can hold. Here we agree with Cowan that the number is somewhere in the region of four and differs from one individual to another, as reflected in differences in working memory span (Cowan, 2005). The buffer is assumed to be accessible through conscious awareness, allowing it to provide the multi-feature binding mechanism that is often suggested as one of the principal functions of consciousness (Baars, 2002).

The episodic buffer concept has been useful in allowing the earlier three-component model to account for a much wider range of data, and to link in fruitful ways to other approaches to working memory (Baddeley, 2000,

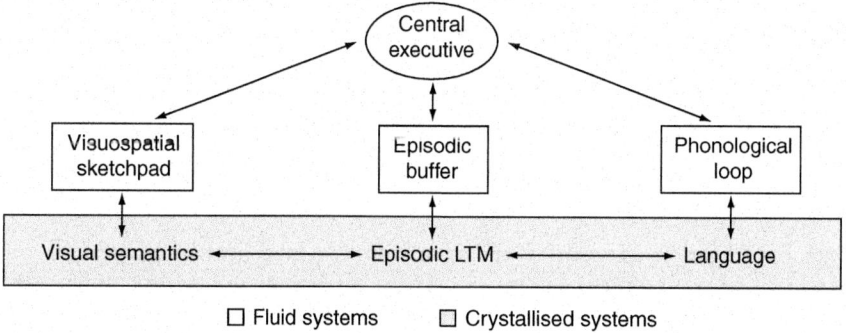

Figure 1 The multicomponent working memory model (Baddeley, 2000).

2007). In particular, our assumption that the proposed phonological and visuo-spatial subsystems feed in to a multidimensional common store, controlled by the central executive, is consistent with a range of other theoretical approaches (Barrouillet, Bernardin, & Camos, 2004; Cowan, 2005; Engle & Kane, 2004). Furthermore, in separating the attentional and storage functions of our earlier concept, it suggests a greater possibility of breaking down and analysing the central executive, thus moving away from an all powerful homunculus that could conveniently provide post hoc explanations, but would not prove productive in generating further understanding. The experiments that follow are part of a programme to tackle this issue. They reflect our decision to focus on the role of the episodic buffer in the binding of features into integrated episodes or chunks. We chose to focus on the interface between the central executive and the episodic buffer, combining dual task methodology with the methods already developed to study binding in visual working memory.

1 Exploring the episodic buffer

In creating the new model, it was tempting to propose a buffer that was directly connected with the whole of the cognitive system. Instead, on the grounds of parsimony, we began by assuming that everything within working memory accessed the buffer via the central executive, which was assumed to provide the processing power that could combine features into a single integrated representation. Consider, for example, a novel concept such as an ice hockey playing elephant, a concept that can certainly be generated, and relevant questions as to the strengths and weaknesses of such a novel player answered. We initially assumed that such processes would be heavily dependent upon the central executive. We set out to test this assumption using the dual task methodology that we have previously found so fruitful in the study of working memory, to systematically disrupt the three other components of the working memory framework with the prediction that blocking the central executive would specifically disrupt binding. We chose to test the generality of our conclusions by studying binding in two distinctive paradigms, namely the binding of features in visual working memory, and the binding of words in the comprehension and retention of prose. In line with the focus of this issue, we will center this review on the visual side of our ongoing research.

2 Binding in visual working memory

Our investigation of visual working memory gained greatly from a line of research on the interface between visual attention and visual short-term memory that was specifically focused on the binding of features into objects. An elegant series of experiments by Luck and colleagues (Luck & Vogel, 1997; Vogel, Woodman, & Luck, 2001) produced evidence for the automatic binding of features into objects. They concluded that storage capacity was limited to about four objects, regardless of the number of features comprising each object. Using a broadly similar approach, Wheeler and Treisman (2002), while accepting the concept of a limited capacity, proposed that the process of combining features into objects and maintaining them in visual working memory was itself attention-demanding. This approach differs principally in emphasizing limited capacity for attention as opposed to storage, and is close to our own initial hypothesis that the combination of features to form integrated representations in the episodic buffer is heavily dependent on executive processes.

A third possibility that we explore below, is that the initial binding of features into objects is automatic, but that subsequent maintenance of the bound features demands attention to prevent disruption from competing stimuli. This would explain why both Vogel et al. and Wheeler and Treisman find little evidence for a binding-specific attentional demand when a single probe is used, in contrast to Wheeler and Treisman's report of poorer retention of binding than of features when participants were required to scan a multi-target array at test, given that such scanning is likely to demand more attention than processing a single probe.

We decided to study the role of attention during encoding using a concurrent task procedure that allowed us to manipulate attentional load. Thus, participants perform feature and binding memory tasks while carrying out either simple or complex verbal tasks at the same time. The logic of this approach is that those conditions relying more on attentional support for successful performance will suffer to a greater extent when this support is disrupted by concurrent complex task performance. As the concurrent tasks are purely verbal in nature, any resulting interference can be attributed to central executive loading, rather than disruption of visuo-spatial processing.

Allen, Baddeley, and Hitch (2006) provided a first attempt to address the question of attentional involvement in visual feature binding, examining the binding of surface features (shape and color). We focused on this form of binding, rather than on feature-location association, because memory for location, either absolute or relative, appears to involve specialized processes and representational levels that are separable from those involved in processing and storing surface features and objects (e.g. Treisman & Gelade, 1980; Treisman & Zhang, 2006).

Allen et al. (2006) used an adaptation of the methodology developed by Wheeler and Treisman (2002) to assess feature and binding memory. Feature memory conditions involved the presentation of arrays of four shapes or colors (in the shape and color conditions respectively), followed by a single recognition probe. This probe shape (or color, depending on the condition) had been present in 50% of trials, while in the remaining 50%, it had not. These conditions were contrasted with an explicit test of binding memory. In this, four colored shapes were presented in each target array (Figure 2a), followed by a colored shape recognition probe. Crucially, while the features of the recognition probe had always been present in the original array, 50% of trials involved a feature recombination, with features drawn from two different target objects (see Figure 2f and g). This condition therefore required a judgment concerning whether the probed feature *combination* had been present in the original set, thus explicitly requiring binding memory for successful performance.

We found that requiring participants to count backwards in threes from a three-digit start number during the encoding phase of each trial did lead to a

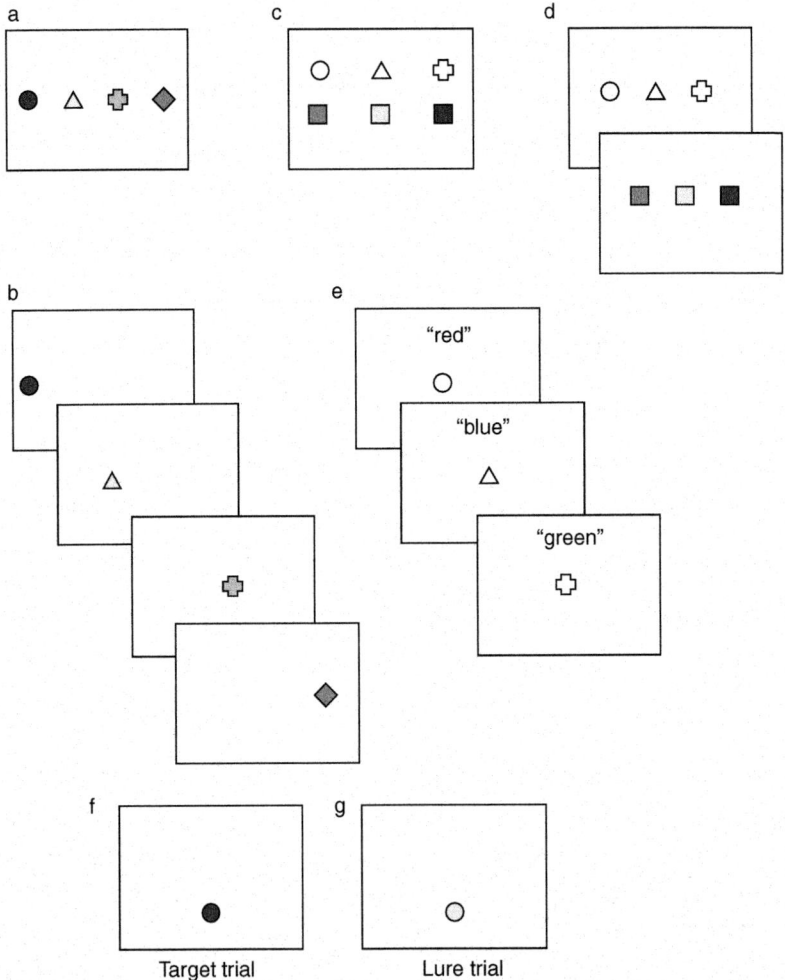

Figure 2 Examples of presentation and test arrays across experiments, showing (a) simultaneous unitized presentation, (b) sequential unitized presentation, (c) spatially separated presentation, (d) temporally separated presentation, (e) cross-modal presentation, (f) target recognition probe, and (g) lure recognition probe.

substantial and significant reduction in recognition accuracy, indicating an important role for central executive resources (cf. Dell'Acqua and Jolicoeur, 2000; Morey & Cowan, 2004, 2005; Stevanovski & Jolicoeur, 2007). However, crucially, this effect was no larger for binding than it was for memory of individual features (Figure 3). We replicated this pattern of findings in experiments using a range of concurrent tasks including recall of 6-digit sequences, and backward counting in decrements of 1. These experiments suggest that the binding of features such as shape and color into integrated object representations in visuo-spatial working memory proceeds automatically.

This series of experiments therefore provided a clear answer to our initial question. The encoding of items in general appears to have a large central executive component, as indicated by the substantial impact of concurrent task performance, but this does not increase when there is a requirement to bind component features into objects.

In apparent contrast to these findings, Brown and Brockmole (2010) have recently observed a larger effect of concurrent load (backward counting in twos) on shape–color binding than on memory for individual features. However, participants in this study performed the concurrent task during the recognition probe test phase as well as during encoding and retention. It is likely that, in requiring a judgment on whether a combination of shape and color were previously presented, the binding condition involves a more complex and demanding decision process at test than that involved in the feature conditions, and it is this stage that is more disrupted by a concurrent task. In contrast, our own consistent findings using an encoding-based executive load that is removed before the test phase (Allen et al., 2006) indicate

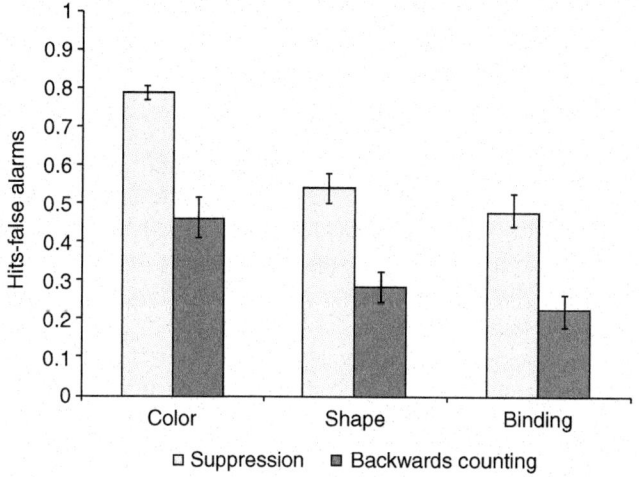

Figure 3 Performance accuracy (and standard error) in feature and binding conditions as a function of concurrent task (from Allen et al., 2006).

that the initial creation of bindings between shape and color does not particularly load on attentional resources.

It is possible however, that binding of features such as shape and color is so basic, rapid and automatic as to be immune to the imposition of a cognitive load. Our next series of experiments aimed to explore how binding might operate when the process is made more demanding.

3 Binding across space, time and modality

We hypothesized that our results might reflect a general tendency for the automatic processing of features linked by general Gestalt principles. For example, Woodman, Vecera, and Luck (2003) have demonstrated an influence of perceptual grouping on visual memory. In addition, visual STM is enhanced by symmetry, an effect that also appears to be automatic, being uninfluenced by concurrent tasks designed to disrupt various components of working memory (Pieroni, Rossi-Arnaud, & Baddeley, in press; Rossi-Arnaud, Pieroni, & Baddeley, 2006).

We ourselves used three methods of disrupting the operation of Gestalt principles, either separating features in space, in time, or across modalities. In each case we studied the impact on performance of concurrent tasks designed to disrupt executive control. Karlsen, Allen, Baddeley, and Hitch (2010) examined binding of spatially or temporally separated features in a series of experiments. Color and shape were presented as features separated in space (Figure 2c) or across a brief delay (Figure 2d), with participants required to bind these together for the purposes of performing the single probe recognition task from Wheeler and Treisman (2002). Karlsen et al. (2010) observed that, while memory for both spatially- and temporally separated conjunctions was less accurate overall than for unitized feature binding, performance was above chance, indicating that binding in these circumstances is possible (cf. Walker & Cuthbert, 1998). However, concurrent executive load did not differentially impact on separated as compared with unitized binding memory. Allen, Hitch, and Baddeley (2009) then extended this investigation beyond the visual modality, examining the potential role of attention in integrating information between visual and verbal modalities (see Figure 2e). In this series of experiments, participants were required to decide whether the features of a visual recognition probe were part of the same visual object or visual-auditory pairing in a preceding sequence. We observed that cross-modal binding was not only as accurate as visually unitized binding, but also was no more disrupted by concurrent backward counting performance. It therefore appears that objects requiring the binding of features across space, time, or modalities are no more disrupted by an attentionally demanding task than are those for which the features are unitized at presentation.

All the experiments described so far are inconsistent with our initial hypothesis, that the episodic buffer is an active binding system depending heavily on the central executive. There is no evidence, from our own studies at least, to suggest that access to the buffer depends crucially on the central executive, or to rule out direct links to the visuo-spatial sketchpad, or indeed the phonological loop. Our negative results were clearly not due to the use of insufficiently demanding secondary tasks, since these consistently resulted in a substantial reduction in overall memory performance.

Our finding that disrupting Gestalt principles does not engage executive processes in binding might seem to imply that the perceptual system binds all features automatically. However, if this were the case it would surely result in perceptual chaos, potentially combining totally unrelated features to create illusory objects. We note that in each of our conditions in which object features were separated at presentation they were nevertheless still close together in space and time. Thus it may be that our conclusions are limited to the binding of perceptual features under the somewhat constrained conditions of our experiments. More generally our results leave open whether executive processes are necessary for binding in the absence of strong perceptual support, as in the case of our earlier hypothetical example of imagining an elephant playing ice hockey.

4 How are bindings maintained?

Like Luck and Vogel (1997), Wheeler and Treisman (2002) obtained results that were consistent with an automatic binding process when they probed recall with a single item. However, when their participants had to scan a stimulus array in order to decide whether a change had occurred or not, their evidence appeared to argue against automaticity. It could be argued however, that this might have resulted from the disruption of binding by the process of scanning the array (Alvarez & Thompson, 2009), with the retention of bindings being more fragile than that of individual features. Such an explanation is consistent with the final study carried out by Allen et al. (2006). This departed from the standard procedure of presenting an array of visual stimuli to be remembered, instead presenting a sequence of single items (Figure 2b), followed by a standard recognition task probing one of the items in the sequence (Figure 2f and g).

The result of this study is shown in Figure 4, from which two conclusions may be drawn. First of all, when the final item was probed, the result was equivalent to earlier studies in showing little difference in performance between individual and bound features. In contrast, earlier items showed a consistent tendency for the retention of bound features to be inferior to single feature recognition. Furthermore, a comparison of simultaneous and sequential presentation conditions revealed that binding memory suffered significantly more than feature memory when items were presented one at a time, relative to simultaneous array presentation. We have recently used concurrent task methodology to explore whether this binding fragility reflects an increased attentional load involved in encoding and maintaining a *sequence* of bound objects. In line with our previous findings however, memory for

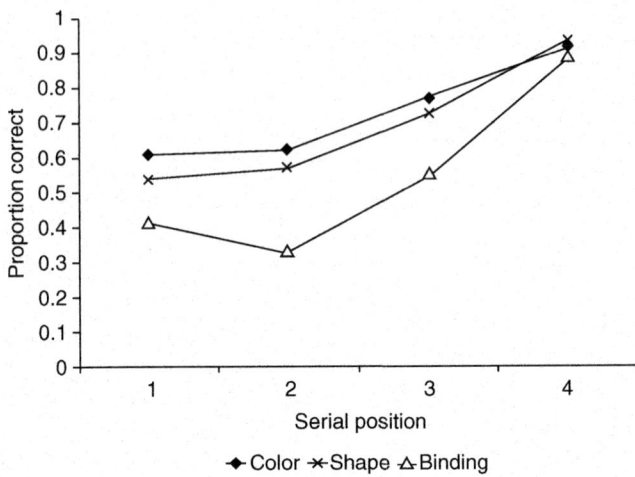

Figure 4 Proportion correct (and standard error) on target trials in feature and binding conditions as a function of serial position with sequential presentation (from Allen et al., 2006).

sequences of shape-color conjunctions was no more affected by backwards counting (in decrements of two) than sequential feature memory, thus indicating a fragility that is independent of attentional demand. These findings complement a growing body of research indicating that retaining arrays of bound objects in working memory does not particularly require attentional support, over and above that required for maintaining featural information (e.g. Delvenne, Cleeremans, & Laloyaux, 2010; Gajewski & Brockmole, 2006; Johnson, Hollingworth, & Luck, 2008; Yeh, Yang, & Chiu, 2005; but see Fougnie & Marois, 2009).

While our sequential presentation experiments produced evidence of some type of interference with memory for feature bindings, its nature is unclear. One possibility is that forgetting reflects the limited capacity of the object storage system; each time a new item is intentionally encoded, it will compete with existing items, resulting in forgetting. Another possibility however, is that forgetting represents an automatic overwriting process that occurs regardless of whether the participant intends to register the presented item. We have started to investigate this using a suffix paradigm, in which a simultaneously presented target array is followed by a single suffix item that participants are instructed to ignore. If memory for feature bindings does possess a greater fragility or susceptibility to overwriting, it should be more disrupted by this potentially interfering stimulus. The suffix method has the advantage of ruling out interference caused by deliberate encoding of further stimuli, as participants know they should try and ignore the suffix. Furthermore, by manipulating the nature of the suffix, important insights can be generated concerning the underlying processes and representations serving visuo-spatial working memory.

Ueno, Allen, Baddeley, Hitch, and Saito (in press) compared memory for color, shape, and color–shape binding using an adaptation of the simultaneous presentation and single probe testing methodology from Allen et al. (2006), although with a longer exposure duration (1000 ms). A first experiment found that presenting a standard suffix (250 ms after target offset, for 250 ms duration in a separate location from the stimuli) containing features that were never part of the experimental set (specifically, a brown hexagon), led to a small but statistically significant decrement that was equivalent for feature and binding memory. Using a suffix that changed on each trial (but still drawing from features outside the experimental set) produced the same pattern of results. Two final experiments examined the effect of suffixes containing features that were drawn from the target set (although not presented on that trial). In this case, such 'plausible' suffixes had a significantly larger disruptive impact on binding than on feature memory (Figure 5), the latter conditions showing a smaller decrement similar to that observed in the initial 'implausible' suffix experiments.

This finding of increased interference with plausible suffixes has recently been replicated in two experiments using cued recall instead of the standard probe recognition paradigm (Ueno, Mate, Allen, Hitch, & Baddeley, 2011). In a key final experiment, Ueno et al. compared the effects of plausible and implausible suffixes with that of suffixes that were 'semi-plausible'. These consisted of one feature dimension that was drawn from the experimental set

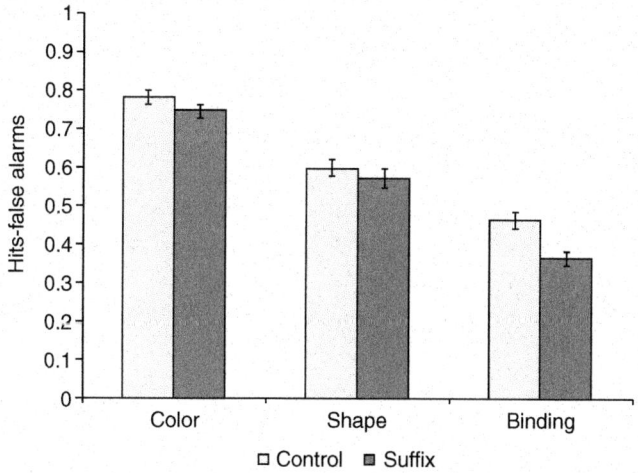

Figure 5 Performance accuracy (and standard error) on target trials in feature and binding conditions as a function of suffix condition (from Ueno et al., in press).

(i.e. plausible), and one that was not. It was found that fully- and semi-plausible suffixes had equivalent effects on cued recall, with both effects being larger than the impact of implausible suffixes.

Our suffix experiments make it clear that disruption of bindings by subsequent items does not depend upon the intention to encode such items, but does depend on their nature. Provided they are clearly not a member of the target set, degree of disruption is minimal, in contrast to the clear disruption from items containing features that might plausibly have been targets. These findings were taken to reflect two distinct processes: a filtering mechanism and a process of overwriting. To-be-ignored stimuli are assumed to require filtering to prevent them from being encoded in visuo-spatial working memory. This process makes a small attentional demand regardless of the nature of the suffix encountered, hence the equivalent but modest effects of implausible suffixes across both feature and binding conditions. Our finding that objects containing only one implausible feature cause as much disruption as suffixes in which both color and shape are plausible suggests that this filtering process operates at the level of individual features, with items containing any plausible feature being more likely to pass through this filter. It should also be noted that, as stimulus-based filtering draws on knowledge of task demands and what has been previously encountered, this process reflects one way in which the products of prior experience can impinge on perceptual input before information is encoded into working memory (see Logie, 1995).

We assume that any item making it through the filter is then automatically represented at the object level. This level is particularly prone to overwriting and loss of existing representations, which would explain why binding memory suffers when further stimuli are encountered (Allen et al., 2006; Alvarez & Thompson, 2009; Logie, Brockmole, & Vandenbroucke, 2009; Wheeler & Treisman, 2002). Any memory task assessing binding (such as probe recognition or cued recall) requires intact representation at the object level. Features, in contrast, have greater redundancy, as they are represented both at the object and feature levels (Treisman & Zhang, 2006). The equivalent effects of semi- and fully-plausible suffixes is also more consistent with the assumption of displacement within a limited capacity store, rather than a feature-based interference effect, where the dual features of a fully-plausible suffix might be thought to be more similar to the target set, and hence more disruptive.

An important element of this account is that visual stimuli are at least initially represented at both the level of constituent features and of objects, as set out by Treisman (e.g. Treisman, 1998), with filtering proceeding on the basis of features and overwriting particularly impacting on object representations.

How might this multi-level approach fit with the working memory model described by Baddeley (2000)? One possibility is that the visuo-spatial sketchpad can be viewed as a hierarchical system containing both feature and object levels of representation (see Figure 6), and it is within this component of working memory that visual features are initially bound together. The results

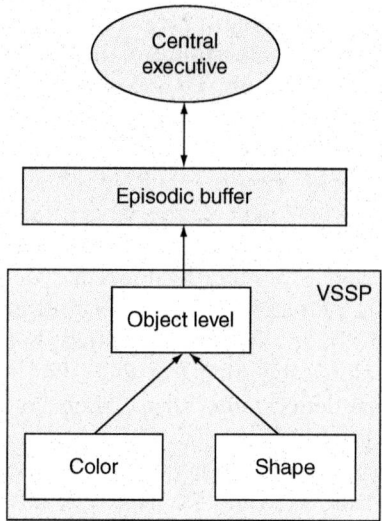

Figure 6 Schematic diagram of the relationship between components of the working memory model.

of this processing are then fed through to the purely object-based episodic buffer, and it is within this storage capacity that objects are consciously retained and manipulated. This tentative outline draws on both object-based (e.g. Vogel et al., 2001) and multi-level (e.g. Treisman, 1998) approaches to visual memory, and places them within the framework of the working memory model.

5 Does binding depend on the hippocampus?

The studies we have described so far are purely behavioral, although they do have implications for the neurobiological basis of binding in both visual working memory, and broadly within the episodic buffer. The buffer is assumed to provide an interface between the verbal and the visuo-spatial subsystems of working memory, and to link both of these to long-term memory. Any system underpinning the episodic buffer therefore needs to be richly connected to a range of brain regions. One obvious candidate for this role is the hippocampus, which appears to play an important role in the process of binding an event to its experienced context, providing the basis of episodic long-term memory (Squire, 2004; Vargha-Khadem et al., 1997; Winocur & Mills, 1970). Given the rich connections between the hippocampus and the structures involved in perception, attention and long-term memory (Suzuki & Amaral, 1994), it is plausible to speculate that it may also play a similar role in the binding of features in working memory. In order to test this, Olson,

Moore, Stark, and Chatterjee (2006) required patients with hippocampal damage to associate line drawings of objects to locations within a 3 × 3 matrix, testing their patients for memory of the object, its location and the binding of object to location after delays of one or eight seconds. Two experiments both showed impairment in memory for binding in the hippocampal patients.

There are however a number of potential objections to this study. First of all, the decision to study binding to spatial location is a limiting factor, since there is considerable evidence for the involvement of the hippocampus in spatial processing (Hartley, Maguire, Spiers, & Burgess, 2003; O'Keefe and Dostrovsky, 1971). Relevant to this point is a study by Piekema, Kessels, Mars, Petersson, and Fernández (2006), who showed that activation of the hippocampus occurred with the binding of an object to a location, but not when a shape had to be bound to a color, suggesting that the Olson et al. (2006) result may have reflected a spatial deficit rather than a problem with binding per se. A second problem with the Olson et al. (2006) study concerns the particular sample of patients tested in both this and subsequent studies (e.g. Ezzyat & Olson, 2008), patients whose damage extends well beyond the hippocampus, making it difficult to attribute any deficits specifically to the hippocampus.

A further study that is cited by Ezzyat and Olson (2008) as a source of evidence for the role of hippocampus in short-term binding was carried out by Hannula, Tranel, and Cohen (2006). This involved associating a single face with a complex visual scene, with results indicating impaired performance in patients with hippocampal damage, even at short delays. However, as Hannula et al. themselves point out, it is possible that this task involves both long- and short-term memory components, with the hippocampal damage reflecting the former. This possibility was investigated by Shrager, Levy, Hopkins, and Squire (2008) in a study based on patients with clear hippocampal damage and moderate or severe amnesia. They used concurrent task methodology to separate out the long and short-term components of performance, studying both the face memory task used by Hannula et al. (2006) and the object location task employed by Olson et al. (2006). In both cases, their evidence indicated a deficit in long-term memory rather than in short-term binding.

So far, the emphasis has been on challenging the interpretation of studies claiming to have demonstrated an association between the short-term binding and the hippocampus in terms of the binding task chosen and the lack of anatomical specificity in the patients studied. Our own studies attempted to test the hippocampal hypothesis using the range of short-term visual binding tests already described, testing a patient with a substantial and relatively precise hippocampal deficit (Baddeley, Allen, & Vargha-Khadem, 2010). The patient in question, Jon, was 28 years old at the time of testing. He was born prematurely, suffered breathing problems during the early weeks of life, and by the age of five began to show memory problems which have continued to be prominent (Gadian et al., 2000). Direct measurement of Jon's MRI scans indicate the reduction of about 50% in both left and right hippocampal regions, with no apparent pathology in the rest of the medial temporal lobe.

Jon has difficulty in reliably finding his way around and, consistent with his hippocampal deficit, is impaired on complex spatial tasks such as judging the appearance of a three-dimensional scene from different orientations (King, Burgess, Hartley, Vargha-Khadem, & O'Keefe, 2002). He makes many prospective memory errors and has great difficulty remembering everyday events. Consistent with this he performs poorly on standardized memory tests, when these involve recall rather than recognition, but is unimpaired on immediate memory tasks, hence his performance on the California Verbal Learning Test: II (Delis, Kramer, Kaplan, & Ober, 1987) was at the first percentile on delayed recall, but the 73rd percentile for immediate. His profile score on the Rivermead Behaviourial Memory Test (Wilson, Cockburn, Baddeley, & Hiorns, 1989) was 3, clearly impaired. However his performance on recognition tests was relatively preserved, showing only a very slight decrement consistent with the potential role of episodic memory in recognition (Baddeley, Vargha-Khadem, & Mishkin, 2001; Gardiner, Brandt, Vargha-Khadem, Baddeley, & Mishkin, 2006). Jon has a full scale IQ of 114 (high average) and performs normally on tests of reading, syntax, semantics and vocabulary (see Baddeley et al., 2001).

We tested Jon on a range of visual binding tasks, as described earlier (Baddeley et al., 2010). In one study, he was presented with a sequence of 2–4 colored shapes, with memory tested by presenting a single probe item. As in our first study, we compared retention of the individual features of color and shape with performance when the binding of shape to color was tested. In a second study, we presented the colors and shapes in separated but adjacent locations, whereas a third study required him to combine information from two modalities, visual and auditory, linking a spoken color name with a visual shape, or vice versa. Figure 7 shows Jon's performance across this range of tasks, together with the performance of six undergraduate students from the University of York who took part as control participants. Jon is clearly not impaired on any of these binding tasks. In conclusion therefore, we find little evidence for a general involvement of the hippocampus in short-term visual binding.

So where anatomically, does binding occur? At this point, we can only speculate, but we suspect that the episodic buffer is likely to reflect a widely distributed system, with different types of binding probably reflecting different subsystems. The observation that visual binding is specifically impaired in Alzheimer's disease as compared to normal aging is intriguing but difficult to interpret in terms of specific neural systems (Parra et al., 2009). Evidence for a separation of the systems involved in visual and verbal binding is provided by an intriguing single case study by Parra, Della Sala, Logie, and Abrahams (2010). Following the removal of a left medial splenoid ridge meningioma, the patient had made an apparently complete recovery. However, completely by chance (she was a voluntary control participant), she was

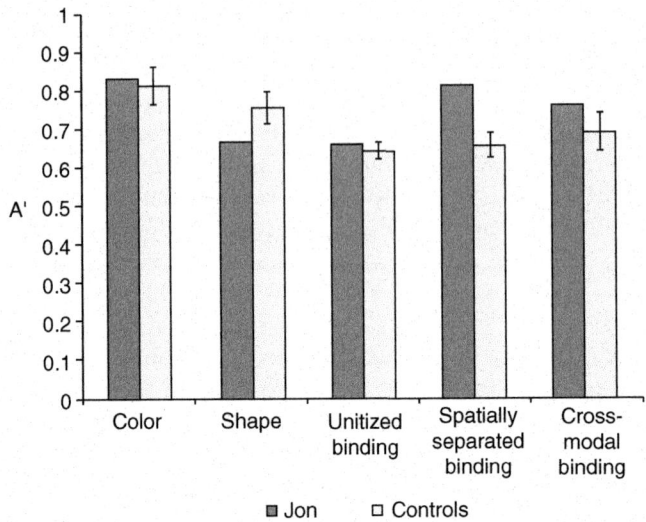

Figure 7 Performance accuracy across feature and binding conditions for patient Jon and controls (from Baddeley et al., 2010).

subsequently discovered to have a very marked deficit in the short-term binding of color to shape, in the complete absence of a deficit in either long-term visual or short-term verbal binding.

6 Binding and the episodic buffer

The study of binding in visual working memory has benefited greatly from the application to working memory of concepts and techniques developed for the study of visual attention. Our own approach has been more top-down, applying techniques that have been developed within a broad multicomponent working memory framework, where many of the ideas were derived from the study of verbal memory. We regard the programme we have just described as being complementary to, and compatible with, attention-based approaches that tend to focus on the neurobiological underpinning of visual working memory, rather than its possible role within a broader multicomponent system. Given our own motivation however, we would like to conclude by discussing the implications of our results for the multicomponent model of working memory that has directed our research.

It may be recalled that our visual binding program was one part of a two pronged attack on binding in working memory, the second prong being concerned with the role of binding in prose recall. Binding in this context involved the binding of words into meaningful chunks or phrases, resulting in a higher immediate memory span for sentences than for the same words in scrambled order. We demonstrated first, that sentential material led to enhanced immediate recall, before going on to study the influence of a range of concurrent tasks on this sentence advantage. Overall performance proved sensitive to concurrent load, with a small decremental effect from a simple visuo-spatial storage task, a larger one from disruption of phonological storage and an even greater one when the central executive was involved. As with the visuo-spatial binding experiments just described, we found no interaction between concurrent load and binding. The binding of words into chunks therefore appears to reflect automatic processes, this time clearly involving long-term memory. It does not appear to depend on the central executive (Allen & Baddeley, 2009; Baddeley, Hitch, & Allen, 2009).

Taking account of the full range of our binding research, we have modified our working memory framework along the lines shown in Figure 8. At the heart of the current model is the episodic buffer, a purely passive system, but one that serves a crucial integrative role because of its capacity to bind information from a number of different dimensions into unitized episodes or chunks. We speculate that smell and taste may also have access to the system, although currently know of no direct evidence on this issue. Our current speculations continue to assume that conscious access to the phonological loop or sketchpad may operate via the buffer. The visuo-spatial and verbal subsystems are themselves assumed to act as lower level buffers allowing, in one case, information from visual, spatial, kinaesthetic and tactile information to be combined. In the case of the phonological loop, language-related information from a number of sources may be combined, including not only speech, but also written, lip read and signed language (see Rönnberg, Rudner, & Ingvar, 2004, for a discussion).

An important feature of the phonological loop is of course, the capacity for vocal rehearsal, providing a system in which sequences of well-learned

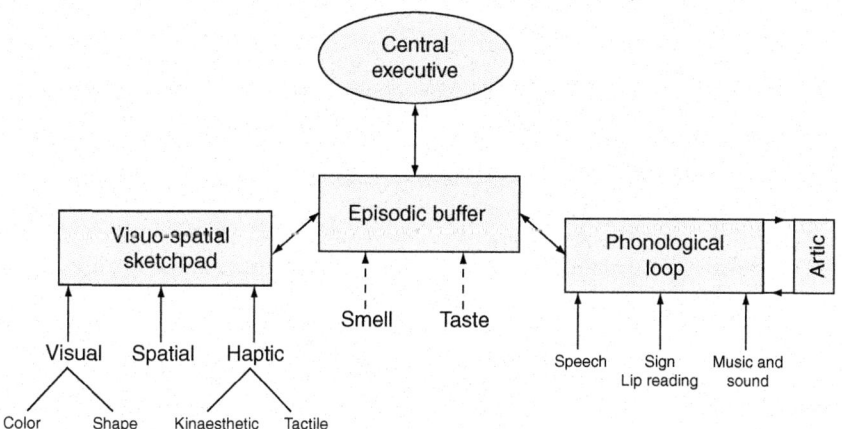

Figure 8 A revised model of working memory.

items such as digits, letters, and words that are readily retrievable from long-term memory may be maintained almost perfectly, provided the sequence is short enough to be repeated before its constituent features are disrupted by decay or interference. The relationship of the phonological loop system to the retention of environmental and musical sounds remains to be fully explored. Our current view however is that there is considerable overlap, but with some potential differences (Williamson, Baddeley, & Hitch, 2010).

We regard the multicomponent approach to working memory as providing a broad theoretical framework, a conceptual map that captures what we already know and encourages further questions that will lead to continued growth, or potentially abandonment of the framework. Our own particular framework reflects its history, which was strongly influenced by its initial focus on verbal short-term memory, and on cognitive neuropsychology. We suggest that it has provided and continues to provide a very productive framework for studying a wide range of problems (see Baddeley, 2007 for an overview).

As described above, our programme of research was driven by a particular theoretical framework, resulting in evidence that caused us to modify that framework, and suggesting further potentially tractable questions. We do not of course wish to argue that our results are incompatible with other conceptual approaches, or to deny the need for more detailed modelling at both the computational and the neurophysiological level. Our results do however place constraints on such models, requiring an explanation of (1) the lack of an effect of attentional load on the capacity to bind the features of shape and color, (2) the fragility of such binding, (3) the capacity to filter out "implausible" but not "plausible" suffixes and (4) the mechanism by which plausible suffixes disrupt the retention of bound objects.

References

Allen, R., Baddeley, A. D., & Hitch, G. J. (2006). Is the binding of visual features in working memory resource-demanding? *Journal of Experimental Psychology: General, 135*, 298–313.

Allen, R. J., & Baddeley, A. D. (2009). Working memory and sentence recall. In A. Thorn, & M. Page (Eds.), *Interactions between short-term and long-term memory in the verbal domain*.

Allen, R. J., Hitch, G. J., & Baddeley, A. D. (2009). Cross-modal binding and working memory. *Visual Cognition, 17*, 83–102.

Alvarez, G. A., & Thompson, T. W. (2009). Overwriting and rebinding: Why feature-switch detection tasks underestimate the binding capacity of visual working memory. *Visual Cognition, 17*, 141–159.

Baars, B. J. (2002). The conscious access hypothesis: Origins and recent evidence. *Trends in Cognitive Sciences, 6*(1), 47–52.

Baddeley, A. D. (2000). The episodic buffer: A new component of working memory? *Trends in Cognitive Sciences, 4*(11), 417–423.

Baddeley, A. D. (2007). *Working memory, thought and action*. Oxford: Oxford University Press.

Baddeley, A. D., Allen, R. J., & Vargha-Khadem, F. (2010). Is the hippocampus necessary for visual and verbal binding in working memory? *Neuropsychologia, 48*, 1089–1095.

Baddeley, A. D., & Hitch, G. J. (1974). Working memory. In G. A. Bower (Ed.), *The psychology of learning and motivation: Advances in research and theory* (pp. 47–89). New York: Academic Press.

Baddeley, A. D., Hitch, G. J., & Allen, R. J. (2009). Working memory and binding in sentence recall. *Journal of Memory and Language, 61*, 438–456.

Baddeley, A. D., & Logie, R. H. (1999). Working memory: The multiple component model. In A. Miyake, & P. Shah (Eds.), *Models of working memory: Mechanisms of active maintenance and executive control* (pp. 28–61). Cambridge University Press.

Baddeley, A. D., Vargha-Khadem, F., & Mishkin, M. (2001). Preserved recognition in a case of developmental amnesia: Implications for the acquisition of semantic memory. *Journal of Cognitive Neuroscience, 13*(3), 357–369.

Barrouillet, P., Bernardin, S., & Camos, V. (2004). Time constraints and resource sharing in adults' working memory spans. *Journal of Experimental Psychology: General, 133*, 83–100.

Brown, L. A., & Brockmole, J. R. (2010). The role of attention in binding visual features in working memory: Evidence from cognitive ageing. *The Quarterly Journal of Experimental Psychology, 63*, 2067–2079.

Cowan, N. (2005). *Working memory capacity*. Hove: Psychology Press.

Daneman, M., & Carpenter, P. A. (1980). Individual differences in working memory and reading. *Journal of Verbal Learning and Verbal Behaviour, 19*, 450–466.

Delis, D. C., Kramer, J. H., Kaplan, E., & Ober, B. A. (1987). *California Verbal Learning Test: Form II*. San Antonio, TX: Psychological Corporation.

Dell'Acqua, R., & Jolicoeur, P. (2000). Visual encoding of patterns is subject to dual-task interference. *Memory and Cognition, 28*, 184–191.

Delvenne, J.-F., Cleeremans, A., & Laloyaux, C. (2010). Feature bindings are maintained in visual short-term memory without sustained focused attention. *Experimental Psychology, 57*(2), 108–116.

Engle, R. W., & Kane, M. J. (2004). Executive attention, working memory capacity and two-factor theory of cognitive control. In B. Ross (Ed.), *The psychology of learning and motivation* (pp. 145–199). New York: Elsevier.

Ezzyat, Y., & Olson, I. R. (2008). The medial temporal lobe and visual working memory: Comparisons across tasks, delays, and visual similarity. *Cognitive, Affective, & Behavioral Neuroscience, 8*, 32–40.

Fougnie, D., & Marois, R. (2009). Attentive tracking disrupts feature binding in visual working memory. *Visual Cognition, 17*, 48–66.

Gadian, D. G., Aicardi, J., Watkins, K. E., Porter, D. A., Mishkin, M., & Vargha-Khadem, F. (2000). Developmental amnesia associated with early hypoxic-ischaemic injury. *Brain, 123*, 499–507.

Gajewski, D. A., & Brockmole, J. R. (2006). Feature bindings endure without attention: Evidence from an explicit recall task. *Psychonomic Bulletin & Review, 13*, 581–587.

Gardiner, J. M., Brandt, K. R., Vargha-Khadem, F., Baddeley, A. D., & Mishkin, M. (2006). Effects of level of processing but not of task enactment on recognition memory in a case of developmental amnesia. *Cognitive Neuropsychology, 23*, 930–948.

Hannula, D. E., Tranel, D., & Cohen, N. J. (2006). The long and the short of it: Relational memory impairments in amnesia, even at short lags. *Journal of Neuroscience, 26*, 8352–8359.

Hartley, T, Maguire, E. A., Spiers, H. J., & Burgess, N. (2003). The well-worn route and the path less travelled: Distinct neural basis of route following and wayfinding in humans. *Neuron, 37*, 877–888.

Johnson, J. S., Hollingworth, A., & Luck, S. J. (2008). The role of attention in the maintenance of feature bindings in visual short-term memory. *Journal of Experimental Psychology: Human Perception & Performance, 34*, 41–55.

Karlsen, P. J., Allen, R. J., Baddeley, A. D., & Hitch, G. J. (2010). Binding across space and time in visual working memory. *Memory and Cognition, 38*, 292–303.

King, J. A., Burgess, N., Hartley, T., Vargha-Khadem, F., & O'Keefe, J. (2002). The human hippocampus and viewpoint dependence in spatial memory. *Hippocampus, 12*, 811–820.

Logie, R. H. (1995). *Visuo-spatial working memory*. Hove, UK: Erlbaum.

Logie, R. H., Brockmole, J. R., & Vandenbroucke, A. R. E. (2009). Bound feature combinations in visual short-term memory are fragile but influence long-term learning. *Visual Cognition, 17*, 160–179.

Luck, S. J., & Vogel, E. K. (1997). The capacity of visual working memory for features and conjunctions. *Nature, 390*, 279–281.

Morey, C. C., & Cowan, N. (2004). When visual and verbal memories compete: Evidence of cross-domain limits in working memory. *Psychonomic Bulletin and Review, 11*, 296–301.

Morey, C. C., & Cowan, N. (2005). When do visual and verbal memory conflict? The importance of working memory load and retrieval. *Journal of Experimental Psychology: Learning, Memory and Cognition, 31*, 703–713.

O'Keefe, J., & Dostrovsky, J. (1971). The hippocampus as a spatial map: Preliminary evidence from unit activity in the freely-moving rat. *Brain Research, 34*, 171–175.

Olson, I. R., Moore, K. S., Stark, M., & Chatterjee, A. (2006). Visual working memory is impaired when the medial temporal lobe is damaged. *Journal of Cognitive Neuroscience, 18*, 1087–1097.

Parra, M. A., Abrahams, S., Logie, R. H., Mendez, L. G., Lopera, F., & Della Sala, S. (2009). Visual short-term memory binding deficits in familial Alzheimer's disease. *Brain, 132*, 1057–1066.

Parra, M. A., Della Sala, S., Logie, R. H., & Abrahams, S. (2010). Selective impairment in visual short-term memory binding. *Cognitive Neuropsychology, 26*, 1–23.

Piekema, C., Kessels, R. P. C., Mars, K. M., Petersson, K. M., & Fernández, G. (2006). The right human hippocampus participates in active maintenance of object-location associations. *NeuroImage, 33*, 374–382.

Pieroni, L., Rossi-Arnaud, C., & Baddeley, A. D. What can symmetry tell us about working memory? In A. Vandierendonck, & A. Szmalec (Eds.), *Spatial working memory*. Hove Psychology Press, in press.

Rönnberg, J., Rudner, M., & Ingvar, M. (2004). Neural correlates of working memory for sign language. *Cognitive Brain Research, 20*, 165–182.

Rossi-Arnaud, C., Pieroni, L., & Baddeley, A. D. (2006). Symmetry and binding in visuo-spatial working memory. *Journal of Cognitive Neuroscience, 139*, 393–400.

Shrager, Y., Levy, D. A., Hopkins, R. O., & Squire, L. R. (2008). Working memory and the organization of brain systems. *Journal of Neuroscience, 28*, 4818–4822.

Squire, L. R. (2004). Memory systems of the brain: A brief history and current perspective. *Neurobiology of Learning and Memory, 82*, 171–177.

Stevanovski, B., & Jolicoeur, P. (2007). Visual short-term memory: Central capacity limitations in short-term consolidation. *Visual Cognition, 15*, 532–563.

Suzuki, W. A., & Amaral, D. G. (1994). Perirhinal and parahippocampal cortices of the macaque monkey: Cortical afferents. *Journal of Comparative Neurology, 350*, 497–533.

Treisman, A. (1998). Feature binding, attention and object perception. *Philosophical Transactions of the Royal Society, London B, 353*, 1295–1306.

Treisman, A., & Gelade, G. (1980). A feature-integration theory of attention. *Cognitive Psychology, 12*, 97–136.

Treisman, A. M., & Zhang, W. (2006). Location and binding in visual working memory. *Memory and Cognition, 34*, 1704–1719.

Ueno, T., Allen, R. J., Baddeley, A. D., Hitch, G. J., & Saito, S. Disruption of binding in visual working memory. *Memory & Cognition*, in press.

Ueno, T., Mate, J., Allen, R. J., Hitch, G. J., & Baddeley, A. D. (2011). What goes through the gate? Exploring interference with visual feature binding. *Neuropsychologia, 29*(23), 1596–1603.

Vargha-Khadem, F., Gadian, D. G., Watkins, K. E., Connelly, A., Van Paesschen, W., & Mishkin, M. (1997). Differential effects of early hippocampal pathology on episodic and semantic memory. *Science, 277*, 376–380.

Vogel, E. K., Woodman, G. F., & Luck, S. J. (2001). Storage of features, conjunctions, and objects in visual working memory. *Journal of Experimental Psychology: Human Perception and Performance, 27*(1), 92–114.

Walker, P, & Cuthbert, L. (1998). Remembering visual feature conjunctions: Visual memory for shape-colour associations is object-based. *Visual Cognition, 5*(4), 409–455.

Wheeler, M. E., & Treisman, A. M. (2002). Binding in short-term visual memory. *Journal of Experimental Psychology: General, 131*, 48–64.

Williamson, V., Baddeley, A., & Hitch, G. (2010). Musicians' and nonmusicians' short-term memory for verbal and musical sequences: Comparing phonological similarity and pitch proximity. *Memory and Cognition, 38*, 163–175.

Wilson, B. A., Cockburn, J., Baddeley, A. D., & Hiorns, R. (1989). The development and validation of a test battery for detecting and monitoring everyday memory problems. *Journal of Clinical and Experimental Neuropsychology, 11*, 855–870.

Winocur, G., & Mills, J. A. (1970). Transfer between related and unrelated problems following hippocampal lesions in rats. *Journal of Comparative & Physiological Psychology, 73*, 162–169.

Woodman, G. F., Vecera, S. P., & Luck, S. J. (2003). Perceptual organization influences visual working memory. *Psychonomic Bulletin & Review, 10*(1), 80–87.

Yeh, Y. Y., Yang, C. T., & Chiu, Y. C. (2005). Binding or prioritization: The role of selective attention in visual short-term memory. *Visual Cognition, 12*(5), 759–799.

19 Working memory
Theories, models, and controversies

A. D. Baddeley

Working memory: theories, models, and controversies

I was honored, pleased, and challenged by the invitation to write this prefatory chapter, pleased because it offered the chance to take a broad and somewhat autobiographical view of my principal area of interest, working memory (WM), but challenged by the potential magnitude of the task. The topic of working memory has increased dramatically in citation counts since the early years, not all of course related to or supportive of my own work, but a recent attempt to review it (Baddeley 2007) ended with more than 50 pages of references. What follows is a partial, as opposed to impartial, account of the origins of the concept of multi-component working memory (M-WM) and of my own views on its subsequent development. My first draft would have filled the chapter page allowance with references; I apologize to all of those whose work should have been cited and is not.

I entered psychology as a student at University College London in 1953, a very exciting time for the field of psychology, which had benefited greatly from developments during the Second World War, where theory was enriched by the need to tackle practical problems. As a result, prewar issues such as the conflict between Gestalt psychology and neobehaviorism began to be challenged by new data and new ideas, some based on cybernetics, the study of control systems, with others influenced by the newly developed digital computers. This in turn led to a renewed interest in the philosophy of science as applied to psychology. Typical questions included, is psychology a science?; if so, is it cumulative or are we doomed to keep on asking the same questions, as appeared to be the case in philosophy? What would a good psychological theory look like?

As students we were offered two answers to this question. The first, championed by Cambridge philosopher Richard Braithwaite (1953), regarded Newton's *Principia* as the model to which scientific theories should aspire, involving as it does postulates, laws, equations, and predictions. Within psychology, the Newtonian model was explicitly copied by Clark Hull in his attempt to produce a general theory of learning, principally based on the study of maze learning in the albino rat.

An alternative model of theorizing came from Oxford, where Stephen Toulmin (1953) argued that theories were like maps, ways of organizing our existing knowledge of the world, providing tools both for interacting with the world and for further exploration. Edward Tolman in Stanford had a view of learning in rats that fitted this model, using it to challenge Hull's neo-behaviorist approach. This raised the crucial question as to how you might decide between the two apparently opposing views. The dominant answer to that question, in the United Kingdom at least, was provided by Karl Popper (1959), a Viennese-trained philosopher who argued strongly that a valid theory should make clear, testable predictions, allowing the rival theories to confront each other in the all-important "crucial experiment" that settles the issue. This approach was closer in spirit to Hull than to Tolman.

My own first published study (Baddeley 1960) attempted just such a crucial experiment, predicting that rats would be smarter than they should be according to Hullian theory, and demonstrating, to my own satisfaction at least, that this was the case. Alas, by the time it was published, the whole field of learning theory seemed to have collapsed. Neither side was able to deliver a knockout blow, and people simply abandoned the research area. I resolved at that point that if I myself were to develop a theory, it would be based very closely on the evidence, which would survive even if the theory proved totally wrong. It is an approach I have followed ever since.

But what is the answer to our original question, should theorists be architects, building elegant structures such as Newton did, or should they be explorers, gradually extending the theory on the basis of more and more evidence, as in the case of Darwin? Clearly both Newton and Darwin got it right, but for fields at a different stage of development. Newton claimed that his success resulted from "standing on the shoulders of giants," who no doubt stood on the shoulders of lesser mortals like ourselves. Darwin had few such giants available. I suggest that any complete theory is likely to require explorers in its initial stages and architects to turn the broad concepts into detailed models. I myself am very much at the explorer end of the continuum, but I fully accept the importance of the skills of the architect if theory is to develop.

My research career really began with my arrival at the Medical Research Council Applied Psychology Unit (APU) in Cambridge. Its role was to form a bridge between psychological theory and practical problems, and the year I arrived, Donald Broadbent, its director, had just published his seminal book, *Perception and Communication*, which provided one of the sparks that ignited what subsequently became known as the cognitive revolution. I was assigned to work on optimizing the design of postal codes, which led me to combine the classic tradition of nonsense syllable learning with new ideas from information theory, resulting in my generating memorable postal codes for each town in the United Kingdom. The Post Office thanked me and went on their way regardless; the code they adopted could, however, have been much worse, as is indeed the case in some countries, but that is another story.

By this time my approach to theory was evolving away from Popper's idea of the need for crucial experiments, largely on the grounds that clear predictions only appeared to be possible in situations that were far narrower than the ones I found interesting. I subsequently discovered that within the philosophy of science, Lakatos (1976), and allegedly Popper himself, had subsequently abandoned the reliance on falsification, arguing instead that the mark of a good theory is that it should be productive, not only giving an account of existing knowledge, but also generating fruitful questions that will increase our knowledge. This more map-like view of theory is the one that I continue to take.

Short-term memory

The term "working memory" evolved from the earlier concept of short-term memory (STM), and the two are still on occasion used interchangeably. I will use STM to refer to the simple temporary storage of information, in contrast to WM, which implies a combination of storage and manipulation.

My interest in STM began during my time at the APU in Cambridge and was prompted by an applied problem, that of finding a way of evaluating the quality of telephone lines that might be more effective than a simple listening test. My PhD supervisor Conrad had recently discovered the acoustic similarity effect. He was studying memory for proposed telephone dialing codes when he noted that even with visual presentation, memory errors resembled acoustic mis-hearing errors (e.g., *v* for *b*), and that memory for similar sequences (*b g t p c*) was poorer than for dissimilar (*k r l q y*), concluding that STM depends on an acoustic code (Conrad & Hull 1964).

I decided to see if the acoustic similarity effect could be used to provide sensitive indirect measure of telephone line quality. It did not; the effects of noise and similarity were simply additive, but I was intrigued by the sheer magnitude of the similarity effect. Similarity was a central variable within the dominant stimulus-response interference theory of verbal learning (see Osgood 1949), but the type of similarity seemed not to be regarded as important. So, would Conrad's effect generalize to other types of similarity in STM?

I tested this, comparing recall of sequences with five phonologically similar words (*man, mat, can, map, cat*), five dissimilar words (e.g., *pit, day, cow, pen, sup*), and five semantically similar sequences (*huge, big, wide, large, tall*) with five dissimilar (*wet, soft, old, late, good*). I found (Baddeley 1966a) a huge effect of phonological similarity[1] (80% sequences correct for dissimilar, 10% for similar) and a small but significant effect for semantic similarity (71% versus 65%). I went on to demonstrate that this pattern reversed when long-term memory (LTM) was required by using ten-word lists and several learning trials; semantic similarity then proved critical (Baddeley 1966b). I concluded that there were two storage systems, a short-term phonological and a long-term semantically based system. My telephony project was passed on to a newly arrived colleague and I was left free to explore this line of basic research.

I saw my work as fitting into a pattern of evidence for separate STM and LTM stores. Other evidence came from amnesic patients who had preserved STM and impaired LTM, while other patients showed the reverse pattern (Shallice & Warrington 1970). A third source of evidence came from two-component memory tasks, which comprised a durable LTM component together with a temporary component. A typical example of this was the recency effect in free recall (Glanzer 1972); the last few words of a list are well recalled on immediate test but not after a brief filled delay, unlike earlier items.

At this point, my simple assumption of two stores, with STM phonologically based and LTM semantically based, led to some clear predictions. Amnesic patients should have semantic coding problems, and recency should be acoustically based. Studies based on amnesic patients suffering from Korsakoff's syndrome did suggest a semantic encoding deficit (Cermak et al. 1974), but our own work showed no evidence of such a deficit (Baddeley & Warrington 1970), and later work (Cermak & Reale 1978) attributed their previously observed deficit to additional executive problems, often found in Korsakoff's syndrome.

In the case of two-component tasks, it became clear that recency did not depend on verbal STM (Baddeley & Hitch 1977) and that the use of semantic or phonological coding was strategy dependent. Phonological coding of verbal material is rapid, attentionally undemanding, and very effective for storing serial order. Semantic coding can be rapid for meaningful sequences such as sentences, but it is much harder to use for storing the order of unrelated words (Baddeley & Levy 1971). We also showed that word sequences can simultaneously be encoded both phonologically and semantically (Baddeley & Ecob 1970) and that standard tasks such as immediate serial recall can reflect both long-term and short-term components, each of which may be influenced by either phonological or semantic factors. In short, STM, retention of material over a brief period, may be based on either phonological or semantic coding. The former is easy to set up but readily forgotten; the latter may take longer to set up but tends to be more durable. Both can operate over brief delays, and the fact that we can learn new words indicates that long-term phonological learning also occurs.

It is worth emphasizing the need to distinguish between STM as a label for a paradigm in which small amounts of information are stored over brief delays and STM as a theoretical storage system. This point was made by Waugh & Norman (1965) and by Atkinson & Shiffrin (1968), but it has often been neglected in subsequent years. Material tested after a brief delay (i.e., an STM task) is likely to reflect both LTM and some form of temporary storage.

Evolution of a multicomponent theory

After nine years at the APU, I moved to Sussex into a new department of experimental psychology, where, in 1972, I was joined by Graham Hitch as a post-doctoral fellow on my first research grant. After a first degree in physics, he had done a psychology MSc in Sussex and a PhD with Broadbent at the APU. We had proposed (perhaps unwisely) to investigate the link between STM and LTM, beginning our grant just when the previously popular field of STM was downsizing itself following criticism of the dominant Atkinson & Shiffrin (1968) model for three reasons. First, the model assumed that merely holding information in STM would guarantee transfer to LTM, whereas Craik & Lockhart (1972) showed that the nature of processing is crucial, with deeper, more elaborate processing leading to better learning. Second, its assumption that the short-term store was essential for access to LTM proved to be inconsistent with neuropsychological evidence. Patients with a digit span of only two items and an absence of recency in free recall should, according to Atkinson and Shiffrin, have a defective short-term store that should lead to impaired LTM. This was not the case. Third, given that Atkinson and Shiffrin assumed their short-term store to be a working memory, playing an important general role in cognition, such patients should have major intellectual deficits. They did not. One patient, for instance, was an efficient secretary, and another ran a shop and a family. Interest in the field began to move from STM to LTM, to semantic memory and levels of processing.

Graham Hitch and I did not have access to these rare but theoretically important STM-deficit patients and instead decided that we would try to manufacture our own "patients" using student volunteers. We did so, not by removing the relevant part of their brain, but by functionally disabling it by requiring participants to do a concurrent task that was likely to occupy the limited-capacity short-term storage system to varying degrees. The concurrent task we chose was serial verbal recall of sequences of spoken digits. As sequence length increased, the digits should occupy more and more of available capacity, with the result that performance on any task relying on WM should be progressively impaired. In one study, participants performed a visually presented grammatical reasoning task while hearing and attempting to recall digit sequences of varying length. Response time increased linearly with concurrent digit load. However, the disruption was far from catastrophic: around 50% for the heaviest load, and perhaps more strikingly, the error rate remained constant at around 5%. Our results therefore suggested a clear involvement of whatever system underpins digit span, but not a crucial one. Performance slows systematically but does not break down. We found broadly similar results in studies investigating both verbal LTM and language comprehension, and on the basis of these, abandoned the assumption that WM comprised a single unitary store, proposing instead the three-component system shown in **Figure 1** (Baddeley & Hitch 1974).

We aimed to keep our proposed system as simple as possible, but at the same time, potentially capable of being applied across a wide range of cognitive activities. We decided to split attentional control from temporary storage, which earlier research suggested might rely on separate verbal and visuospatial short-term systems, all of which were limited in capacity. We labeled

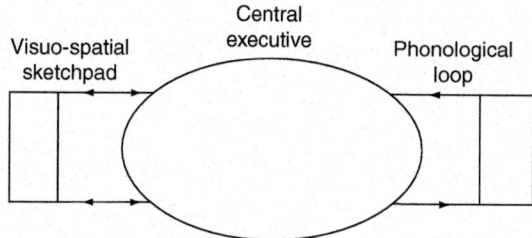

Figure 1 The original Baddeley & Hitch (1974) working memory model.

the central controller as a "central executive" (CE), initially referring to the verbal system as the "articulatory loop," after the subvocal rehearsal assumed to be necessary to maintain information, and later adopting the term "phonological loop" to emphasize storage rather than rehearsal. We termed the third component the "visuo-spatial sketchpad," leaving open the issue of whether it was basically visual, spatial, or both.

We began by focusing on the phonological loop on the grounds that it seemed the most tractable system to investigate, given the very extensive earlier research on verbal STM. At this point, I unexpectedly received an invitation from Gordon Bower to contribute a chapter to an influential annual publication presenting recent advances in the area of learning and memory. We hesitated; our model was far from complete, should we perhaps wait? We went ahead anyhow (Baddeley & Hitch 1974), presenting a model that is still not complete nearly 40 years and many publications later.

Over the next decade we continued to explore the model and its potential for application beyond the cognitive laboratory. At this point I agreed to summarize our progress in a monograph (Baddeley 1986). This was approaching completion when I realized that I had said nothing about the CE, very much a case of Hamlet without the prince. My reluctance to tackle the executive stemmed from two sources: first, its probable complexity, and second, because of the crucial importance of its attentional capacity. Although there were a number of highly developed and sophisticated theories of attention, most were concerned with the role of attention in perception, whereas the principal role of the CE was the attentional control of action. The one directly relevant article I could find (Norman & Shallice 1986) appeared as a chapter because of the difficulty of persuading a journal to accept it (Shallice 2010, personal communication), alas, all too common with papers presenting new ideas.

Norman and Shallice proposed that action is controlled in two rather separate ways. One is based on well-learned habits or schemata, demanding little in the way of attentional control. An example of this might be the activity of driving a well-learned route to your office. This source of control can be overridden by a second process, the supervisory attentional system (SAS), which responds to situations that are not capable of being handled by habit-based processes, for example, coping with the closure of a road on your normal route.

With some relief, I incorporated the Norman and Shallice model into my own concept of a CE, producing a book (Baddeley 1986) that attempted to pull together developments in WM that had occurred in the previous decade and then apply them to data from the literature in three areas: fluent reading, the development of WM in children, and the effects of aging. Although I tended to refer to our proposals as a model, using the criteria proposed earlier, it might better be regarded as a simple theory, in the sense of Toulmin's idea of theories as maps, linking together existing knowledge and encouraging further investigation. If so, it was a map with many blank areas that I hoped would be filled by myself and others, leading in due course to more detailed modeling.

What then are the essentials of the broad theory? The basis is the assumption that it is useful to postulate a hypothetical limited-capacity system that provides the temporary storage and manipulation of information that is necessary for performing a wide range of cognitive activities. A second assumption is that this system is not unitary but can be split into an executive component and at least two temporary storage systems, one concerning speech and sound while the other is visuo-spatial. These three components could be regarded as modules in the sense that they comprise processes and storage systems that are tightly interlinked within the module and more loosely linked across modules, with somewhat more remote connections to other systems such as perception and LTM. I regard the very rigid definition of modularity by Fodor (1983) as unhelpful and neuropsychologically implausible. A consequence of my rejection of Fodorian simplicity is the assumption that each of these systems can be fractionated into subsystems and that these will be linked to perceptual and LTM processes in ways that require further investigation.

My overall view of WM therefore comprised, and still comprises, a relatively loose theoretical framework rather than a precise model that allows specific predictions. The success of such a framework should be based, as suggested by Lakatos (1976), not only on its capacity to explain existing data but also on its productivity in generating good, tractable questions linked to empirical methods that can be widely applied. The proposed components of WM are discussed in turn, beginning with the phonological loop.

Characteristics of the phonological loop

We saw the phonological loop as a relatively modular system comprising a brief store together with a means of maintaining information by vocal or sub-vocal rehearsal. In the 1960s, a number of studies attempted to decide whether forgetting in the STM system was based on trace decay or interference (see Baddeley 1976). None of these studies proved to be conclusive, a state of affairs that remains true, in my own opinion. We opted to assume a process of trace decay, partly on the basis of our results and partly because it avoided the need to become involved in the many controversies surrounding traditional approaches to interference theory at the time (see Baddeley 1976, chapter 5), although we did assume a limited-capacity store, which in turn implies some unspecified form of interference, either by displacement or by overwriting. We used existing results, together with our own subsequent studies, to create a simple model that is based on the method of converging operations. This involves combining evidence from a range of different phenomena, each consistent with the model, but each individually explicable in other ways. If none of the competing interpretations are able to explain the whole pattern, whereas the phonological loop model can, then this provides valuable support. This approach has the advantage of potentially producing a robust model, but it has the disadvantage of being required to confront a range of different possible alternative explanations for each individual phenomenon.

The phonological similarity effect

As described above, this is regarded as an indication that phonological storage is involved. Its effect is principally on the storage of order information. Indeed, item information may be helped by similarity since it places constraints on possible responses. For this reason, studies that specifically attempt to investigate the loop tend to minimize the need to retain item information by repeatedly using the same limited set, for example, consonants. Studies using open sets, for instance, different words for each sequence, are more likely to reflect loss of item information and to show semantic and other LTM-based effects.

The word length effect

We assumed that vocal or subvocal rehearsal was likely to occur in real time, with longer words taking longer and hence allowing more time for trace decay, thus leading to poorer performance. We studied the immediate recall of sequences of five words ranging in length from one syllable (e.g., *pen day hot cow tub*) to five syllables (e.g., *university, tuberculosis, opportunity, hippopotamus, refrigerator*) and found that performance declined systematically with word length. As expected, when participants were required to read out words of different lengths as rapidly as possible, there was a close correspondence between word length and articulation time. The simple way of expressing our results was to note that people are able to remember as many words as they can articulate in two seconds (Baddeley et al. 1975b).

We interpreted our data by assuming that longer words take longer to rehearse, resulting in more trace decay and poorer recall. Such decay is also likely to continue during the slower spoken recall of longer words. We presented evidence for time-based decay, which has since faced challenge and counter-challenge (see Baddeley 2007, pp. 43–49). Fortunately, however, the general hypothesis of a phonological loop will function equally well with either a decay or interference interpretation of short-term forgetting, illustrating the value of combining a broad theoretical map while leaving more detailed modeling to be decided by further experimentation.

Articulatory suppression

If the word length effect is dependent on subvocalization, then preventing it should eliminate the effect. This is indeed the case (Baddeley et al. 1975b). When participants are required to continuously utter a single word such as "the," performance drops and is equivalent for long and short words. Suppression also removes the phonological similarity effect for visually presented materials but not when presentation is auditory (Baddeley et al. 1984). We interpret this as suggesting that spoken material gains obligatory access to the phonological store, whereas written material needs to be subvocalized if it is to register.

The claim that auditory presentation allows a phonological trace to be laid down despite suppression has recently been challenged. Jones et al. (2006) have suggested that the effect is limited to the recency component of immediate serial recall, suggesting that it is better regarded as a perceptual effect. However, although this may be true for long lists, shorter lists show an effect that operates throughout the serial position curve (Baddeley & Larsen 2007).

Irrelevant sound effects

Colle & Welsh (1976) required their participants to recall sequences of visually presented digits presented either in silence or accompanied by white noise or by speech in an unfamiliar language that they were told to ignore. Only the spoken material disrupted performance on the visually presented digits, an effect that was independent of the loudness of the irrelevant sound sources. Pierre Salame, a French visitor to Cambridge, and I followed up and extended Colle's work, demonstrating that visual STM was disrupted to the same extent by irrelevant words and nonsense syllables; indeed, irrelevant digits had no more effect on digit recall than did nondigit words containing the same phonemes (e.g., *one two* replaced by *tun woo*), suggesting that interference was operating at a prelexical level. We did, however, find slightly less disruption of our monosyllabic digits from bisyllabic words than from monosyllabic words, concluding rather too hastily that this suggested that interference was dependent on phonological similarity (Salame & Baddeley 1986). Like Colle and Welsh, we suggested an interpretation in terms of some form of mnemonic masking. This proved to be something of an embarrassment when it was clearly demonstrated that irrelevant items that were phonemically similar to the remembered sequence were no more disruptive than dissimilar items (Jones & Macken 1995, Larsen et al. 2000). Unfortunately, our initial hypothesis came to be regarded as central to WM, despite our subsequent withdrawal, a salutary lesson in premature theorizing.

Meanwhile Dylan Jones and colleagues in Wales were developing a very extended program of research on irrelevant sound. They showed that STM was disrupted not only by irrelevant speech, but also by a range of other sounds, including, for example, fluctuating tones (Jones & Macken 1993). In order to account for their results they proposed the "changing state" hypothesis, whereby the crucial feature was that the irrelevant sound needed to fluctuate. Jones (1993) coupled this with the object-orientated episodic record (OOE-R) hypothesis, which assumes that both digits and irrelevant sounds are represented as potentially competing paths on a multidimensional surface. The OOE-R hypothesis is not spelled out in detail but would appear to assume that serial order is based on chaining, whereby each item acts as a stimulus for the response that follows, which in turn acts as a further stimulus.

Retaining serial order

A typical memory span is around six or seven digits, not because the digits themselves are forgotten, but rather because their order is lost. Retaining serial order is a crucial demand for a wide range of activities, notably including language, in which sequences of sounds within words and words within sentences must be maintained, and skilled motor performance such as striking a ball with a bat or playing the piano. However, as Lashley (1951) points out, it is far from easy to explain how this is achieved. The most obvious hypothesis is through the previously described mechanism of chaining through sequential associations. However, this has some major potential problems; if one item is lost, then the chain is broken and subsequent recall should fail, and yet it is often the case that despite errors in the middle of a sequence, the latter part is reproduced correctly. Similarly, if an item is repeated within the chain (e.g., 7 5 3 5 9 6), then the chain should be disrupted, but this disruption, when it occurs, is typically far from dramatic.

A third phenomenon appears to be even more problematic. This again is an effect that was discovered when trying to solve a practical problem, that of trying to reduce the negative impact of phonological similarity on the recall of postal codes. It seemed plausible to me to assume that the principal effect of similarity would come from having two or more similar items bunched together, in which case it might prove possible to greatly minimize the effect by alternating similar and dissimilar items (e.g., *dfvkpl*). The results were disappointing; the similar items appeared to be just as liable to be forgotten when sandwiched between dissimilar items as when they were adjacent, so we put the experiment to one side. It was only later, when I was attempting to pin down the nature of the phonological loop effect, that I realized that our result had clear implications for theories of serial order retrieval in general (Baddeley 1968) and were in particular inconsistent with hypotheses that depended upon chaining. The argument goes as follows: If one considers a sequence of six letters as a series of pairs, then we know that the principal source of interference comes from similarity at the stimulus level, which then gives rise to errors on the subsequent response (Osgood 1949). We would therefore expect errors to follow the similar items, whereas in fact the similar items themselves were the main source of error (Baddeley 1968). This result has continued to present a challenge to models of serial order.

The past decade has seen considerable activity in the attempt to produce clearly specified computational or mathematical models of serial order retention, with a number located within the phonological loop tradition. Very briefly, approaches fall into two categories. One class of models assumes that items are associated with a series of internal markers, which may be temporal oscillators as in Brown et al.'s (2000) OSCAR hypothesis, or other forms of ordinal marking, as in the case of the model and its subsequent refinement by Burgess & Hitch (1999, 2006). A second approach is typified by the primacy hypothesis of Page & Norris (1998), which assumes a limited capacity of excitation that is shared among the sequence of items. The first item is the most strongly activated, the second slightly less, and so forth. At recall, the strongest item is retrieved first and then inhibited to avoid further repetition before going on to the next strongest. Both of these approaches can handle the similarity sandwich effect, as they do not depend upon chaining. Furthermore, they require two stages, a store and a serial order link, offering an interpretation of the irrelevant sound effect in terms of adding noise to this additional stage (Page & Norris 2003), an explanation as to why similarity between irrelevant and remembered items is not important.

Modeling serial order continues to be a very lively field with considerable interaction between proponents of the different models, which are now starting to become more ambitious. Burgess and Hitch are now attempting to model the link between the phonological loop and long-term phonological learning (Burgess & Hitch 2006, Hitch et al. 2009), while a further challenge being addressed lies in the interpretation of chunking, the effect that makes sentences so much more readily recalled than scrambled words (Baddeley et al. 2009). Can models of serial order in verbal STM be generalized to visual STM? The answer seems to be that they can (Hurlstone 2010). If so, do they reflect a single common system? I myself think it more likely that evolution has applied the same solution to a problem, maintaining serial order, that crops up in a range of different domains.

The phonological loop and LTM

What function might the phonological loop (PL) serve, other than making telephoning easier (an unlikely target for Mother Nature)? The opportunity to investigate this question cropped up when an Italian colleague, Giuseppe Vallar, invited me to help him to investigate a patient, PV, with a very pure and specific deficit in phonological STM. Her intellect was preserved, but her auditory digit span was only two items. She had fluent language production and comprehension, except for long, highly artificial sentences in which ambiguity could only be resolved by retaining the initial part of a long sentence until the end, again not a great evolutionary gain. We then came up with the idea that her phonological loop might be necessary for new long-term phonological learning. We tested this by requiring her to learn Russian vocabulary (e.g., *flower-svieti*), comparing this with her capacity for learning to pair unrelated Italian words, for example (*castle-table*). When compared to a group of matched controls, her capacity to learn native language pairs was normal, whereas she failed to learn a single Russian word after ten successive trials, a point at which all the normal participants had perfect performance (Baddeley et al. 1988). We had found a function for the phonological loop.

Although the work with PV had a major influence on my theoretical views, of much greater practical importance was my collaboration with Susan Gathercole, in which we explored the role of the phonological loop in vocabulary learning, both in children with specific language impairment and in normal children. A series of studies showed that WM plays a significant role in the initial stages of vocabulary acquisition and is also linked to reading skills (see Baddeley et al. 1998 for a review). It formed the basis of an extensive and successful application of the M-WM theory to the identification and treatment of WM deficits in school-age children (Gathercole & Alloway 2008; Gathercole et al. 2004a, b).

At a theoretical level, work with PV led to a major development. I had previously tended to treat WM and LTM as separate though interrelated systems. The fact that the loop specifically facilitates new phonological learning implies a direct link from the loop to LTM. Gathercole (1995) showed that existing language habits influence immediate nonword recall, making the nonwords that have a similar letter structure to English, such as *contramponist*, easier than less familiar sounding nonwords such as *loddenapish* (Gathercole 1995). This suggests that information flows from LTM to the loop, as well as the reverse. Furthermore, it seemed reasonable to assume that a similar state of affairs would occur for the visuo-spatial sketchpad, leading to a revision of the original model along the lines indicated in **Figure 2**. Here, a crucial distinction is made between WM, represented by a series of fluid systems that require only temporary activation, and LTM, representing more permanent crystallized skills and knowledge.

The phonological loop: master or slave?

In formulating our model, we referred to the loop and sketchpad as slave systems, borrowing the term from control engineering. It is, however, becoming increasingly clear that the loop can also provide a means of action control. In my own case, this first became obvious during a series of studies of the CE, in this case concentrating on its capacity for task switching. We used a very simple task in which participants were given a column of single

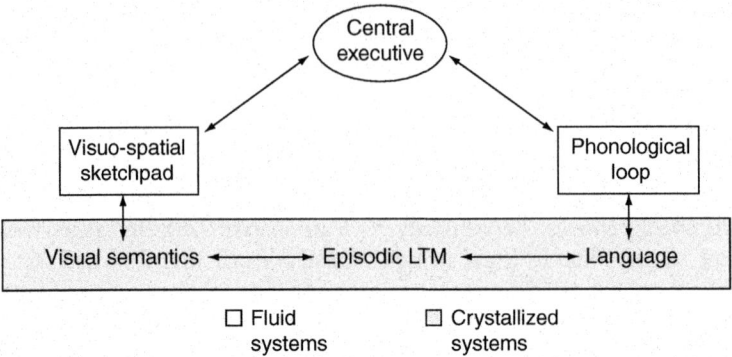

Figure 2 A modification of the original model to take account of the evidence of links between working memory and long-term memory (LTM).

digits and required in one condition to add 1 and write down a total, and in another condition, to subtract 1, or in the switching condition, to alternate addition and subtraction. Switching leads to a substantial slowing, and we wanted to know why. We used dual task methods, disrupting the CE with an attentionally demanding verbal task and a task involving simple verbal repetition. To our surprise, switching was disrupted almost as much by articulatory suppression as by the much more demanding executive task. It became clear that people were using a simple subvocal code of "plus–minus–plus," etc., to cue their responses. When the relevant plus and minus signs were provided on the response sheet, the suppression effect disappeared (Baddeley et al. 2001). Similar results have been obtained and further developed by Emerson & Miyake (2003).

The importance of self-instruction had of course already been beautifully demonstrated by the great Russian psychologist Alexander Luria, who showed that children gradually learn to control their actions using overt self-instruction, a process that later becomes sub-vocal. He went on to demonstrate the value of self-instructions in neuropsychological rehabilitation (Luria 1959).

The phonological loop: critique

The loop is probably the best-developed and most widely investigated component of WM, possibly because of the availability of a few simple tools such as the phonological similarity, word length, and suppression effects. It is, however, only one very limited component of WM. When its use in digit span is prevented by combining visual presentation with articulatory suppression, the cost is something in the region of two digits (Larsen & Baddeley 2003). Its strength is that it can provide temporary sequential storage, using a process that is rapid, and requires minimal attention. It is a system that is extremely useful, widespread, and one that, as experimenters, we ignore at our peril. The analogy that comes to mind is that of the role of the thumb in our motor behavior: small, not essential, but very useful. There is, however, a danger of exaggerating its importance. It appears to be this that Nairne criticized under the label "the standard hypothesis" (Nairne 2002), by which he appears to refer to attempts to account for a range of time-specified STM effects purely in terms of the loop. This hypothesis seemed to be attributed to myself, although as discussed elsewhere (Baddeley 2007, pp. 35–38), Nairne's criticisms do not apply to WM more generally. I agree that what Nairne describes as the standard hypothesis is far from adequate as a theory of WM or even as a general account of STM.

I have discussed the phonological loop thus far as if it were limited to the storage of heard and spoken speech. It is important to note, however, that the same system, operating under broadly similar constraints, appears to underpin memory for both lip-read and signed material (see Rönnberg et al. 2004 for a review). All of these are language related, which raises the question of whether the same system is used for nonlinguistic auditory information such as environmental sounds and music. Neither of these topics is well explored, although there is growing interest in comparing language and music and some indication of overlap (Williamson et al. 2010).

Visuo-spatial sketchpad

Interest in visuo-spatial memory developed during the 1960s, when Posner & Konick (1966) showed that memory for a point on a line was well retained over a period ranging up to 30 seconds, but it was disrupted by an interpolated information-processing task, suggesting some form of active rehearsal. Dale (1973) obtained a similar result for remembering a point located in an open field. In contrast to these spatial memory tasks, Posner & Keele (1967) produced evidence suggesting a visual store lasting for only two seconds. However, their method was based on speed of processing letters, in which a visual letter code appeared to be superseded by a phonological code after two seconds. Although this could reflect the duration of the visual trace, it could equally well reflect a more slowly developing phonological code that then overrides the visual.

Visual STM

A colleague, Bill Phillips, and I decided to test this using material that would not be readily nameable. We chose 5×5 matrices in which approximately half the cells would be filled at random on any given trial. We tested retention over intervals ranging from 0.3 to 9 seconds, by presenting either an identical stimulus or one in which a single cell was changed, with participants making a same/different judgment. We found a steady decline over time, regardless of whether we measured performance in terms of accuracy or reaction time (Phillips & Baddeley 1971). A range of studies by Kroll et al. (1970), using articulatory suppression to disrupt the use of a name code in letter judgments, came to a similar conclusion, that the Posner and Keele result was based on switching from a visual to a phonological code, perhaps because of easier maintenance by subvocal rehearsal. Meanwhile, Phillips went on to investigate the visual memory store using matrix stimuli, demonstrating that accuracy declines systematically with number of cells to be remembered (Phillips 1974), suggesting limited visual STM capacity. It was this work that influenced our initial concept of the visuo-spatial sketchpad.

Spatial STM

The most frequently used clinical test of visuo-spatial memory is the Corsi block-tapping test (Milner 1971), which is spatially based and involves sequential presentation and recall. The participant views an array of nine blocks scattered across a test board. The tester taps a sequence of blocks, and the participant attempts to imitate this. The number of blocks tapped is increased until performance breaks down, with Corsi span typically being around five, about two less than digit span. Della Sala et al. (1999), using a modified version of the Phillips matrix task, showed that visual pattern span is dissociable from spatial Corsi span, with some patients being impaired on one while the other is preserved, and vice versa. Furthermore, pattern span can be disrupted by concurrent visual processing, whereas Corsi span is more susceptible to spatial disruption (Della Sala et al. 1999). I return to the visual-spatial distinction at a later point.

Visuo-spatial WM

During the 1970s, research moved from visual STM to its role in visual imagery. Our own studies used a technique developed by Brooks (1968), in which participants are required to remember and repeat back a sequence of spoken sentences. In half of the cases the sentences can be encoded as a path through a visually presented matrix. The other half of the instructions were not readily encodable spatially. We found that recall of the visuo-spatially codable sentences was differentially disrupted by pursuit tracking (Baddeley et al. 1975a). We interpreted this result in terms of the sketchpad, leading to the question of whether the underlying store was visual or spatial. This we tested using a task in which blindfolded participants tracked a sound source (spatial but not visual) or detected the brightening of their visual field (visual but not spatial), again while performing the Brooks task. We found that the tracking still disrupted the spatial but did not interfere with the verbal task, whereas the brightness judgment showed a slight tendency in the opposite direction, leading us to conclude that the system was spatial rather than visual (Baddeley & Lieberman 1980).

Although these results convinced me that the system was essentially spatial, Robert Logie, who was working with me at the time, disagreed and set out to show that I was wrong. He succeeded, demonstrating that some imagery tasks were visual rather than spatial. He used a visual imagery mnemonic whereby two unrelated items are associated by forming an image of them interacting; for example, *cow* and *chair* could be remembered as a cow sitting on a chair. Logie (1986) showed that this process can be disrupted by visual stimuli such as irrelevant line drawings or indeed by simple patches of color. There are now multiple demonstrations of the dissociation of visual and spatial WM. Klauer & Zhao (2004) critically review this literature before performing a very thorough series of investigations controlling for potential artifacts; their results support the distinction between visual and spatial STM, a distinction that is also supported by neuroimaging evidence (Smith & Jonides 1997).

Yet further fractionation of the sketchpad seems likely. Research by Smyth and colleagues has suggested a kinesthetic or movement-based system used in gesture and dance (Smyth & Pendleton 1990). Another possible channel of information into the sketchpad comes from haptic coding as used in grasping and holding objects, which in turn is likely to involve a tactile component. Touch itself depends on a number of different receptor cells capable of detecting pressure, vibration, heat, cold, and pain. We currently know very little about these aspects of STM, and my assumption that information from all of these sources converges on the sketchpad is far from clearly established.

The nature of rehearsal in the sketchpad is also uncertain. Logie (1995, 2011) suggests a distinction between a "visual cache," a temporary visual store, and a spatial manipulation and rehearsal system, the "inner scribe," although the precise nature of visuo-spatial rehearsal remains unclear.

The central executive

The executive as homunculus

The CE is the most complex component of WM. Within the original model it was assumed to be capable of attentional focus, storage, and decision making, virtually a homunculus, a little man in the head, capable of doing all the clever things that were outside the competence of the two subsystems. Although our model tended to be criticized for taking this approach, like Attneave (1960) I regard homunculi as potentially useful if used appropriately. It is important that they are not seen as providing an explanation, but rather as a marker of issues requiring explanation. Provided the various jobs performed by the homunculus are identified, they can be tackled one at a time, hopefully in due course allowing the homunculus to be pensioned off.

Much of our work has used concurrent tasks to disrupt the various components of WM, with the assumption typically being that attentionally demanding tasks will place specific demands on the CE, in contrast to tasks that require simple maintenance. For example, counting backward in threes from a number such as 271 is assumed to load the executive, whereas simply repeating 271 would not. This and related tasks have proved to be a successful strategy for separating out contributions of the three initially proposed WM subcomponents (e.g., Baddeley et al. 2011).

Fractionating the executive

In an attempt to specify the functions of the CE, I speculated as to what these might be; what would any adequate executive need to be able to do? I came up with four suggestions (Baddeley 1996). First it would need to be able to focus attention; evidence of this came from the impact of reducing attention on complex tasks such as chess (Robbins et al. 1996). A second desirable characteristic would be the capacity to divide attention between two important targets or stimulus streams. I had been studying this in collaboration with Italian colleagues for a number of years, focusing on Alzheimer's disease. We selected two tasks involving separate modalities: one verbal, involving recall of digit sequences, and the other requiring visuospatial tracking. We titrated the level of difficulty for each of these to a point at which our patients were performing at the same level as both young and elderly controls. We then required tracking and digit recall to operate simultaneously. There was a marked deficit in the performance of the patients when compared to either of the two control groups. Perhaps surprisingly, age did not disrupt this specific executive capacity, provided the level of difficulty is equated in the first place (Logie et al. 2004). In the absence of titration of level of difficulty, however, performance tends to decline with age on the tasks when performed singly, with the deficit even greater when the two tasks are performed at the same time (Riby et al. 2004).

The third executive capacity we investigated involved switching between tasks, for which we felt there might be a specific control system. As mentioned earlier, we chose to study a task involving alternating between simple addition and subtraction, using a demanding concurrent verbal executive task and articulatory suppression as its nondemanding equivalent. We found a large effect of articulatory suppression coupled with a rather small additional effect when an executive load accompanied suppression. The study of task switching has expanded very substantially in recent years (Monsell 2005), becoming theoretically rather complex, and in my view at least, arguing against a unitary executive capacity for task switching. I should point out that there are many other suggestions as to the basic set of executive capacities that are too numerous to discuss in this context (see, for example, Engle & Kane 2004, Miyake et al. 2000, Shallice 2002).

Interfacing with LTM

The fourth executive task that I assigned to our homunculus was the capacity to interface with LTM. In an attempt to constrain our WM model, we had made the assumption that the CE was a purely attentional system with no storage capacity (Baddeley & Logie 1999). However, this created a number of problems. One concerned the question of how subsystems using different codes could be integrated without some form of common storage. Participants do not simply use either one code or another, but rather combine them, with both visual and phonological codes being usable simultaneously (Logie et al. 2000). This capacity is particularly marked in the case of language processing, where a single phrase can show the influence of phonological coding at short delays and semantic coding at longer intervals (Baddeley & Ecob 1970). Memory span for unrelated words is around 5, increasing to 15 when the words make up a sentence. This enhanced span for sentence-based sequences seems to reflect an interaction between phonological and semantic systems rather than a simple additive effect (Baddeley et al. 1987), a conclusion that is consistent with later dual-task studies (Baddeley et al. 2009). But how might this interaction occur?

A further challenge to the concept of a purely attentional executive came from the very extensive work on individual differences in WM stemming from the initial demonstration by Daneman & Carpenter (1980) of a correlation between a measure they termed "WM span" and capacity for prose comprehension. Their measure required participants to read out a sequence of sentences and then recall the final word of each. This and similar tests that require the combination of temporary storage and processing have proved enormously successful in predicting performance on cognitive tasks ranging from comprehension to complex reasoning and from learning a programming language to resisting distraction (see Daneman & Merikle 1996 and Engle et al. 1999 for reviews). Such results were gratifying in demonstrating the practical significance of WM, but embarrassing for a model that had no potential for storage other than the limited capacities of the visuo-spatial and phonological subsystems. In response to these and related issues, I decided to add a fourth component, the episodic buffer (Baddeley 2000). Although I was reluctant to add further systems to the multicomponent theory, I felt that one in 25 years was perhaps acceptable.

The episodic buffer

The characteristics of the new system are indicated by its name; it is episodic in that it is assumed to hold integrated episodes or chunks in a multidimensional code. In doing so, it acts as a buffer store, not only between the components of WM, but also linking WM to perception and LTM. It is able to do this because it can hold multidimensional representations, but like most buffer stores it has a limited capacity. On this point we agree with Cowan (2005) in assuming a capacity in the region of four chunks. I made the further assumption that retrieval from the buffer occurred through conscious awareness, providing a link with our earlier research on the vividness of visual and auditory imagery (Baddeley & Andrade 2000). This results in a theory of consciousness that resembles that proposed by Baars (1988), which assumes that consciousness serves as a mechanism for binding stimulus features into perceived objects. He uses the metaphor of a stage on which the products of preconscious processes, the actors, become available to conscious awareness, the audience.

Our new component could be regarded as a fractionation of our initial 1974 version of the CE into separate attentional and storage systems. It had a number of advantages in addition to providing a possible answer to the question of the interaction between LTM and WM. At a theoretical level it formed a bridge between our own bottom-up approach based on attempting to understand the peripheral systems first, and the more top-down approaches predominant in North America, which were more concerned with analyzing the executive and attentional aspects of WM (e.g., Cowan 2005, Engle et al. 1999). Perhaps for this reason, the concept appears to have been welcomed and is frequently cited. However, although that suggests that people find it useful, if it is to be theoretically productive, there is a need to use it to ask interesting and tractable questions, a challenge that has kept Graham Hitch, Richard Allen, and myself busy over recent years.

WM and binding

Like Baars (1988), we assume that a central role of the buffer is to provide a multidimensional medium that allows features from different sources to be bound into chunks or episodes, not only perceptually but also creatively, allowing us to imagine something new, for example, an ice-hockey-playing elephant. We could then reflect on this new concept and decide, for example, whether our elephant would be better doing a mean defensive body check or keeping goal. This all seemed likely to be an attention-demanding process, so we speculated that the buffer would depend heavily on the CE. In the initial (Baddeley 2000) model (see **Figure 3**), I intentionally required all access to go through the executive, arguing that we could then investigate empirically whether other links were needed.

We studied the role in binding played by each of the three initial components of WM, using our well-tried concurrent task strategy to disrupt each in turn. If, as our initial hypothesis proposed, the CE controls access to and from the buffer, then an attentionally demanding concurrent task should have a very substantial effect on the capacity to bind information, in contrast to minor effects from disrupting the other subsystems. We decided to examine binding in two very different modalities, namely the binding of visual features into perceived objects on one hand, and the binding of words into sentences on the other.

Visual binding and WM

Our work on visual binding was strongly influenced by some new developments that were beginning to extend the methods applied to the study of visual attention to the subsequent short-term storage of perceived items. A central question

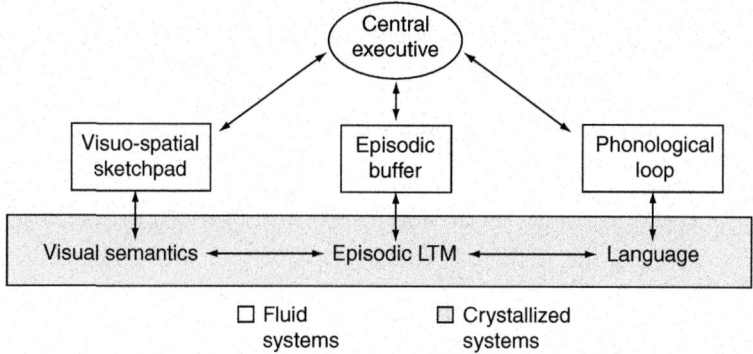

Figure 3 The model following the introduction of a fourth component, the episodic buffer, a system for integrating information from a range of sources into a multidimensional code (Baddeley 2000).

of this approach concerned the factors that determine the conditions under which features such as color and shape are integrated and bound into perceived and remembered objects. The basic experimental paradigm was developed by Luck & Vogel (1997, Vogel et al. 2001). As in the work of Phillips (1974), it involved presenting an array of visual stimuli, followed (after a brief delay) by a probe stimulus, with participants deciding whether or not the probe had been in the array. A number of important results emerged, notably including the observation that capacity was limited to about four objects and was approximately the same, regardless of whether participants were remembering only a single feature, for example, color or shape, or were required to bind the two features and remember not only that a red stimulus had been presented, or a square, but also that the two had been bound together as a red square (Vogel et al. 2001). A subsequent study by Wheeler & Treisman (2002) obtained the same result when testing involved a single probe item. However, they found a binding impairment when the memory test required searching through an array of stimuli in order to find a target match, a result they interpreted as suggesting that maintaining the binding of features was attentionally demanding.

We ourselves tested the attentional hypothesis using our concurrent task procedure. Presentation of the stimulus array was accompanied by a demanding task such as counting backward by threes. If the CE is heavily involved in binding, then the concurrent task should prove more detrimental to the binding condition (e.g., remembering a red square) than to either of the single-feature probe tasks (e.g., red or square). We compared the backward counting condition to one involving articulatory suppression. As expected, we found an overall impairment in performance when accompanied by backward counting. However, this was just as great for the single features as for the binding condition.

A series of further studies explored this finding, using other concurrent tasks and more demanding binding conditions. In one case, for example, shapes and the color patches to which each shape should be bound were presented in separate locations. In another study, the features to be bound were separated in time, while a third experiment presented one feature visually (e.g., a patch of red) and the associated shape verbally. Although some of these activities led to a lower overall level of performance, in no case did we obtain a differential disruption of binding (see Baddeley et al. 2011 for a review).

The final experiment of the Allen et al. (2006) paper did, however, obtain a differential effect. In this study, colored shapes were presented sequentially, followed by a probe. When the final item was probed, the results were as before: no additional binding deficit. However, earlier items did show poorer retention of bound stimuli. We interpreted this as suggesting that binding did not demand extra attention but that maintaining it against distraction did. We explored this disruption effect further, again using simultaneous presentation, but this time inserting a single additional item that participants were instructed to ignore between presentation and test. Binding was differentially impaired even though participants were told to ignore the suffix, which suggests that although visual binding per se is not attention demanding, maintaining bindings against distraction is (see Baddeley et al. 2011 for an overview).

Binding in verbal WM

Although it appears that attention may be useful for maintaining visual bindings, our data indicate that the simple binding of color and shape is not itself attention demanding. It could, of course, be argued that perceptual binding is atypical in not requiring central resources. Fortunately, however, as part of our converging operations approach to theory, we had pursued a parallel series of experiments investigating the role of executive processes in the binding of words into chunks during retention of spoken sentences.

We carried out a series of experiments, the results of which can be summarized quite simply (Baddeley et al. 2009). Concurrent tasks involving the visuo-spatial sketchpad had a small but significant effect on recall that increased when they also had a visually based executive component. Simple articulatory suppression had a greater effect that was further amplified when both suppression and attentional load were required. Most importantly, however, none of these tasks differentially disrupted the binding of words into chunks as reflected in magnitude of the advantage in recalling sentences over unrelated word sequences. Hence, just as with visual binding, although concurrent tasks impair overall performance, they do not appear to interfere with the binding process itself, which in the case of sentences, we assume operates relatively automatically in LTM.

The evidence from both visual and verbal binding is thus inconsistent with the original proposal that the process of binding involves the active manipulation of information within the episodic buffer, which we now regard as being an important but essentially passive structure on which bindings achieved elsewhere can be displayed. It remains important in that it allows executive processes to carry out further manipulation. This may in turn lead to further bindings involving, for example, the binding of phrases into integrated sentences or objects into complex scenes.

In conclusion, although binding is sometimes discussed as if it were a unitary function, we suggest that it differs depending on the specific type of binding involved. For example, binding may be perceptual or linguistic, and it may be temporary, as required to perform WM tasks, or durable, as in the binding of new information to its context in LTM, a capacity that is disrupted in amnesic patients, who may nonetheless show normal binding in WM (Baddeley et al. 2010). All of these types of binding may, however, result in bound representations accessible through the episodic buffer.

Linking long-term and working memory

Is WM just activated LTM?

A number of approaches describe WM as activated LTM (e.g., Cowan 2005, Ruchkin et al. 2003). My view on this issue is that working memory involves the activation of many areas of the brain that involve LTM. This is also true of language, for which activated LTM is not taken as an explanation. I assume that in the case of Cowan's (2005) model, it is a way of referring to those aspects of WM that are not his current principal concern and not a denial of a need for further explanation. He and I would, I think, agree that the phonological loop, the simplest component of WM, is likely to depend on phonological and lexical representations within LTM as well as procedurally based language habits for rehearsal.

Long-term WM

Ericsson & Kintsch (1995) proposed this concept in explaining the superior performance of expert mnemonists, going on to extend it to the use of semantic and linguistic knowledge to boost memory performance. They argue that these and other situations utilize previously developed structures in LTM as a means of boosting WM performance. I agree, but I cannot see any advantage in treating this as a different kind of WM rather than a particularly clear example of the way in which WM and LTM interact.

LTM and the multicomponent model

It seems likely that some of the misunderstandings confronting M-WM stem from the rather limited links with LTM shown in **Figures 2** and **3**. This was also reflected in a disagreement between myself and Robert Logie, who insisted that all information entered the sketchpad via LTM. It was only when

I tried to represent my views in the simple model shown in **Figure 4** that we found we agreed. Incoming information is processed by systems that themselves are influenced by LTM. I see WM as a complex interactive system that is able to provide an interface between cognition and action, an interface that is capable of handling information in a range of modalities and stages of processing.

Neurobiological approaches to working memory

The development of my own views on WM has been strongly influenced by the study of patients with neuropsychological deficits, and particularly by patients with specific impairment in the absence of general cognitive deficits. Brain damage can be seen scientifically as producing a series of unfortunate experiments of nature. Nature is not usually a good experimenter: Patients typically have a range of different deficits, but just occasionally "pure" deficits occur that potentially, given careful and thorough investigation, allow clear theoretical conclusions to be drawn. These can then be extended to help diagnose and treat patients with related but more complex disabilities.

There are two aspects of such research: the behavioral, linking the performance of the patient to cognitive psychology, and the neurobiological, linking it to its anatomical and neurophysiological basis. Both are important and will ultimately be combined. However, my own expertise and current concern is for the behavioral and the extent to which neurobiological research has so far contributed to the cognitive understanding of M-WM.

There is no doubt that the popularity of the concept of WM owes a great deal to neurobiological studies that appear to suggest that WM may depend on one or more specific anatomical locations. Single-unit recording of the

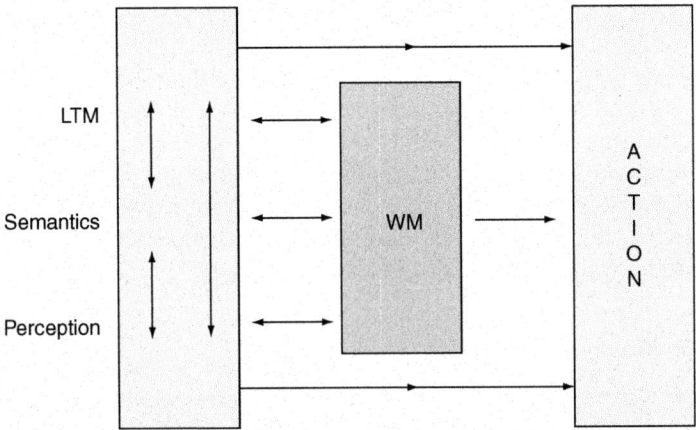

Figure 4 My current view of the complex and multiple links between working memory (WM) and long-term memory (LTM).

brains of awake monkeys performing a visual STM task found that continued activation of cells in the frontal cortex was associated with successful recall, and disrupted activation was associated with failure (Goldman-Rakic 1988). This led some to conclude that it reflected a specific frontal location of WM. My own view was that it probably formed part of a complex circuit underpinning the visuo-spatial sketchpad. Subsequent discovery of similar cells elsewhere in the brain is consistent with this view.

A second source of apparent support for the concept of WM came from neuroimaging when the various subsystems of M-WM appeared to be relatively closely localizable (for reviews, see Henson 2001, Smith & Jonides 1997). This led to a large number of further neuroimaging investigations from many different laboratories, producing a range of different results, which when fed into a meta-analysis often failed to show a consistent pattern (see Baddeley 2007, chapter 7). My own view is that this simply reflects the unreliability of such results and the complexity of WM, as well as the need to modify paradigms to fit the constraints of neuroimaging, for example, avoiding overt speech.

A different interpretation is offered by Jonides et al. (2008), who comment on the "near-revolutionary changes in psychological theories about STM, with similarly great advances in the neurosciences" that have occurred in the past decade. Their very ambitious interpretation of the "mind and brain of STM" is discussed below. More generally, although I think it is very important to understand the neurobiological basis of WM, I am not yet convinced that it has made a major contribution to psychological theories of WM. This does not reflect a general rejection of neuroimaging, which offers an essential and potentially powerful tool for understanding cognition and its neural basis. There are some important areas, such as those investigating conscious awareness, in which neuroimaging provides a crucial component in testing and potentially validating distinctions such as that between "remembering" and "knowing" (Düzel et al. 2001). In the case of WM, however, I have two major sources of doubt. The first concerns the lack of apparent replicability in the field. The second more basic concern is the validity of the assumption that anatomical localization will provide a firm theoretical basis for a system as complex as WM in the absence of a much better understanding of the temporal structure of activation than is typically available at present.

Some alternative approaches

Theories of STM

A good deal of the controversy surrounding M-WM concerns those aspects associated with STM rather than WM, and in particular with the phonological loop model. Whereas some are concerned with the specifics of the model, such as trace decay, for example, other concerns stem from a difference in purpose. Much of our own work has focused on analyzing one source of short-term storage, that based on phonological coding. That means using tasks that minimize semantic and other longer-term factors. Other theorists focus on the whole range of codes that may be contributing, as in the case of Nairne's (1990) feature model. These will inevitably include long-term factors such as semantic coding and will emphasize the similarities between STM and LTM. I regard these two approaches as complementary.

Theories of WM

There are a number of ambitious models of WM that I regard as broadly consistent with the multicomponent framework, although each has a different emphasis and terminology.

Cowan's embedded processes theory

Cowan defines WM as "cognitive processes that are maintained in an unusually accessible state" (Cowan 1999, p. 62). His theory involves a limited-capacity attentional focus that operates across areas of activated LTM. A central issue for Cowan over recent years has been to specify the capacity of this attentional focus and hence the capacity of WM. He produces extensive evidence to suggest that, unlike an earlier suggestion of seven items, the capacity is much closer to four. Importantly, however, this is four chunks or episodes, each of which may contain more than a single item (Cowan 2005).

At a superficial level, Cowan's theories might seem to be totally different from my own. In practice, however, we agree on most issues but differ in our terminology and areas of current focus. I see Cowan's model as principally concerned, in my terminology, with the link between the CE and the episodic buffer. Cowan refers to the material on which his system works as "activated LTM" but does not treat this as providing an adequate explanation, accepting the need for a more detailed analysis of the processes operating beyond attentional focus as reflected in his extensive and influential work on verbal STM, research that interacts with and is complementary to the phonological loop hypothesis of verbal STM (e.g., Cowan et al. 1992). I regard our differences as principally ones of emphasis and terminology.

Individual difference-based theories

The demonstration by Daneman & Carpenter (1980) that WM-span measures can predict comprehension has provided a major focus of research on WM over the past 30 years, involving multiple replications and extensions (Daneman & Merikle 1996). At a theoretical level, there has been considerable interest in identifying the feature of such complex span measures that allows them to predict cognitive performance so effectively. Purely correlational approaches to this issue have a number of limitations, and in my view, the most promising work in this area comes from combining experimental and correlational methods to tackle the question of why some people are better able to sustain material under these complex conditions. Some explanations focus on the capacity to utilize gaps between the processing operations of the span task in order to maintain a fading memory trace (Barrouillet et al. 2004). Others also assume the need to resist time-based decay but emphasize efficiency at switching between the various tasks involved in span (Towse et al. 2000) or they emphasize the role of interference rather than decay (Saito & Miyake 2004).

However, the most extensively developed theoretical account of the mechanisms underpinning WM capacity is that proposed by Engle and colleagues (Engle et al. 1999, Engle & Kane 2004). They emphasize the importance of inhibitory processes, which they argue are crucial to shielding the memory content from potential disruption. Much of their work involves a combination of individual difference and experimental approaches, typically initially testing a large group of participants and then selecting two subgroups, those with very high and those with very low WM span. They have demonstrated that such groups differ not only in WM performance but also in susceptibility to interference across tasks ranging from recall from episodic LTM, through the capacity to generate items from a semantic category, to performance on an antisaccade task of eye movement control (for a review, see Engle & Kane 2004). Although this is an impressive program of work, I suspect that a theory of executive processing based entirely on inhibitory control may be a little narrow. Control is clearly important, but I suspect people also differ in more positive aspects of attentional capacity.

In general, I would see most of the models of WM based on individual differences as consistent with the broad M-WM framework typically focusing on executive control but accepting the contribution of separate visual and verbal STM components (see Alloway et al. 2006 for an example of such a model). Once again, overall similarities may be obscured by terminological differences. Engle and colleagues (Unsworth & Engle 2007) have recently reverted to an earlier distinction between primary and secondary memory, which I would interpret in terms of the distinction between the fluid and crystallized systems that reflect temporary structural representations in the M-WM model (see **Figure 2**).

Jonides and the mind and brain of STM

This approach (Jonides et al. 2008) is strongly influenced by neuroimaging in assuming, for each of a range of modalities, that perception, STM, and LTM are all performed in the same anatomical locations. They also cite evidence from neuropsychology, suggesting that amnesic patients have a general difficulty in binding features together (Hannula et al. 2006, Olson et al. 2006). However, this evidence has been criticized on two grounds: first, that the measures used comprise both long- and short-term components (Shrager et al. 2008), and second, that the conclusions are based on spatial binding. There is strong evidence that both spatial processing and episodic LTM depend on the hippocampus. Nonspatial binding such as that of color to shape was not found to be impaired in a hippocampally compromised amnesic patient (Baddeley et al. 2010), whereas classic amnesic patients do not appear to show evidence of a WM deficit (Baddeley & Warrington 1970, Squire 2004). Furthermore, developmental amnesic patient Jon, who has greatly reduced hippocampal volume, performs well on a range of complex WM tasks (Baddeley et al. 2011).

The major source of evidence cited by Jonides et al. comes from neuroimaging, where STM tasks often activate areas of the brain that also are involved in LTM (e.g., Ruchkin et al. 2003). However, as Jonides et al. note in their discussion of the single-unit studies, the fact that an area becomes active during a given task does not mean that it is essential for performance on that task. Presenting a word is likely to activate regions responsible for its phonological, articulatory, lexical, and semantic dimensions, but that does not mean that all these are necessary in order to repeat that word. Potentially more powerful evidence exists based on lesions in neuropsychological patients (Olson et al. 2006), but the interpretation of this has been questioned (Baddeley et al. 2010); the classic neuropsychological literature typically reports a dissociation between perceptual and memory deficits (Shallice 1988).

The "mind and brain" model proposed by Jonides et al. is somewhat complex, involving five psychological assumptions and six assumptions about neurobiological processing levels. They go on to illustrate their model, using the case of remembering three visual items over a two-second delay, resulting in a figure that involves 13 psychological processes operating across 10 neural levels. I remain somewhat skeptical as to how productive such a model will prove to be. This reflects a difference between us in theoretical style, with my own preference for the gradual development of detailed modes within a broad theoretical framework, whereas Jonides et al. are rather more ambitious.

Computational models of WM

The WM theorists discussed so far have all taken a broad-based approach to theory. There are, however, theorists who attempt a much more detailed account of WM, typically accompanied by computer simulation. This is a very flexible approach, giving rise to a range of different models of WM, which can on occasion result in subcomponents resembling aspects of M-WM including the sketch pad (Anderson et al. 2004, p. 1037) and the loop (Anderson et al. 1996).

Barnard's (1985) ambitious computationally based "interacting cognitive subsystems" model can also be mapped directly onto M-WM. It was initially developed to account for language processing but was subsequently used extensively by Barnard to analyze situations involving human-computer interaction

(Barnard 1987). The model can simulate most aspects of WM while linking it to motor control, emotion, and levels of awareness as part of a broad, ambitious, and insightful model, which in Barnard's hands has been applied with success to an impressive range of situations from choreography to theories of depression (Teasdale & Barnard 1993). However, the sheer complexity of the model makes it difficult for others to use. It is also unclear how important the computational detail really is, and indeed whether it gives an adequate account of what is happening within the more peripheral subcomponents. In discussing his attempt to produce a full simulation of the model, Howard Bowman (2011), a computer scientist who had worked with Barnard in a simulation, now advocates a hierarchical decomposition of the model using components that can be built in isolation, avoiding unnecessary detail such as premature attempts to specify at a neural level.

I suspect that undue complexity may in due course also prove to be a problem for an ambitious new model proposed by Oberauer (2010), who attempts to provide a blueprint for the whole WM system. He sees the main focus of WM as being "to serve as a blackboard for information processing on which we can construct new representations with little interference from old memories." He proposes six requirements for a WM system, namely, (a) maintaining structural representations by dynamic bindings, (b) manipulating them, (c) flexibly reconfiguring them, (d) partially decoupling these from LTM, (e) controlling LTM retrieval, and (f) encoding new structures into LTM. He postulates mechanisms for achieving each of these, hence attempting to put flesh on the previously vague concept of "activated LTM."

A crucial feature of Oberauer's model is the distinction he makes between declarative and procedural WM. Declarative WM is the aspect of WM of which we are aware, comprising most of the current work in the area, whereas procedural WM is concerned with the nondeclarative processes that underpin such operations: I assume that an example would be the process controlling subvocal rehearsal. However, he also considers a higher level of procedural control through what he refers to as the "bridge," as in the bridge of a ship, and what I myself would call the central executive. Consider the following: A participant in my experiment is instructed to press the red button when the number 1 appears, press the green for number 2, and neither for 3. We would expect this simple instruction to be followed throughout the experiment. It is as if some mini-program is set up and then runs, but we currently know very little about how this is achieved. I think the investigation of this aspect of procedural working memory, sometimes referred to as "task set," will become increasingly influential.

This is certainly a very ambitious program, and as Oberauer points out, the evidence at present is rather sparse, but it could be an exciting development. However, the sheer complexity of the model may make it difficult to evaluate experimentally. But then, I am a theoretical mapmaker and temperamentally skeptical of complex theoretical architectures. Time will tell.

What next?

A speculative model and some questions

I have described my attempts to turn a broad theoretical framework into a more detailed model by a process of speculation followed by empirical exploration. It is therefore perhaps appropriate to end on my own current speculations and some of the many questions they raise.

As **Figure 5** shows, my current views are not dramatically different from our original speculation, apart from the episodic buffer, and the attempt to provide considerably more speculative detail. In each case this suggests questions that will not be easily answered but that potentially offer a way forward. I consider the various components in turn.

Central executive

This is an attentional system; how does it differ from the limited-capacity component of Cowan's (2005) model? I assume that it comprises a number of executive functions, but how many, and how are they organized and inter-related? Just how far can one take attempts to explain executive control in terms of a single factor such as that of inhibition? Do we need to worry about precisely what is being inhibited and whether this differs between individuals? Do we need a concept of cognitive energy?

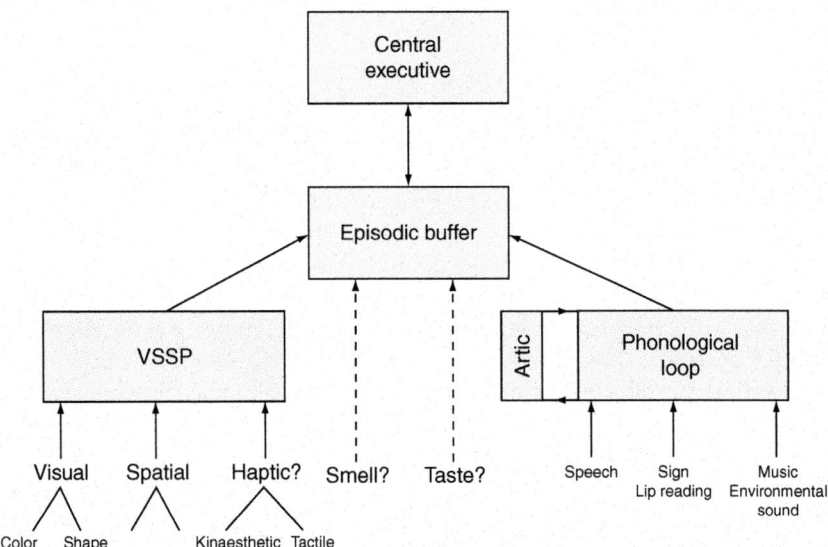

Figure 5 A speculative view of the flow of information from perception to working memory. VSSP, visuo-spatial sketchpad.

Episodic buffer

How should we measure its capacity? To what extent is this limited by number of chunks and to what extent by similarity between chunks? If similarity is important, can other modalities such as smell and taste be added without impacting visual or verbal capacity? Are there separate subsystems for smell and taste? How is rehearsal maintained? My current speculation is that it operates according to the principle of attentionally based refreshing, as discussed by Johnson et al. (2002). What about emotion? Elsewhere (Baddeley 2007) I have suggested that it impacts WM via a hedonic detector system; where within the M–WM system is this located?

I assume that the buffer provides access to conscious awareness; does this mean that we are not directly aware of the other subsystems but only of their products when registered in the buffer?

Phonological loop

Can we reach a conclusion on the ancient trace decay/interference controversy? Is subvocal rehearsal atypical of other types of rehearsal, as I suspect? To what extent is the loop used for remembering non-verbal material such as music or environmental sounds?

Visuo-spatial sketchpad

The visual and spatial aspects appear to be clearly separable but linked within the sketchpad; is this true of haptic, tactile, and kinesthetic memory? What is the mechanism of visuo-spatial rehearsal? Is it a spatial analogue to the phonological loop, as Logie (1995) suggests, or is it more like attentionally based refreshing? Finally, given that our attempt to link the loop with LTM through language acquisition proved very fruitful, would pursuing the link between the sketchpad and LTM prove equally useful?

Integration

Finally, how do these increasingly detailed accounts fit together to provide an interactive unitary system that mediates between perception, LTM, and action?

In praise of negative results

It is of course very easy to raise questions, but much more difficult to answer them. This can lead to a program that cautiously seeks easy confirmation of what we are pretty sure we already know, resulting in confirmation bias, and an avoidance of too much risk of negative results. Negative results are a pain for a number of reasons. First of all, they are hard to interpret. They could result from a poor design or sloppy experimentation. They also raise the question of whether the experiment has sufficient power and indeed whether the question is worth asking in the first place, with all these factors making negative results harder to publish. If we are to understand WM, however, it is important to know what it does not do, and this is likely to involve negative results, as has often proved to be the case in the various stages of developing the current M-WM model. Publication is justifiably more difficult, and there needs to be a good justification for the question. Negative results can, however, be very important and publishable, provided the problem of sensitivity is addressed through inclusion of other conditions showing positive effects together with clear evidence of replication. This was the case with our original 1974 studies, where the effect of concurrent tasks was much less than anticipated, and even more so in our recent exploration of the episodic buffer (Baddeley et al. 2009, 2011).

So what does WM not do? My own conclusion after surveying the experimental literature and its implications for clinical and social psychology (Baddeley 2007) is that we have evolved an overall cognitive system that attempts to minimize the demands made on WM while allowing it to intervene where necessary. A very basic example is that of breathing, far too important to be left to working memory. However, as any diver or singer will know, we clearly do have considerable, though limited, control. Suicide by breath holding is not an option.

Applications

A central requirement of our original framework was that it should be applicable outside the laboratory. Although I have not discussed this aspect of M-WM, it does appear to have had success in achieving this, at two levels. First, through direct application of the M-WM framework to specific practical problems, Gathercole's extensive development of a WM measure applied to school-aged children has been successful in identifying children at risk and providing methods of helping teachers identify and help children with WM problems (Gathercole & Alloway 2008). Another instance is the development and validation of a dual-task performance measure for the early detection of Alzheimer's disease (Kaschel et al. 2009, Logie et al. 2004).

A second aspect of theoretical application is the use of the M-WM theory as a tool for investigating and understanding other research areas. Here the applications are very extensive, ranging from human factors to psychiatry, neuropharmacology to language therapy, and even to paleoanthropology, where the development of working memory is proposed as an explanation of the differences between Neanderthal man and homo sapiens, suggested by a study of surviving artifacts (Wynn & Coolidge 2010).

My own view is that this breadth of application has reflected the simplicity of the theoretical framework together with the availability of a few basic methodologies, such as the use of similarity effects as an indication of coding dimension and of dual-task performance as a way of controlling processing. Such techniques are easily learned, and while not guaranteeing fruitful answers, do at least provide conceptual and practical tools for investigating a wide range of problems. From a theoretical viewpoint, such practical applications can be extremely valuable both in helping explore the boundaries of the laboratory-based effects and in highlighting theoretical anomalies that have the potential to become future growing points.

Conclusion

So where does this leave our early question of what makes a good theory? Clearly, my own preference has been for Toulmin's view of theories as maps, coupled with the Lakatos criterion of judging success by productiveness rather than predictive accuracy. However, as we begin to fill in the empty spaces on the theoretical map, it hopefully will be increasingly possible to develop interlinked and more detailed models of the components of WM and their mode of interaction.

Acknowledgments

I am grateful to Graham Hitch, Richard Allen, Robert Logie, Susan Gathercole, and Christopher Jarrold for many stimulating discussions and for commenting on an earlier draft.

Note

1 I subsequently abandoned the term "acoustic similarity" because it suggested an input modality-based system, which is not the case; I mistakenly assumed that phonological was a more neutral term. It was not intended as a statement of the linguistic basis of the memory system, which remains an open question.

References

Allen R, Baddeley AD, Hitch GJ. 2006. Is the binding of visual features in working memory resource-demanding? *J. Exp. Psychol.: Gen.* 135:298–313

Alloway TP, Gathercole SE, Pickering SJ. 2006. Verbal and visuospatial short-term and working memory in children: Are they separable? *Child Dev.* 77:1698–716

Anderson JR, Bothell D, Byrne MD, Douglass S, Lebiere C, Qin Y. 2004. An integrated theory of the mind. *Psychol. Rev.* 111:1036–60

Anderson JR, Reder LM, Lebiere C. 1996. Working memory: activation limitations on retrieval. *Cogn. Psychol.* 30:221–56

Atkinson RC, Shiffrin RM. 1968. Human memory: a proposed system and its control processes. In *The Psychology of Learning and Motivation: Advances in Research and Theory*, ed. KW Spence, JT Spence, pp. 89–195. New York: Academic

Attneave F. 1960. In defense of homunculi. In *Sensory Communication*, ed. W Rosenblith, pp. 777–82. Cambridge, MA: MIT Press

Baars BJ. 1988. *A Cognitive Theory of Consciousness*. Cambridge, UK: Cambridge Univ. Press

Baddeley AD. 1960. Enhanced learning of a position habit with secondary reinforcement for the wrong response. *Am. J. Psychol.* 73:454–57

Baddeley AD. 1966a. Short-term memory for word sequences as a function of acoustic, semantic and formal similarity. *Q.J. Exp. Psychol.* 18:362–65

Baddeley AD. 1966b. The influence of acoustic and semantic similarity on long-term memory for word sequences. *Q. J. Exp. Psychol.* 18:302–9

Baddeley AD. 1968. How does acoustic similarity influence short-term memory? *Q.J. Exp. Psychol.* 20:249–64

Baddeley AD. 1976. *The Psychology of Memory*. New York: Basic Books

Baddeley AD. 1986. *Working Memory*. Oxford, UK: Oxford Univ. Press

Baddeley AD. 1996. Exploring the central executive. *Q. J. Exp. Psychol. A* 49:5–28

Baddeley AD. 2000. The episodic buffer: a new component of working memory? *Trends. Cogn. Sci.* 4:417–23

Baddeley AD. 2007. *Working Memory, Thought and Action*. Oxford, UK: Oxford Univ. Press

Baddeley AD, Allen RJ, Hitch GJ. 2011. Binding in visual working memory: the role of the episodic buffer. *Neuropsychologia* 49:1393–400

Baddeley AD, Allen RJ, Vargha-Khadem F. 2010. Is the hippocampus necessary for visual and verbal binding in working memory? *Neuropsychologia* 48:1089–95

Baddeley AD, Andrade J. 2000. Working memory and the vividness of imagery. *J. Exp. Psychol.: Gen.* 129:126–45

Baddeley AD, Ecob JR. 1970. Simultaneous acoustic and semantic coding in short-term memory. *Nature* 277:288–89

Baddeley AD, Gathercole SE, Papagno C. 1998. The phonological loop as a language learning device. *Psychol. Rev.* 105:158–73

Baddeley AD, Grant W, Wight E, Thomson N. 1975a. Imagery and visual working memory. In *Attention and Performance V*, ed. PMA Rabbitt, S Dornic, pp. 205–17. London: Academic

Baddeley AD, Hitch GJ. 1974. Working memory. In *The Psychology of Learning and Motivation: Advances in Research and Theory*, ed. GA Bower, pp. 47–89. New York: Academic

Baddeley AD, Hitch GJ. 1977. Recency re-examined. In *Attention and Performance VI*, ed. S Dornic, pp. 647–67. Hillsdale, NJ: Erlbaum

Baddeley AD, Hitch GJ, Allen RJ. 2009. Working memory and binding in sentence recall. *J. Mem. Lang.* 61:438–56

Baddeley AD, Larsen JD. 2007. The phonological loop unmasked? A comment on the evidence for a "perceptual-gestural" alternative. *Q. J. Exp. Psychol.* 60:497–504

Baddeley AD, Levy BA. 1971. Semantic coding and short-term memory. *J. Exp. Psychol.* 89:132–36

Baddeley AD, Lewis V, Vallar G. 1984. Exploring the articulatory loop. *Q. J. Exp. Psychol. A* 36:233–52

Baddeley AD, Lieberman K. 1980. Spatial working memory. In *Attention and Performance VIII*, ed. R Nickerson, pp. 521–39. Hillsdale, NJ: Erlbaum

Baddeley AD, Logie RH. 1999. Working memory: the multiple component model. In *Models of Working Memory: Mechanisms of Active Maintenance and Executive Control*, ed. A Miyake, P Shah, pp. 28–61. Cambridge, UK: Cambridge Univ. Press

Baddeley AD, Papagno C, Vallar G. 1988. When long-term learning depends on short-term storage. *J. Mem. Lang.* 27:586–95

Baddeley AD, Thomson N, Buchanan M. 1975b. Word length and the structure of short-term memory. *J. Verbal Learn. Verbal Behav.* 14:575–89

Baddeley AD, Vallar G, Wilson BA. 1987. Sentence comprehension and phonological memory: some neuropsychological evidence. In *Attention and Performance XII: The Psychology of Reading*, ed. M Coltheart, pp. 509–29. London: Erlbaum

Baddeley AD, Vargha-Khadem F, Mishkin M. 2001. Preserved recognition in a case of developmental amnesia: implications for the acquisition of semantic memory. *J. Cogn. Neurosci.* 13:357–69

Baddeley AD, Warrington EK. 1970. Amnesia and the distinction between long- and short-term memory. *J. Verbal Learn. Verbal Behav.* 9:176–89

Barnard PJ. 1985. Interactive cognitive subsystems: a psycholinguistic approach to short-term memory. In *Progress in the Psychology of Language*, ed. A Ellis, pp. 197–258. London: Erlbaum

Barnard PJ. 1987. Cognitive resources and the learning of human-computer dialogs. In *Interfacing Thought: Cognitive Aspects of Human-Computer Interaction*, ed. JM Carroll, pp. 112–58. Cambridge, MA: MIT Press

Barrouillet P, Bernardin S, Camos V. 2004. Time constraints and resource sharing in adults' working memory spans. *J. Exp. Psychol.: Gen.* 133:83–100

Bowman H. 2011. *Is there a future in architectural modeling of mind?* Paper presented at Philip Barnard retirement symp., MRC Appl. Psychol. Unit, Cambridge, UK

Braithwaite RB. 1953. *Scientific Explanation.* Cambridge, UK: Cambridge Univ. Press

Brooks LR. 1968. Spatial and verbal components in the act of recall. *Can. J. Psychol.* 22:349–68

Brown GDA, Preece T, Hulme C. 2000. Oscillator-based memory for serial order. *Psychol. Rev.* 107:127–81

Burgess N, Hitch GJ. 1999. Memory for serial order: a network model of the phonological loop and its timing. *Psychol. Rev.* 106:551–81

Burgess N, Hitch GJ. 2006. A revised model of short-term memory and long-term learning of verbal sequences. *J. Mem. Lang.* 55:627–52

Cermak LS, Butters N, Moreines J. 1974. Some analyses of the verbal encoding deficit of alcoholic Korsakoff patients. *Brain Lang.* 1:141–50

Cermak LS, Reale L. 1978. Depth of processing and retention of words by alcoholic Korsakoff patients. *J. Exp. Psychol. Hum. Learn.* 4:165–74

Colle HA, Welsh A. 1976. Acoustic masking in primary memory. *J. Verbal Learn. Verbal Behav.* 15:17–32

Conrad R, Hull AJ. 1964. Information, acoustic confusion and memory span. *Br. J. Psychol.* 55:429–32

Cowan N. 1999. An embedded-processes model of working memory. In *Models of Working Memory*, ed. A Miyake, P Shah, pp. 62–101. Cambridge, UK: Cambridge Univ. Press

Cowan N. 2005. *Working Memory Capacity.* Hove, UK: Psychol. Press

Cowan N, Day L, Saults JS, Keller TA, Johnson T, Flores L. 1992. The role of verbal output time and the effects of word-length on immediate memory. *J. Mem. Lang.* 31:1–17

Craik FIM, Lockhart RS. 1972. Levels of processing: a framework for memory research. *J. Verbal Learn. Verbal Behav.* 11:671–84

Dale HCA. 1973. Short-term memory for visual information. *Br. J. Psychol.* 64:1–8

Daneman M, Carpenter PA. 1980. Individual differences in working memory and reading. *J. Verbal Learn. Verbal Behav.* 19:450–66

Daneman M, Merikle PM. 1996. Working memory and language comprehension: a meta-analysis. *Psychonom. Bull. Rev.* 3:422–33

Della Sala S, Gray C, Baddeley A, Allamano N, Wilson L. 1999. Pattern span: a tool for unwelding visuo-spatial memory. *Neuropsychologia* 37:1189–99

Düzel E, Vargha-Khadem F, Heinze HJ, Mishkin M. 2001. Brain activity evidence for recognition without recollection after early hippocampal damage. *Proc. Natl. Acad. Sci. USA* 98:8101–6

Emerson MJ, Miyake A. 2003. The role of inner speech in task switching: a dual-task investigation. *J. Mem. Lang.* 48:148–68

Engle RW, Kane MJ. 2004. Executive attention, working memory capacity and two-factor theory of cognitive control. In *The Psychology of Learning and Motivation*, ed. B Ross, pp. 145–99. New York: Elsevier

Engle RW, Kane MJ, Tuholski SW. 1999. Individual differences in working memory capacity and what they tell us about controlled attention, general fluid intelligence, and functions of the prefrontal cortex. In *Models of Working Memory: Mechanisms of Active Maintenance and Executive Control*, ed. A Miyake, P Shah, pp. 102–34. New York: Cambridge Univ. Press

Ericsson KA, Kintsch W. 1995. Long-term working memory. *Psychol. Rev.* 102:211–45

Fodor JA. 1983. *The Modularity of Mind*. Cambridge, MA: MIT Press

Gathercole SE. 1995. Is nonword repetition a test of phonological memory or long-term knowledge? It all depends on the nonwords. *Mem. Cognit.* 23:83–94

Gathercole SE, Alloway TP. 2008. *Working Memory and Learning: A Practical Guide*. London: Sage

Gathercole SE, Pickering SJ, Ambridge B, Wearing H. 2004a. The structure of working memory from 4 to 15 years of age. *Dev. Psychol.* 40:177–90

Gathercole SE, Pickering SJ, Knight C, Stegmann Z. 2004b. Working memory skills and educational attainment: evidence from National Curriculum assessments at 7 and 14 years of age. *Appl. Cogn. Psychol.* 40:1–16

Glanzer M. 1972. Storage mechanisms in recall. In *The Psychology of Learning and Motivation: Advances in Research and Theory*, ed. GH Bower, pp. 129–93. New York: Academic

Goldman-Rakic PW. 1988. Topography of cognition: parallel distributed networks in primate association cortex. *Annu. Rev. Neurosci.* 11:137–56

Hannula DE, Tranel D, Cohen NJ. 2006. The long and the short of it: relational memory impairments in amnesia, even at short lags. *J. Neurosci.* 26:8352–59

Henson R. 2001. Neural working memory. In *Working Memory in Perspective*, ed. J Andrade, pp. 151–74. Hove, UK: Psychol. Press

Hitch GJ, Flude B, Burgess N. 2009. Slave to the rhythm: experimental tests of a model for verbal short-term memory and long-term sequence learning. *J. Mem. Lang.* 61:97–111

Hurlstone MJ. 2010. *The problem of serial order in visuospatial short-term memory*. Unpubl. PhD dissert., Univ. York, UK

Johnson MK, Reeder JA, Raye CL, Mitchell KJ. 2002. Second thoughts versus second looks: an age-related deficit in reflectively refreshing just-activated information. *Psychol. Sci.* 13:64–67

Jones D, Hughes RW, Macken WJ. 2006. Perceptual organization masquerading as phonological storage: further support for a perceptual-gestural view of short-term memory. *J. Mem. Lang.* 54:265–81

Jones DM. 1993. Objects, streams and threads of auditory attention. In *Attention: Selection, Awareness and Control*, ed. AD Baddeley, L Weiskrantz, pp. 87–104. Oxford, UK: Clarendon

Jones DM, Macken WJ. 1993. Irrelevant tones produce an irrelevant speech effect: implications for phonological coding in working memory. *J. Exp. Psychol.: Learn. Mem. Cogn.* 19:369–81

Jones DM, Macken WJ. 1995. Phonological similarity in the irrelevant speech effect: within- or between-stream similarity? *J. Exp. Psychol.: Learn. Mem. Cogn.* 21:103–15

Jonides J, Lewis RL, Nee DE, Lustig CA, Berman MG, Moore KS. 2008. The mind and brain of short-term memory. *Annu. Rev. Psychol.* 59:193–224

Kaschel R, Logie RH, Kazén M, Della Sala S. 2009. Alzheimer's disease, but not ageing or depression, affects dual-tasking. *J. Neurol.* 256:1860–68

Klauer KC, Zhao Z. 2004. Double dissociations in visual and spatial short-term memory. *J. Exp. Psychol.: Gen.* 133:355–81

Kroll NE, Parks T, Parkinson SR, Bieber SL, Johnson AL. 1970. Short-term memory while shadowing: recall of visually and aurally presented letters. *J. Exp. Psychol.* 85:220–24

Lakatos I. 1976. *Proofs and Reputations*. Cambridge, UK: Cambridge Univ. Press

Larsen J, Baddeley AD. 2003. Disruption of verbal STM by irrelevant speech, articulatory suppression and manual tapping: Do they have a common source? *Q. J. Exp. Psychol. A* 56:1249–68

Larsen JD, Baddeley AD, Andrade J. 2000. Phonological similarity and the irrelevant speech effect: implications for models of short-term verbal memory. *Memory* 8:145–57

Lashley KS. 1951. The problem of serial order in behavior. In *Cerebral Mechanisms in Behavior: The Hixon Symposium*, ed. LA Jeffress, pp. 112–36. New York: John Wiley

Logie RH. 1986. Visuo-spatial processing in working memory. *Q. J. Exp. Psychol.* 38A:229–47

Logie RH. 1995. *Visuo-Spatial Working Memory*. Hove, UK: Erlbaum

Logie RH. 2011. The functional organisation and the capacity limits of working memory. *Curr. Dir. Psychol. Sci.* 20:240–45

Logie RH, Cocchini G, Della Sala S, Baddeley A. 2004. Is there a specific capacity for dual task co-ordination? Evidence from Alzheimer's disease. *Neuropsychology* 18:504–13

Logie RH, Della Sala S, Wynn V, Baddeley AD. 2000. Visual similarity effects in immediate serial recall. *Q. J. Exp. Psychol.* 53A:626–46

Luck SJ, Vogel EK. 1997. The capacity of visual working memory for features and conjunctions. *Nature* 390:279–81

Luria AR. 1959. The directive function of speech in development and dissolution, part 1. *Word* 15:341–52

Milner B. 1971. Interhemispheric differences in the localization of psychological processes in man. *Br. Med. Bull.* 27:272–77

Miyake A, Friedman NP, Emerson MJ, Witzki AH, Howenter A, Wager TD. 2000. The unity and diversity of executive functions and their contributions to complex "frontal lobe" tasks: a latent variable analysis. *Cogn. Psychol.* 41:49–100

Monsell S. 2005. The chronometrics of task-set control. In *Measuring the Mind: Speed, Control, and Age*, ed. J Duncan, L Phillips, P McLeod, pp. 161–90. Oxford, UK: Oxford Univ. Press

Nairne JS. 1990. A feature model of immediate memory. *Mem. Cognit.* 18:251–69
Nairne JS. 2002. Remembering over the short-term: the case against the standard model. *Annu. Rev. Psychol.* 53:53–81
Norman DA, Shallice T. 1986. Attention to action: willed and automatic control of behaviour. In *Consciousness and Self-Regulation. Advances in Research and Theory*, ed. RJ Davidson, GE Schwartz, D Shapiro, pp. 1–18. New York: Plenum
Oberauer K. 2010. Design for a working memory. *Psychol. Learn. Motiv.* 51:45–100
Olson IR, Moore KS, Stark M, Chatterjee A. 2006. Visual working memory is impaired when the medial temporal lobe is damaged. *J. Cogn. Neurosci.* 18:1087–97
Osgood CE. 1949. The similarity paradox in human learning: a resolution. *Psychol. Rev.* 56:132–43
Page MPA, Norris D. 1998. The primacy model: a new model of immediate serial recall. *Psychol. Rev.* 105:761–81
Page MPA, Norris DG. 2003. The irrelevant sound effect: what needs modeling, and a tentative model. *Q. J. Exp. Psychol.* 56A:1289–300
Phillips WA. 1974. On the distinction between sensory storage and short-term visual memory. *Percept. Psychophys.* 16:283–90
Phillips WA, Baddeley AD. 1971. Reaction time and short-term visual memory. *Psychon. Sci.* 22:73–4
Popper K. 1959. *The Logic of Scientific Discovery*. London: Hutchison
Posner MI, Keele SW. 1967. Decay of visual information from a single letter. *Science* 158:137–39
Posner MI, Konick AF. 1966. Short-term retention of visual and kinesthetic information. *Organ. Behav. Hum. Perform.* 1:71–86
Riby LM, Perfect TJ, Stollery B. 2004. The effects of age and task domain on dual task performance: a meta-analysis. *Eur. J. Cogn. Psychol.* 16:863–91
Robbins T, Henderson E, Barker D, Bradley A, Fearneyhough C, et al. 1996. Working memory in chess. *Mem. Cognit.* 24:83–93
Rönnberg J, Rudner M, Ingvar M. 2004. Neural correlates of working memory for sign language. *Cogn. Brain Res.* 20:165–82
Ruchkin DS, Grafman J, Cameron K, Berndt RS. 2003. Working memory retention systems: a state of activated long-term memory. *Behav. Brain Sci.* 26:709–77
Saito S, Miyake A. 2004. On the nature of forgetting and the processing-storage relationship in reading span performance. *J. Mem. Lang.* 50:425–43
Salame P, Baddeley AD. 1986. Phonological factors in STM: similarity and the unattended speech effect. *Bull. Psychonom. Soc.* 24:263–65
Shallice T. 1988. *From Neuropsychology to Mental Structure*. Cambridge, UK: Cambridge Univ. Press
Shallice T. 2002. Fractionation of the supervisory system. In *Principles of Frontal Lobe Function*, ed. DT Stuss, RT Knight, pp. 261–77. New York: Oxford Univ. Press
Shallice T, Warrington EK. 1970. Independent functioning of verbal memory stores: a neuropsychological study. *Q. J. Exp. Psychol.* 22:261–73
Shrager Y, Levy DA, Hopkins RO, Squire LR. 2008. Working memory and the organization of brain systems. *J. Neurosci.* 28:4818–22
Smith EE, Jonides J. 1997. Working memory: a view from neuroimaging. *Cognit. Psychol.* 33:5–42
Smyth MM, Pendleton LR. 1990. Space and movement in working memory. *Q. J. Exp. Psychol.* 42A:291–304
Squire LR. 2004. Memory systems of the brain: a brief history and current perspective. *Neurobiol. Learn. Mem.* 82:171–77

Teasdale JD, Barnard PJ. 1993. *Affect, Cognition and Change: Remodeling Depressive Thought*. Hove, UK: Erlbaum

Toulmin S. 1953. *The Philosophy of Science*. London: Hutchison

Towse JN, Hitch GJ, Hutton U. 2000. On the interpretation of working memory span in adults. *Mem. Cognit.* 28:341–48

Unsworth N, Engle RW. 2007. The nature of individual differences in working memory capacity: active maintenance in primary memory and controlled search from secondary memory. *Psychol. Rev.* 114:104–32

Vogel EK, Woodman GF, Luck SJ. 2001. Storage of features, conjunctions, and objects in visual working memory. *J. Exp. Psychol.: Hum. Percept. Perform.* 27:92–114

Waugh NC, Norman DA. 1965. Primary memory. *Psychol. Rev.* 72:89–104

Wheeler ME, Treisman AM. 2002. Binding in short-term visual memory. *J. Exp. Psychol.: Gen.* 131:48–64

Williamson V, Baddeley A, Hitch G. 2010. Musicians' and nonmusicians' short-term memory for verbal and musical sequences: comparing phonological similarity and pitch proximity. *Mem. Cognit.* 38:163–75

Wynn T, Coolidge FL. 2010. Beyond symbolism and language: an introduction to Supplement 1, *Working Memory*. *Curr. Anthropol.* 51:5–16

Index

Page numbers in *italics* denote tables, those in **bold** denote figures.

acoustic code 4–5, 15, 108n1, 130, 301, 334
acoustic coding 5, 16, 19, 24, 28, 36, 43, 52, 173; recoding 39; short lived effects 15; system 13
acoustic confusions *25, 26*; intrusions *27, 29*; mis-hearing errors 334; similarity 10
acoustic confusions 29, 36; cues 4, 9; information 299; intrusion errors 19, 23; LTM 36; nature 108n1, 117; recoding 39; signal 140; similarity effect on STM 9, 11–13, 15, *16*, 18–19, 28, 43, 64, 70, 151, 173, 334, 363n1; store 4, 70, 107, 127, 130; system 132
acoustic store 107, 130; precategorical 70, 127
acoustic-semantic difference in verbal STM 5
Adams, A.-M. 166, *167*, 183–6, 272
Alderman, N. 262–3
Allen, R.J. 295–6, 315, **317**, 318–19, **320**, 321–2, 324, 327, 352
Alloway, T.P. 343, 357, 362
Allport, A. (D.A) 187, 270–1
Alvarez, G.A. 319, 322
Alzheimer's disease (AD) 244, 251, 257, 259, 280–2, 325, 348, 362; patients 251, 257–60, 263; studies 263
amnesia 324; alcoholic Korsakoff syndrome 6; clinically diagnosed 20; intact performance in 88
amnesic 6; group 25, 34–5; Korsakoff patients 19, 21, 35; memory dependence 24; patients 22, 281; syndrome 87, 255, 272

amnesic patients 5, 18–21, 35, 304; acoustic coding 36; developmental 358; difficulty in binding features 357; digit spans unimpaired 30; free-recall performance 34; frontal 91; hippocampally compromised 358; LTM 18, 34, 272, 306, 308, 353; memory for location 199; preserved implicit learning 6; priming effects 88; prose recall 303; recency effect 81, 88; semantic coding 335; STM 19, 25, 29; temporal lobe damage 19; WM 358
amnesic subjects (Ss) 18–21, **22**, 23, 24, 26–8, 30–2, 34–6; delayed recall *23,* **24**; Hebb effect **32**; immediate memory for digits **31**; impairment in learning digit sequences 36; paired-associates *29,* **30**; rate of development of PI *27,* **28**; retention of word triads **26**; ST forgetting curves 35
amnesics 20, **22**, *23,* **24**, *25,* **26**, *27,* **28**, *29,* **30–3**, *34*
Anderson, J.R. 274, 358
Anderson, N. 269
Anon. 280
articulation 49, 102, 126; impaired 185; irrelevant 171; maximum 145; memory span-articulation rate ratio 118; motor codes necessary 302; preventing 101, 132–3, 147; process of 139; rate 107, 117, *118*, 122, 141, 185–6; subvocal **300**; subvocal and overt 127, 85; suppression 109, **122**, 131, 134, 137, 142, 147; time 110, 125; *see also* concurrent articulation
articulatory rehearsal 126, 299; component 191, 301; loop 126;

prevent 147; process 130, 139, 153, 186, 302
articulatory similarity 18; system 49–50, 131, **190**
articulatory suppression 65–6, 101–2, 110, 126, 131–2, 135, 137, 139, 151, 187, 258, 263, 265, 280, 290–1, 304, 306, 340, 351–2; concurrent **288**; impairs performance 127; influence on immediate memory 130, *133–4, 136*; procedure 49; recall of long and short words 143, **144**, 145, 147; as secondary task 283, 286; technique 64, 70; and word length effect **122**, **124**, 125
articulatory suppression effects 51, 53, 67, 72, 171–3, 185, 281, 298, 302, 348; deleterious 66; disruption of switching 344; disruption in use of name code 346; dual task 291; impairs retention 65; negative 185; in phonological coding 146; in reasoning task 64; on serial recall 304, 306; in subvocal rehearsal 107, 121, 124
Atkinson, R.C. 39–40, 43, 72, 76, 81, 83, 88, 130, 150, 159, 162, 297, 299, 335–6
attentional control 107, 243, 250, 254, 271, 281, 336–7; system 299, 303
attentional controller 257, 264, **300**
Attneave, F. 243, 347
Atwood, G.E. 102, 221
Avons, S. 107, 236

Baars, B.J. 313, 349–50
Baddeley, A.D. 4, 6, 9, 13, 19, 23, 28, 43, 45–6, 51–3, 64, **66, 69**, 70–1, 73, **74**, 80–5, 87–9, 91, 93–4, 99, 101–2, 110, 126, 130–2, 138–40, 142, 145–6, 151–3, 160–1, 164–6, *167*, 168–76, 178, 181–2, 184–7, 191, 199, 206–8, 211, 213–14, 218, 220, 221, 224–5, 227, 233, 236, 238, 243–4, 247–51, 253–5, 257–66, 268–9, 272–3, 276, 281–3, 287, 296, 297–9, **300**, 305, 312, **313**, 315, 318, 321–2, 324–5, **326**, 327–8, 332–6, **327**, 328–46, 348–50, **351**, 352–3, 355, 358, 361–2
BADS test 249–50
Barnard, P.J. 358–9
Barrouillet, P. 314, 357
Bartlett, F.C. 303; Lecture 244
Basso, A. 152–3, 157, 161

Bellugi, U. 178–9
Benton, A.L. 284–5
Besner, D. 102, 145, 187, 238
binding 318, **326**, 352; automatic 309, 319; colorshape **321**; disruption of 319, 321–2, 352; of features 312, 314–15, 317–18, 323, 351; fragility 319–20, 328; information 296–7, **307**; impairment 351; mechanism 313, 350; neurobiological basis of 323; nonspatial 358; object to location 324; overwriting 322; problem 305–6; in prose recall 326; retention of 315, 319; role of hippocampus 324; shape-color 317–18, 325–6; short-term 226, 324, 326; verbal 226, 325, 352; visual 324–6, 350, 352; visuo-spatial 327; of words 314, 326–7, 350, 352; *see also* feature binding
binding conditions **317, 320–1**, 322, **326**, 351–2; simultaneous/sequential presentation 319
binding memory 315–16, 318–19, 321–2; visual working 314, 326; working 326, 350, 353
Bishop, D.V.M. 180–1; Test for the Reception of Grammar 179
Bjork, R.A. 73, 82, 84, 87
Blick, K.I. 201
Bortolini, U. 154–5
Bower, G.H. 211, 218, 337
brain damage 175, 354
brain lesions 191, 282
British Abilities Scales 166
Broadbent, D.E. 1, 40, 100, 113, 224, 226–7, 232, 238, 264, 297, 333, 336
Brockmole, J.R. 317, 320, 322
Brooks, D.N. 6, 88
Brooks, L.R. 71, 207–9, 218, 224–5, 232–5, 237, 246
Brown, G.D.A. 41, 177, 187, **188**, 189, 342
Brown, J. 4, 19, 24
Brown, R. 36, 165, 183
Brown–Peterson technique 27, 34
Bruce, D. 28, 64
Burgess, N. 85, 191, 324–5, 342
Burgess, P. 247, 254, 256, 261
Butterworth, B. 165

Carpenter, P.A. 99, 218, 257, 262, 273–4, 276, 295, 313, 349, 356
Cattell, R.B. 270; Culture Fair Test 259, 270

372 *Index*

central executive (CE) 100, 103–4, 107, 126, 243–4, 247–8, 250, 254–5, 259, 266, 282, 292, **300**, 309, 312–13, 327, 347; access to episodic buffer **307**, **313**, 314, **323**, **351**; activation of LTM 272–3; attentional controller 257; blocking 314; deficit 281; fractionable 299; general-purpose 218; impairment 251, 303; involvement of 145; lacks storage capacity 301; light demands on 206; loading 315; operation disrupted 268; resources 317; role in integration of information 306; served by slave systems 220, 225, 227, 281; single 220; subcomponents 262; as unitary system 249, 276; in working memory 227, 237, 255, 269, 283, **327**, **344**, **360**
central executive (CE) component 69, 71, 76, 206, 281, 308, 317; controlling 131; of working memory 253, 265, 280
Cermak, L.S. 6, 335
children: congenitally deaf 130–1; dyslexic 152, 161; with impaired articulation 185; McCarthy Scales *167*; older 88, 179, 188; overt self-instruction 344; passages suitable for 55; repeating unfamiliar non-words 187; school-aged 343, 362; young 170, 181, 183, 185–6
children's digit span 170; development in children 131, 166; reduced in children 161;
children's language development 181; delay 181; developmental dyslexia 151–2; impaired 180; learning foreign language 168; rapid learning of syntactic rules 183; with specific impairment (SLI) 180–1, 186, 343; word learning 164, 169, 172
children's memory: poor 180; recognition 88; WM deficits 343, 362; WM development 338
children's memory span 101, 151, 185; impaired long-term non-word matching 186; increase with age 101; reduced digit span 161
children's phonological learning 185; long-term 186
children's phonological loop 166, 168, 343; ability 170, 180, 182; performance 299
children's phonological memory 179, 186; ability 183
children's phonological store 184; impaired storage capacity 152
children's rehearsal 304; auditorily presented words 147
children's vocabulary knowledge 171, 174, 177; acquisition 150, 166, 178; development 187, 300; learning 343
chunking 70, 109, 125, 302–3, 305, 309, 342
chunks of information 81, 109, 125, 303, 313–14, 326–7, 349–50, 352, 356, 361
Clarke, H.H. 138
Cloze technique 53–4
Cocchini, G. 244
coding 173, 356; acoustic 16, 23–4, 36; articulatory 64; episodic 273; lexical-semantic 173; phonemic 45, 52, 64, 67, 101, 125; phonological 131, 145–6, 159, 162, 293, 335, 349, 356; semantic 4–5, 16, 19, 23, 28–9, 43, 161–2, 173, 175, 236, 349, 356; short-term phonological 159; speech 110, 130–1; visuo-spatial 216
coding in amnesic patients acoustic 5, 18–21, 35, 304; semantic 335
coding role in long-term learning: articulatory 64; short-term phonological 159
Cohen, R.L. 94, 100, 324
Colle, H.A. 139, 146, 340
Coltheart, V. 43, 67
concurrent articulation 50, 102; disrupting effect 146; effect on free recall **66**; effect on reasoning times and error rates *50*
Conrad, R. 3–4, 9, 18, 24, 43, 52, 101, 107, 110, 113–14, 130–1, 171, 206, 236, 238, 302, 334
Conway, A.R.A. 274–5, 295
Corsi Blocks Test 178, 218–19, 281; Corsi span 346
Cowan, N. 184, 313–14, 317, 349–50, 353, 356
Craik, F.I.M. 19, 39, 64, 70, 74, 88–9, 109–10, 125, 137, 150, 159, 162, 258, 268–9, 293, 336
Crowder, R.G. 70, 81, 83, 92, 127, 142, 272
crystallized 179; skills and knowledge 343; systems **300**, **307**, 357

Dale, H.C.A. 4, 9, 13, 19, 28, 201–2, 345
Dalezman, J.J. 73, 91

Dalrymple-Alford, J.C. 263
Daneman, M. 99, 183, 262, 273–4, 276, 295, 313, 349, 356
Daneman–Carpenter WM span measure 99, 183, 262, 273, 276, 349, 356
Davis, R. 142, 147
Della Sala, S. 244, 250–1, 256–7, 259–60, 281, 283–4, 325, 346
dementia 21, 280, 282, 289, 291; early detection 280–1; effects of 283, 292–3; normotensive hydrocephalus 282; progressive 260, 284; senile 145, **288**; *see also* Alzheimer's disease, Huntingdon's disease
dementia of Alzheimer type (DAT) 280–2, 293; patients 281, 283–5, 287, **288**, *289–90*, 291–2
De Renzi, E. 218–19, 284
D'Esposito, M. 249, 251
digit span 5, 29, 34, 40, 59, 65, 83, *167*, 168, 186, 191, 218, 301, 346; auditory 166, 182–3, 280, **288**, 343; in children 170; concurrent 67, 258, 266, 283, 287; data 35; development in children 131, 166; impaired 66, 76, 126, 160, 178, 281; increase in 70; low 176; measure 166–7; normal in amnesics 18–19, 30; norms 101; reduced in children 161; task 50, 64, 67, 72, 254, 258, 260–1, 263, 291; tracking and 290; traditional technique 31; of two items 44, 151, 336; use of phonological loop prevented 344; of woman with Down's syndrome 179
disruption 40, 102, 225, 336, 341; in autobiographical memory 262; of binding 319, 322, 352; by concurrent digit span task 67; by concurrent spatial activity 206; of context 84; degree of 67, 211, 232, 265, 322; differential 236–7, 240, 251, 292; due to secondary tasks 235, 237–8, 287; effect 211, 352; enhancement of 287; of information processing 146; irrelevant sound sources 340; of memory for patterns 226; minimized 209; in performance 267, 292; by phonemic similarity 58; of phonological coding 146; of phonological storage 327; potential 357; preventing 315; of recall 211; resistant to 219; sensitive to 232; significant 224; spatial 346; of strategic

thought 264; of verbal STM 227; visual 221; of visuo-spatial processing 315
distraction 15; attentional hypothesis 146; effects of 24 limit 271; maintaining bindings against 352; resisting 349
Doors and People Test 176
Down's syndrome 178–9
Drachman, D.A. 19, 29–31, 36
dual task performance 249–50; control by frontal lobes 251; primary and secondary *253*; sensitive to age 259
Duncan, J. 254, 256, 264–6, 269
Dunn, L.M. *167*
dyslexia 255; developmental in children 151
dyslexic children 152, 161
dysarthric patients 302

Ellis, N.C. 101, 131, 151–2, 161, 174, 185
Engle, R.W. 257, 261–2, 273–6, 295, 313–14, 348–50, 357
episodic buffer 295, 297, 306, **307**, 308–9, 312, **313**, 314–15, 318, **323**, 325–6, **327**, 349, **351**, 352–3, 356, **360**, 361–2
episodic memory 87, 189–90, 306, 309; coding 273; long-term 88–9; role in recognition 325
Ericsson, K.A. 272, 303, 353
executive processes 172, 247–51, 254–6, 260, 262, 264, 267–9, 304, 315, 319, 357; central 255; operation of 307; role in chunking 309, 352–3; search and retrieval 261
explicit retrieval 87; strategy 80, 91
Expressive One-Word Picture Vocabulary Scale *167*
Ezzyat, Y. 324

Farah, M.J. 239
feature binding: memory for 320; multi- 313; unitized 318; visual 312, 315
fluid capacities 300, **307**; intelligence 266, 270–1; IQ test 270; systems **307**, **313**, 343, **344**, **351**, 357
foreign language 182, 189; learning 174, 176–7; learning English as 168; vocabulary acquisition 108, 152, 159, 168, 172–3, 176, 300
formal similarity 9, 11–13
Fowler, A.E. 101, 165

free recall 4, 21, 25, 40, *67*, 73; absence of recency 336; additional memory load 68; backward counting interval 73, 81–2; concurrent articulation **66**; delaying 59, 61; depends on semantic coding 19; immediate 80, 89, 91–3, 95; immediate and delayed 18, **60**, 61, **63**; intrusions in *25*; learning 66, 68, 126; memory preload 58, 60; phonemic similarity 64; priming paradigms 93; recency effect 34, 45, 59, 65, 70, 72, 76, 80, 88, 109–10, 125–6, 335; of unrelated words 65; visuo-spatial coding less appropriate 216
free recall components; long-term 68, 159; LTS 72–3; prerecency 159; recency 89; secondary memory 60
free recall task 6, 58, 64; concurrent digit-span 82; concurrent memory load 61, **63**, 64

Gathercole, S. 161, 164, 166, *167*, 168–72, 179, 181, 183–9, 253, 272, 299, 343, 362
Gick, M.L. 258, 269
Glanzer, M. 5, 19, 21, 45, 64–5, 73–4, 81, 88–9, 110, 125, 335
Glenberg, A.M. 82–5, 92
Grant, J. 179
Grant, S. 71, 199, 206
Greene, J. 258, 263
Greene, R.L. 81–2, 89

Hammerton, M. 49, 51
Hannula, D.E. 324, 357
Hartley, T. 191, 324–5
Hasher, L. 269, 271, 276
Haynes, C. 181
Hebb, D.O. 6, 31
Hebb effect 31, **32–3**, 34–6; repeated digit-sequence technique 18–19
Henson, R.N.A. 191, 355
Hinton, G.E. 162, 189
Hintzman, D.L. 89
hippocampal 324; damage 324; deficit 324–5; reduced volume 358; syndrome 255
hippocampus 188, 312, 323–5, 358
Hitch, G.J. 40, 81–4, 91–2, 99–101, 110, 126, 131, 147, 151, 160, 164–6, *167*, 169, 171–2, 184, 186, 191, 206, 220, 224–5, 236, 247, 253, 255, 269, 273, 276, 281, 283, 295–300, 312, 315, 318, 321, 327–8, 335–6, **337**, 342, 350
homunculus 243–4, 248, 250, 253–4, 256–7, 276, 313–14, 347–8
Horowitz, L.M. 9, 11
Hull, A.J. 52, 130–1, 135, 142, 171, 332–4
Hulme, C. 177, 180, 185, 187, **188**, 189
Hunt, E. 102, 240
Huntingdon's disease 282

iconic storage 201–2
implicit learning 80, 89, 95; application of explicit retrieval strategy 80, 91; preserved in amnesic patients 6; recency 87
implicit memory 83, 87, 90–1

Janssen, W.H. 102, 221
Jarrold, C. 178
Johnson-Laird, P.N. 46, 50, 272
Jolicoeur, P. 317
Jones, D. 181, 340–1
Jonides, J. 191, 347, 355, 357–8
Just, M.A. 257, 273

Kail, R. 152, 161, 184
Karlsen, P.J. 318
Katz, J. 94
Keppel, G. 27
Korsakoff's syndrome 335; alcoholic 6, *20*, 21; amnesic 19, 21, 35, 335
Kosslyn, S.M. 239
Kroll, N.E.A. 71, 346
Kyllonen, P.C. 256, 273

Lakatos, I. 334, 338, 363
language acquisition 165, 182–4, 299, 361
Larsen, J. 340–1, 344
Leonard, L.B. 152, 161, 181
lesions 5; brain 191, 282; clinical studies 299; frontal 249; frontal lobe 257, 260; neuropsychological patients 358; peripheral nerve *20*, 21; posterior 218; posterior right-hemisphere 219
Levy, B.A. 19, 49, 64–5, 89, 101, 110, 123, 126, 146, 335
Lezak, M.D. 261, 267
Liberman, I.Y. 101–2
Loess, H. 19, 27, 35
Logie, R.H. 200, 224–5, 235, 237–40, 240n1, 244, 253, 257, 259, 298, 213, 322, 325, 347–9, 353, 361–2

long-term learning 34, 39, 43, 89, 151–2, 156; ability unimpaired 76; capacity 175; deficit 176; difficulty with disyllabic nonwords 160; of digit sequences 70; disrupted 40; efficient 35; impaired 19, 159, 162, 175, 186; maintenance of incoming material 161; normal 6, 36, 76, 151, 153, 160; phonological loop function 169, 171–2, 174, 189; poor 281; role of articulatory coding in 64; role of short-term phonological coding 159; role of working memory in 58; temporary short-term storage 150
long-term memory (LTM) 4–5, 9, 13, *23*, 24–5, 31, 35, 40, 87, 175, 226, 299, 308, 334, 338–9, 354; access to 160; acoustic confusions 36; activated 356, 359; activation of 186, 272–4, 303; additional information from 244; in amnesic patients 81; capacity to hold and manipulate information 253, 257; contribution to WM 305; dating of events 84; defective 18, 30, 34, 36, 81; deficit in 324; impact of STM deficit 107–8; impaired 6, 19–23, 28, 34, 36, 304, 335–6; lexical 188; not responsible for Hebb effect 34; preload 58; propositional codes 102; recency effects 41, 81; representations constructed 183; retention of items 18; role in WM 273; semantic coding 19, 175; store 335; structures within 187; tasks 35, 153; temporary activation of 303; transfer of information to 64; verbal 336; visual images generated from 239; vocal rehearsal 327–8
long-term memory (LTM) binding process 352; words into chunks 327
long-term memory (LTM) control of 249; differentially controlled 309
long-term memory (LTM), effects of similarity in 9; acoustic 43; semantic 334
long-term memory (LTM), episodic 257, **300**, 306, **307**, 308, **313**, 323, **344**, **351**, 357–8; buffer 297, 313
long-term memory (LTM) interfacing 348; central executive 250; slave systems 307
long-term memory (LTM) link 295; central executive 244; with WM **344**, 349–50, 353, **354**

long-term memory (LTM), phonological 176, 188–9; learning **190**; loop (PL) 342–3, 361
long-term memory (LTM) retrieval 261, 359; WM role 100, 276
long-term store (LTS) 39, 44, 65, *67*, 68, 150; component 61, 72–3; registration 73; transfer 72, 76; working memory-LTS system 76
LTS-STS 76
Luck, S.J. 199, 314, 318–20, 351
Luria, A.R. 344

McCarthy Scales for Children *167*
McClelland, J.L. 188
McKenzie, W.A. 90
McLeod, P. 142, 147, 240
Mandler, G. 35, 88
Martin, A.J. 43, 166, *167*, 169, 171–2, 186–7
Masur, E.F. 174
Melton, A.W. 4, 18–19, 34
Melzack, R. 94
memory span 24, 53, 70, 110, 127, 135; capacity 111; in children 185; correlation to reading rate 118–19, **120**, 121, 125; Corsi non-verbal 281; cultural differences 131; for digits 341; deficit 152; effect of articulatory suppression, 187; evidence of speech coding 110; immediate 19, 101, 109, 159, 281, 326; increase with age in children 101; for letters 52, *136*; limited 72, 76, 113, 125; noise interference 138; phonemic coding 125; phonological similarity *136*; preload 47; role of working memory 68; for sentences 5, 326; spatial 218–19; temporal duration of item 113, 116; and tracking *290*; for unrelated words 349; verbal 51–2, 180, 218, 281, 295; visual 157, 228; for visual matrix patterns 224; Williams and Down's syndromes 178; word length effect 131; working 262, 273, 275, 295, 313
memory span, auditory 298; impairment 152, 160
memory span, effects of word length 101, 109, 111, **112**; sensitive to 125; varying 69
memory span impairment 150–1; auditory 152, 160; by phonemic similarity 126

memory span for nonwords 153, 272; for sequences of 187
memory span tasks 51, 63–4, 67, 125; concurrent 73, 110; Corsi non-verbal 281; impairment on 151; performance on 265; phonological loop function 301; role of working memory 68, 126; traditional 45
Michas, I.C. 166, *167*, 170
Miles, T.R. 151–2, 161
Miller, E. 280–1
Miller, G.A. 70, 109–10, 115, 125, 302
Milner, B. 5, 19–20, 24, 218, 346
Miyake, A. 295, 344, 348, 357
mnemonics capacity 151; experts 272; first-letter alphabetic 216, 225; imagery 206–7, 211–13, 215–16, 224–5, 347; location 214, **215**, 216–17; masking 340; pegword 212, 214, 225; spatial 221, 225; spelling 176
Moar, I.T. 222n1
modalities/modality 67, 71–2, 82–3, 123, 270, 306, 361; based input system 363n1; binding 350; combining information from a range of 325, 348, 354, 357; disrupted 318; effect 70, 110, 115; manipulation of 65; multi- 273; presentation **124**, 147; rehearsal 304; specific 211, 236; switching 270–1; visual 159, 318; visual and auditory 312
Monsell, S. 90, 348
Moray, N. 142, 240
Morris, R. 145, 251, 258, 269, 281
Morton, J. 70, 127, 142
multicomponent working memory (M-WM) 40, 80, 95, 131, 151, 160, 164, 199, 247, **307**, 312, **313**, 326, 328, 332, 335, 343, 349, 353–5, 356–8, 361–2
Murdock, B.B. 19, 28, 43, 68, 72, 74, 91
Murray, D.J. 49, 121, 131–2, 139, 146, 171–2

Nairne, J.S. 345
Navon, D. 239
Neale, M.D. 55
Neale Analysis of Reading Ability 55
Nelson, H.E. 261
Nelson, R. 64, 110
neuropsychological patients 150, 164, 256, 268, 308, 358
Nicholson, R. 101, 131
nonphonological 153, 169, 180

nonwords 154, 181; ability to repeat 168–9; capacity for reading 157; disyllabic 156, 160; immediate recall *154*, 176; less familiar sounding 343; maintenance of 161; memory span for 153, 159, 173; multisyllabic 176; polysyllabic 161; poor repetition 185; sequences of 186–7; similar letter structure to English 343; trisyllabic 156; with unusual sound patterns 189; wordlike 170
Norman, D.A. 18, 43, 53, 83, 89, 130, 243–4, 247, 254, 257, 264, 281–2, 235, 337–8

Oberauer, K. 359
O'Keefe, J. 324–5
Olson, I.R. 323–4, 357–8
Osgood, C.E. 334, 342

Page, M.P.A. 191, 342
paired-associate learning 4, 9, 172, 182, 216; minimal learning 18, 28, 91; nonword learning 154, 156–7; recall **30**, 35; retention 29; short-term *29*; short-term retention 34; tasks 173, 207; tests 179; verbal learning 153; word learning 153
Paivio, A. 29, 207, 212, 216
Papagno, C. 93, 159, 172–3, 175, 179, 182, 185, 260
Parkin, A.J. 88, 247–50
Parkinson's disease 263
Parra, M.A. 325
Patterson, K.A. 44, 53, 76
PET studies 251, 268
Peterson, L.R. 4, 6, 18–19, 24, 26, 28, 35–6, 44, 49, 51, 123, 199, 281
Peterson forgetting curve 36
Peterson short-term forgetting task 6, 18, 27, 44
Phillips, W.A. 199, 218, 220–1, 224–7, 232, 235–6, 238, 345–6, 351
Phillips matrix task 346
phonemic 108n1; code 123–4, 127; coding 45, 52, 64, 67, 101, 125; component 68, 70; discrimination task 160; loop 69; processing demands 71; rehearsal buffer 76; rehearsal routines 69; response buffer 68–70; similarity effects 51, *52*, 53, 56–8, 64–5, *67*, 70, 101–2, 110, 126, 206; trace 70
phonological 363n1; deficit 308; development 187; effect of articulatory

suppression 146; experience of the remember 272; factors 177, 335; information 301; knowledge 188, 190; material 108, 165, 173, 182–3; pool 169; processing 88, 160, 187; recording of material 172; rehearsal 185; sequences 188; structures 173–4, 187, 191; trace 139, 340; unit 152; working memory 183
phonological code 5, 102, 137, 159, 301, 345–6, 349; coding 131, 145–6, 159, 162, 293, 335, 349, 356; encoding 159–60
phonological forms 164, 169–71, 174–5, 177–8, 183–4, 186, 236; unfamiliar 166, 172–3
phonological learning, long-term 162, 170–2, 184, 189, **190**, 191, 300, 335, 342–3; in young children 185; deficit in 175–6, 299; new words 164, 169, 174, 186; unfamiliar material 172, 182
phonological long-term system 189; LTM 176, 188–9, **190**; representations within LTM 353
phonological loop 107–8, 108n1, 164–7, 172, 174, 243, 253, 272–3, **300**, 301–2, **307**, **313**, 319, **327**, 328, **337**, 339–40, **344**, 345, **351**, 353, 355–6, **360**, 361; below-average function 182; capacity 165–6, 170–1, 175, 177, 303, 305; characteristics of 273, 338; in children 168; component of WM 191; conscious access to 327; constraints 170–1, 186; deficit 180–1, 187, 299; effect 341; evolutionary significance 254; function 167, 169–70, 175, 180–1, 184, 191, 343; impaired 176–7, 179, 185; in language acquisition 183–4; limited-capacity resource 191; limits and limitations 176, 298; skills 169, 182; temporary storage system 312; vocal rehearsal 327
phonological loop operation 174, 187; impaired 173; influenced 171
phonological loop-based learning process 173; long-term 170–2, 189, 342; vocabulary 343
phonological memory *167*, 170, 179, 308; ability in young children 183; deficits 181; link with vocabulary acquisition 184–6; long-term 5; performance 182, 185; problems 180; skills 169, 172, 180; tests 176, 181–3; 186

phonological memory, short-term 165, 179; deficit 175; impaired 298; tests 176
phonological representation 162, 172; acquisition of novel long-term 186; integrity of 183; quality of 185; stable long-term 173, 182
phonological short-term system 334; deficits 299; input store 153, 161; STM 171, 188, **190**, 250; STM deficit 302, 342; STM skills 179; store 150–1, 153, 161, 184, 188
phonological similarity 101, 107, *133–4, 136*, 344; degree of 173; effect 130–5, 137–9, 145–6, 171, 238, 281, 301, 334, 339; effect removed 172, 302, 340; effect of suppression on 131; errors and sequence length 136; negative impact 341
phonological storage 171, 184, 339; component impairment 161; disruption of 327; impaired capacity 151–2; long-term 162; short-term 150–1, 162; short-term component 159, 161
phonological store 93, 130, 139, 146–7, 151–3, 156, 160–2, 164, 172, 175, 184–6, 189–91, 272, 300–3, 340; access to 146–7, 153, 172, 340; activated 190, 272; buffering system 156; decaying representations 164, 184; defective 302; impaired 151, 161; input 139, 146, 161; location in left hemisphere 191; long-term 162; maximum load on 175; memory trace 300; registration of auditory items 302; registration of material in 147; role in learning new words 186; short-term 150–2, 160, 162, 184, 188; short-term input 153, 161; subvocal rehearsal 184; temporal capacity 185; temporary 301
phonological system 162, 165, 189, 200, 349; long-term 189; short-term 334; short-term deficits 299; slave 249, 305; subsystems 314, 349
Pieroni, L. 318
Pinker, S. 165, 183
Pinto, A. da C. 82, 85
Plunkett, K. 183–4
Popper, K. 333–4
Posner, M.I. 76, 142, 147, 199, 201–4, 283, 345–6
Prabhakaran, V. 307–9

primary memory 80–3, 94, 110, 126, 357
priming 41, 83, 93; effects 80, 85, 90–1; mechanism 87, 273; negative 271; perceptual 87–8; presentation 90
proactive interference (PI) 18, 27; effects 25, 27, 35; rate of development **28**
psychoacoustics 3

Rabbitt, P. 100, 138, 216, 256, 259, 269–70
Raven, J. 284
Raven's Matrices *167*, 179, 284
reaction-time (RT) *50*, *57*, 201–4, 264, 268, **288**, *289*; concurrent task 258, 280, 289–91
recency effect 5, 21, 40, 58, 62, 73, 84, 90–1, 281; backward-counting destroys 75; buffer-storage account 64; concurrent load 63–4; constant in number of items 110; delaying free recall abolishes 59; demonstration in LTM 81; disrupted 95; in free recall 34, 45, 59, 65, 70, 72, 76, 80, 88, 109–10, 125–6, 335; limited size 75; long-term 82, 92–4; in long-term memory 41, 81; marked 74, 82; negative 89; one-item 226–7, 232; phonemic similarity 64; preload influence on 59–61; preserved in amnesic patients 81; priming 85–8; remained intact 226; short-term 82; specialist form of retrieval 273; temporary buffer store 72, 74; unaffected by concurrent task 83; visual 226
rehearsal 4, 26, 81, 108n1, 124, 186, 305; buffer 76; central code 302; continuous 24; impact of length of nonwords 173; of information 225; longer 162; loop 126; maintenance 162; memory items 49, 51, 61; phonemic 67, 69; prevented 19, 21, 24, 121, 123, 139; problem of 304; process 130, 153, 164, 171, 185; rate 117; rote verbal 174; SLI group 181; of visual aspect 204; visuo-spatial 347
rehearsal, articulatory 147; component 191, 301; process 139, 153, 186, 302; system 126, 299
rehearsal strategies 61, 116, 185, 203; rote 213; subvocal 304
rehearsal, subvocal 100–1, 107, 130, 139, 145, 173, 184–5, 301, 304, 337–9,
346, 359, 361; process 184–5; strategy 304
rehearsal system: articulatory 126, 299; phonemic 67; spatial 347
Rejman, M.H. 83, 87
retrieval 273; from buffer store 72–3, 297, 349; checking 254; cue 75, 84; difficulty in initiating 262; explicit mode 95; habitual schemata 264; of information 36; implicit conditions 90; limited-capacity processor 268; from LTM 261, 359; mechanism 85; plans 44, 267–9; role of attentional processes 268; role of WM in 76, 99–100, 276; rules 69–70; serial order 341; targets 86
retrieval process 80, 86, 92; automaticity in 268, 275; explicit 87
retrieval strategy 75; capacity to switch 253, 269; explicit 80, 91; last-in-first-out 41, 91–2
Richardson, J.T.E. 64, **66**, 110, 221
Richardson-Klavehn, A. 87
Rivermead Behaviourial Memory Test 325
Robbins, T.W. 264, 348
Robertson, I.H. 263, 269
Roediger, H.L. 267, 272
Rönnberg, J. 327, 345
Rossi-Arnaud, C. 318
Ruchkin, D.S. 353, 358
Rumelhart, D.E. 43–4

Saito, S. 321, 357
Salamé, P. 139, 146, 238, 340
Salthouse, T.A. 256, 258, 269
Schacter, D.L. 6, 80, 85, 87–91
Schneider, G.E. 220, 264
secondary memory 60, 110, 357
semantic 45, 57, 325; associations 160, 173; attributes 186; categories 87, 118, 140, 216, 259, 266, 274, 357; characteristics 214; compatibility 15, *16*; complexity 272; confusions 26; dimensions 358; effects 15; impossibility 58; intrusions 23, 29; knowledge 300, 353; processing 39, 74, 236, 336; representation 137; system 349; visual **300**, **307**, **313**, **344**, **351**; visuospatial 299
semantic code 5, 159; coding 4–5, 16, 28–9, 43, 161–2, 173, 236, 349, 356; deficit 6; encoding 5, 88, 137, 335; free recall 19; LTM 19, 175; problems in amnesic patients 335; STM 335

semantic factors 335; minimized 356; in STM performance 19; in text recall 187
semantic memory 40, 190, 207, 211, 273; LTM 334–5; LTM-based effects 339; system 221
semantic similarity 4, 130; effect on concept formation 43; effect on LTM 19, 334; effect on STM 9, 11; lack of effect 28, 301; manipulated 89; unimportance 13
serial recall 9, 183, 187, 214; back-up store 301; effect of articulatory suppression 304, 306; imagery mnemonic 216; immediate 3, 5, 145, 181, 190, 335, 340; lists of long and short words 140; phonological store 302; short-term **68**, 72; unsuitability of visuospatial sketch-pad 216, 298, 303
Shallice, T. 5, 39, 44, 53, 76, 81, 103, 126, 150–1, 159–60, 165, 175, 187, 236, 238, 243–4, 247, 254, 256–7, 261, 264, 269, 281–2, 335, 337–8, 358
Shepard, R.N. 102, 218
Shiffrin, R.M. 39–40, 43, 72, 76, 81, 83, 130, 150, 159, 162, 264, 267, 297, 299, 335–6
short-term memory (STM) 9, 13, 18–20, 23, 44, 76, 80, 92, 99, 110, 125, 164, 169, 272, 297, 299, 334–6, 355–7; acoustic coding technique 24; in amnesics 30–1, 34, 36; capacity 109, 114–17, 121, 166, 346; components 324; comprehension tasks 56; contribution to free-recall 23; defective 139, 299; disorder 19; experiments 9; holding language material 53; individual differences 26; labile effects 35; limitations 176; for matrix patterns 236; nonverbal 182; nonverbal skills 182; normal 18–20, 22–3, 25, 30, 34, 36; patients 151–2, 175, 187, 281; performance 166–7; PI effect 27; processes 187; profile 176; recency effect 61, 72; separate vocabulary acquisition system 188; spatial 346–7; speech coding in 130–1; storage **190**; store 150; system 338; tasks 18, 35, 41, 43, 126, 151, 206, 335, 355, 358; time-specified 345; trace 4, 189; unitary 247; for word sequences 9, 11; as working memory 99, 164–5
short-term memory deficit 40, 150, 175, 177, 186, 281, 306; patients 336; phonological 302, 342

short-term memory disruption 341; by phonemic similarity 43
short-term memory impairment 5, 39, 81, 146; in performance 150; reasoning 50
short-term memory load 50; additional **60**
short-term memory, phonological 171, 188, **190**; deficit in 302, 342; superior skills 179
short-term memory system 4; articulatory 116; fragile 5; speech-based 111; time-based 117; unitary 131, 151; verbal, specialized 236; visual 226, 237, 239
short-term memory, verbal 5, 18, 107, 166, 168, 178, 191, 328, 335, 337, 342, 356–7; deficits 238; disruption by mental arithmetic 227; specialized system 236
short-term memory, visual 201–4, 227, 318, 235, 340, 342, 345–7, 355; durability of 199; function 240; independent of verbal temporary storage 239; interface with visual attention 314; limited by pattern complexity 238; self-contained system 226; specialized system 237
short-term memory, visuo-spatial 227, 235, 250; span 182; storage of information 224
short-term retention 191; of familiar verbal material 166; task 19; of word triads **26**
short-term store (STS) 43–5, 52–3, 58–9, 61, 64, 72–3, 76, 81; defective 336; limited capacity 39, 150; phonological 150–1, 153, 161, 184, 188; specialized 226; unitary 160; visual 238
Shrager, Y. 324, 357–8
Shute, V.J. 273–4
simultaneous 93, 259; access to codes 306; array presentation 319–21, 352; availability of components 157; encoding 335, 349; monitoring 262; operation 267; performance of tasks 232, 263, 280; processing and storing of information 273, 281; recency effects 82; semantic and acoustic coding 16, 23; storage and manipulation of material 255; storing of digits 55, 62, 126; subsidiary task 43; tracking and digit recall 348; unitized presentation **316**

380 *Index*

slave system(s) 131, 249; articulatory loop 151, 225, 343; implicit priming mechanism 273; phonological 305; serving central executive 220, 227, 281; simpler more tractable 250, 253–4; subsidiary 100, 256, 299; temporary interface between **307**; temporary storage provided 313; visual and verbal 306; visuo-spatial scratch pad (VSSP) 206, 225, 305, 343; WM 257, 272, 304, 308
Smith, E.E. 191, 347, 355
Smith, J.C. 23, 73
Smyth, M.M. 347
Snowling, M. 185
spatial working memory 206–7, 211–13, 218–21; *see also* visuo-spatial representations
Speidel, G.E. 183
Sperling, G.A. 52, 113–14, 127, 130
Spinnler, H. 152, 251, 256–7, 259–60, 281, 283–4
Squire, L. 87–8, 323–4, 358
standardized ability test batteries 166
Sternberg, S. 274–6
Street, R.F. 284
Street Gestalt Completion Test 284
stroke 152, 219, 282
STS-LTS 73, 76
subsidiary task 43, 73
subvocal rehearsal 100–1, 107, 130, 139, 145, 173, 184–5, 301, 304, 337–9, 346, 359, 361
subvocalization 126, 206, 304, 340
switching 84, 267; attention 49, 93, 249–50; codes 5, 346; demands 267; disrupted 344; modalities 270–1; retrieval plans 267–9; retrieval strategies 253; from one state to another 94; tasks 343, 348, 357

Talland, G. 19
Taylor, R.L. 201, 203
temporary storage 99, 127, 191, 300, **307**, 335–6, 349; of information 297–8, 312, 334, 338; limited-capacity system 306; of new words 182; provided by slave systems 313; speech-based material 131; of unfamiliar phonological forms 166; unitary system 299; verbal 239
Thorn, A.S.C. 187, 189
Thorndike, E.L. 10

Thorndike-Lorge frequency 10–11, 21, 212
tracking 210, 238, 260–1, 289, *290*; adaptive 285–6; AD patients 258; digit recall 348; and digit span 290–1; disrupted recall 209; effect on learning using imagery 215; influence on mnemonic 214; interferes with memory organization 216–17; pursuit 71, 207, **215**, 216, 258, 283, 346; visual STS involvement 238
tracking, concurrent 212–13, 217; effects of *289*; visuo spatial 225
tracking performance 43, 207, 210, 212–13, 215, 287, 289, 292; DAT patients **288**, 291
tracking task 199, 207, 213–15, 263, 280; auditory 208–10, 222n1; concurrent 212–13; primary 286–7; secondary disruption 287; spatial 213; visuo-spatial 216
tracking, visuo-spatial 216: concurrent 225
Treisman, A. 142, 315, 318–19, 322–3, 351
Trojano, L. 176
Tulving, E. 6, 35, 74, 80, 83, 85, 87–91, 125, 304, 306, 309
Tzeng, O.J.L. 73, 75, 82

Ueno, T. **321**
Underwood, B.J. 1, 9, 13, 27

Vallar, G. 93, 101, 139, 146, 151–3, 159–61, 165, 172–3, 179, 182, 187, 236, 238, 342
Vargha-Khadem, F. 323–5
visual attention 314, 326, 350
visual memory span 157, 228; for matrix patterns 224
visual short-term memory (STM) 199, 226–7, 235, 237–40, 314, 342, 345, 355; decay of 203; disruption 340; duration of 201–3; enhanced by symmetry 318; limited capacity 346; rehearsal 204
visuo-spatial 258; memory deficits 308; slave system 249, 305; STM 182, 350; tracking 348
visuo-spatial representations 317, 321; filtering of to-be-ignored stimuli 322
visuo-spatial sketch pad (VSSP) 199, 200, 200n1, 224–5, 237, 243, 253, 273, 283, 298–9, **300**, 303–5, **307**, 312,

313, 319, 322, **327**, **337**, 343, **344**, 345–7, **351**, 352–3, 355, 358, **360**, 361
vocabulary acquisition 165–6, 168, 170–1, 174, 184, 191; natural 175, 177–8; poor 182; problems 177; STM system role 188; WM role 343
vocabulary learning 169, 172, 180, 185; experimental simulations 175; imitation 174; memory problems in 178; role of phonological loop 343
Vogel, E.K. 199, 314–15, 319, 323, 351

Wang, P.P. 178
Warrington, E.K. 5–6, 23, 34, 39, 44, 53, 64, 76, 81, 88, 126, 150–1, 159–60, 175, 199, 220, 236, 238, 281, 335, 358
Wason, P.C. 46, 50
Waters, G.F. 256, 273
Watkins, M.J. 70, 82, 84, 87, 93, 110, 115, 127, 150, 159, 162, 239
Waugh, N.C. 18, 27, 35, 43, 83, 89, 130, 335
Wechsler, D. 166, 272
Wechsler Adult Intelligence Scale 166
Wechsler Intelligence Scale for Children 166
Wechsler Memory Scale 272

Weiskrantz, L. 23, 34, 44, 220
Welford, A.T. 258, 269
Wheeler, M.E. 315, 318–19, 322, 351
Whitten, W.B. 73, 82, 84, 92
Wickelgren, W.A. 18–19, 52, 64
Wickens, C.D. 236, 239
Williams syndrome 178–80
Williamson, V. 328, 345
Wilson, B.A. 88, 91, 187, 227, 248–50, 255, 262, 266, 272, 281, 325
Wilson, J.T.L. 227
Wisconsin Card Sorting Test 261
Woodman, G.F. 314, 318
word length 57, 59, 72, 118, 143, 151, 173; varied, varying 69–70
word length effect 107, 109, 113–17, 119, 121, **122**, 123–7, 130–2, 139–41, 145–7, 171, 238, 302, 339–40, 344; on memory span 101, 109–10, **112**, *113*, **120**, **141**, 142, **144**; recency 70; on short-term serial recall **69**
words: acoustically different 10, 12; encoded 15–16, 36; acoustically similar 9–13, 15–16, 26, 135
word sequences 9, 11, 64; encoded phonologically and semantically 335; five-word 11–12, 133, 143; four-word 143; immediate memory for *133–4*; random 35; unrelated 352